INDUSTRIAL FIRE PROTECTION ENGINEERING

TH 30449100145484
9445 Zalosh, Robert G.
.M4   Industrial fire protection
Z35 engineering
c.2

INDUSTRIAL FIRE PROTECTION ENGINEERING

Robert G. Zalosh
*Center for Firesafety Studies, Worcester Polytechnic Institute,
Worcester, MA, USA*

LIBRARY
WEST GEORGIA TECHNICAL COLLEGE
303 FORT DRIVE
LAGRANGE, GA 30240

WILEY

Copyright © 2003   John Wiley & Sons Ltd, The Atrium, Southern Gate, Chichester,
West Sussex PO19 8SQ, England

Telephone (+44) 1243 779777

Email (for orders and customer service enquiries): cs-books@wiley.co.uk
Visit our Home Page on www.wileyeurope.com or www.wiley.com

All Rights Reserved. No part of this publication may be reproduced, stored in a retrieval system or transmitted in any form or by any means, electronic, mechanical, photocopying, recording, scanning or otherwise, except under the terms of the Copyright, Designs and Patents Act 1988 or under the terms of a licence issued by the Copyright Licensing Agency Ltd, 90 Tottenham Court Road, London W1T 4LP, UK, without the permission in writing of the Publisher. Requests to the Publisher should be addressed to the Permissions Department, John Wiley & Sons Ltd, The Atrium, Southern Gate, Chichester, West Sussex PO19 8SQ, England, or emailed to permreq@wiley.co.uk, or faxed to (+44) 1243 770620.

This publication is designed to provide accurate and authoritative information in regard to the subject matter covered. It is sold on the understanding that the Publisher is not engaged in rendering professional services. If professional advice or other expert assistance is required, the services of a competent professional should be sought.

Neither the author nor John Wiley & Sons Ltd accept any responsibility or liability for loss or damage occasioned to any person or property through using the materials, instructions, methods or ideas contained herein, or acting or refraining from acting as a result of such use.

Other Wiley Editorial Offices

John Wiley & Sons Inc., 111 River Street, Hoboken, NJ 07030, USA

Jossey-Bass, 989 Market Street, San Francisco, CA 94103-1741, USA

Wiley-VCH Verlag GmbH, Boschstr. 12, D-69469 Weinheim, Germany

John Wiley & Sons Australia Ltd, 33 Park Road, Milton, Queensland 4064, Australia

John Wiley & Sons (Asia) Pte Ltd, 2 Clementi Loop #02-01, Jin Xing Distripark, Singapore 129809

John Wiley & Sons Canada Ltd, 22 Worcester Road, Etobicoke, Ontario, Canada M9W 1L1

Wiley also publishes its books in a variety of electronic formats. Some content that appears in print may not be available in electronic books.

Library of Congress Cataloging-in-Publication Data

Zalosh, Robert G.
    Industrial fire protection engineering / Robert G. Zalosh.
        p. cm.
    Includes bibliographical references and index.
    ISBN 0-471-49677-4 (alk. paper)
    1. Industrial buildings – Fires and fire prevention.   2. Fire protection engineering. I. Title.

TH9445.M4 Z35 2002
628.9'22 – dc21   2002069032

British Library Cataloguing in Publication Data

A catalogue record for this book is available from the British Library

ISBN 0-471-49677-4

Typeset in 9.5/11.5pt Times by Laserwords Private Limited, Chennai, India
Printed and bound in Great Britain by TJ International, Padstow, Cornwall
This book is printed on acid-free paper responsibly manufactured from sustainable forestry in which at least two trees are planted for each one used for paper production.

CONTENTS

Preface	xi
1 Introduction and perspective	1
1.1 Engineering approach to industrial fire protection	1
1.1.1 Fire/explosion scenario identification	2
1.1.2 Consequence analysis	6
1.1.3 Alternative protection evaluation	8
1.2 Statistical overview of industrial fires and explosions	10
1.2.1 Industrial occupancies in large loss fires	10
1.2.2 Types of fires/explosions in the largest losses	14
1.2.3 Facilities involved in multiple fatality fires and explosions	14
1.2.4 Ignition sources	17
1.2.5 Need for automatic detection and suppression	18
1.3 Historic industrial fires and explosions	20
1.3.1 Fire protection lessons learned	21
1.3.2 Lessons not learned	23
References	24
2 Plant siting and layout	27
2.1 Fire protection siting considerations	27
2.1.1 Safe separation distances	27
2.1.2 Water supplies	36
2.1.3 Local firefighting organizations	41
2.1.4 Local codes and attitudes	42
2.1.5 Local environmental effects	42
2.2 Plant layout for fire/explosion protection	43
2.2.1 General principles and procedures	43
2.2.2 Hazard segregation and isolation	43
2.2.3 Ignition source isolation	46
2.2.4 Passive barriers	51
2.2.5 Sprinkler system layout	51
2.2.6 Accessibility for manual firefighting	52
2.2.7 Emergency exits	52
2.2.8 Computer aided plant layout	54
References	53
3 Fire resistant construction	57
3.1 Construction materials	57
3.1.1 Steel	57

3.1.2 Steel insulation	61
3.1.3 Concrete	61
3.2 Fire resistance calculations	61
3.3 Fire resistance tests	67
3.3.1 Furnace exposure tests	67
3.3.2 Empirical correlations	69
3.3.3 High intensity fire resistance tests	72
3.4 Fire walls	73
3.4.1 General criteria for fire walls	73
3.4.2 Fire wall design	73
3.4.3 Fire wall loss experience	78
3.5 Fire doors	78
3.5.1 Types of fire doors	78
3.5.2 Fusible links and detectors	81
3.5.3 Reliability issues	81
3.6 Insulated metal deck roofing	83
3.6.1 Description	83
3.6.2 White house tests	84
3.6.3 Small-scale tests and classifications	85
3.7 Water spray protection of exposed structures	86
References	87
4 Smoke isolation and venting	**91**
4.1 Isolation and halon suppression within ventilated equipment	91
4.2 Isolation within rooms–building smoke control	96
4.2.1 Buoyancy pressure differences	96
4.2.2 Volumetric expansion pressures	99
4.2.3 Isolation via ventilation exhaust	100
4.2.4 Upstream smoke propagation	104
4.2.5 Door and damper smoke leakage	107
4.3 Heat and smoke roof venting	107
4.4 Heat and smoke venting in sprinklered buildings	112
4.4.1 Testing	112
4.4.2 Loss experience	113
4.4.3 Mathematical modeling	113
4.4.4 Closing remarks	114
References	114
5 Warehouse storage	**117**
5.1 Warehouse fire losses	117
5.2 Storage configurations	118
5.3 Effect of storage height, flue space, and aisle width	124
5.4 Commodity effects	128
5.4.1 Generic commodity classification	128
5.4.2 Laboratory flammability testing	132
5.4.3 Small array tests	135
5.4.4 Large array sprinklered fire tests	145

5.5 Sprinkler flow rate requirements	148
5.5.1 Ceiling spray sprinklers	149
5.5.2 In-rack sprinklers	157
5.5.3 Early suppression fast response (ESFR) sprinklers	158
5.6 Sprinklered warehouse fire modeling	159
5.6.1 Conceptual model overview	159
5.6.2 Free burn heat release rates and flame spread rates	159
5.6.3 Warehouse fire plumes and ceiling jets	159
5.6.4 Sprinkler actuation model	162
5.6.5 Spray-plume penetration model	163
5.6.6 Reduction in heat release due to actual delivered density	164
5.6.7 Fire control criteria: can wetted commodity be ignited?	165
5.6.8 Fire suppression criteria	166
5.7 Cold storage warehouse fire protection	167
References	168
6 Storage of special commodities and bulk materials	**171**
6.1 Roll paper	171
6.1.1 Commodity description	171
6.1.2 Loss experience	173
6.1.3 Roll paper fire tests	173
6.1.4 Roll paper protection requirements	177
6.2 Nonwoven roll goods	178
6.2.1 Commodity description	178
6.2.2 Loss experience	179
6.2.3 Fire tests	179
6.2.4 Sprinkler protection requirements for nonwovens	181
6.3 Rubber tire storage	181
6.4 Aerosol products	184
6.4.1 Product description	184
6.4.2 Aerosol warehouse fires	185
6.4.3 Aerosol product formulation effects	186
6.4.4 Sprinkler protection guidelines	188
6.5 Solid oxidizers	188
6.6 Bulk storage	191
6.6.1 General description	191
6.6.2 Spontaneous ignition testing	192
6.6.3 Spontaneous ignition theory	192
6.6.4 Detection and suppression of bulk storage fires	196
References	198
7 Flammable liquid ignitability and extinguishability	**201**
7.1 Incident data	201
7.2 Ignitability temperatures	202
7.2.1 Flash points and fire points	202
7.2.2 Autoignition temperatures	205
7.2.3 Time to reach fire point	205
7.3 Electrostatic ignitability	209

- 7.4 Pool and spill fire heat release rates — 215
  - 7.4.1 Confined pool fires — 215
  - 7.4.2 Unconfined spill fires — 217
- 7.5 Spray fires — 219
- 7.6 Water spray extinguishment — 222
  - 7.6.1 High flash point liquids — 224
  - 7.6.2 Water miscible liquids — 226
  - 7.6.3 Low flashpoint liquids — 227
  - 7.6.4 Spray fires — 228
- 7.7 Foam extinguishment — 230
  - 7.7.1 Low Expansion Foam — 230
  - 7.7.2 Medium and high expansion foam — 234
- 7.8 Dry chemical and twin agent extinguishment — 234
- 7.9 Carbon dioxide suppression — 236
- 7.10 Halon replacement suppression agents — 237
- References — 238

8 Flammable liquid storage — 243

- 8.1 Storage tanks — 243
  - 8.1.1 Generic tank designs — 243
  - 8.1.2 Storage tank loss history and fire scenarios — 247
  - 8.1.3 Tank burning rates and spacing criteria — 251
  - 8.1.4 Tank emergency venting — 256
  - 8.1.5 Tank fire suppression — 266
  - 8.1.6 Portable tanks and intermediate bulk containers — 267
- 8.2 Drum storage — 268
  - 8.2.1 Drum designs and storage modes — 268
  - 8.2.2 Loss experience and fire scenarios — 270
  - 8.2.3 Drum failure times and failure modes — 271
  - 8.2.4 Fire suppression systems for drum storage — 276
- 8.3 Flammable liquids in small containers — 279
  - 8.3.1 Container types — 279
  - 8.3.2 Loss experience — 281
  - 8.3.3 Container failure times and failure modes — 282
  - 8.3.4 Sprinkler protection for flammable liquids in small containers — 285
- References — 293

9 Electrical cables and equipment — 297

- 9.1 Electrical cables: generic description — 297
- 9.2 Cable fire incidents — 300
- 9.3 Cable flammability testing and classifications — 304
- 9.4 Vertical cable tray fire test data — 309
- 9.5 Horizontal cable tray fire test data — 311
- 9.6 Cable fire suppression tests — 314
  - 9.6.1 Sprinkler and water spray suppression tests — 314
  - 9.6.2 Gaseous suppression system tests — 316
- 9.7 Passive protection: coatings and wraps — 317
- 9.8 Protection guidelines and practices — 319

9.9 Electronic equipment flammability and vulnerability	322
9.9.1 Electronic component flammability	322
9.9.2 Electronic cabinet flammability	323
9.9.3 Electronic equipment vulnerability	324
9.9.4 Detection and suppression of electronic equipment fires	326
9.10 Transformer fire protection	327
9.10.1 Transformer generic description	327
9.10.2 Transformer fire scenarios	328
9.10.3 Transformer fire incidents	329
9.10.4 Installation and fire protection guidelines	332
9.10.5 Water spray protection of transformers	332
References	334
Appendix A: Flame Radiation Review	337
A.1 Flame emissive power	337
A.2 Flame height	341
A.3 Configuration factor	342
A.4 Atmospheric transmissivity	342
A.5 Point source approximation	343
References	346
Appendix B: Historic industrial fires	347
B.1 General Motors Livonia fire – August 12 1953	347
B.2 McCormick Place fire – January 16 1967	350
B.3 K MART fire – June 21 1982	351
B.4 New York Telephone Exchange fire – February 27 1975	354
B.5 Ford Cologne, Germany Warehouse fire – October 20 1977	358
B.6 Triangle Shirtwaist Company fire, N.Y.C. – March 25 1911	361
B.7 Hinsdale, Illinois Telephone Central Office Fire	363
B.8 Sandoz Basel fire	369
References	373
Appendix C: Blast Waves	375
C.1 Ideal blast waves	376
C.2 Pressure vessel ruptures	378
C.3 Vapor cloud explosions	379
C.4 Vented gas and dust explosions	379
References	380
Index	381

PREFACE

This work was made possible in part by a grant from the Society of Fire Protection Engineers Educational and Scientific Foundation.

The material in this text was compiled and presented while the author was teaching a course in Industrial Fire Protection at the Worcester Polytechnic Institute (WPI) center for Firesafety Studies. The course is intended for graduate students who have an undergraduate education in engineering or the physical sciences, and who have already studied combustion chemistry, fire dynamics, and the basics of automatic fire suppression systems. However, a keen interest in industrial fire protection and an inquisitive, analytical psyche are perfectly acceptable substitute prerequisites for readers of this text.

Neither this nor any other textbook can replace consensus codes and standards for the majority of industrial fire protection applications. On the other hand, codes and standards often do not suffice for the probing practitioner or pathfinder in industrial fire protection. Many of the author's students and his current and former colleagues at WPI and at Factory Mutual Research Corporation fall into this category. Their work and encouragement have made this text possible.

Publisher's Note: Whilst all efforts have been made to identify and contact holders of copyrighted material it is possible one or two items may not be acknowledged. If anyone is aware of an item in this book so affected the publisher would welcome their comments.

1 INTRODUCTION AND PERSPECTIVE

1.1 Engineering approach to industrial fire protection

The prevailing impression among many employees, regulators, and 'captains of industry' has been that industrial firesafety can be achieved through common sense, enforcement of prescriptive codes and standards, and guidance from local fire chiefs. Indeed, these methods should suffice in a simple workplace producing simple and unchanging products or services. However, today's industrial facilities are rarely simple and unchanging. A more effective approach to industrial firesafety is needed to deal with the complexities and changes that exist in modern industrial facilities.

The approach to industrial fire protection espoused in this text is an application and elaboration of the performance-based approach espoused by the Society of Fire Protection Engineers (SFPE, 2000). It involves engineering analyses of generic industrial hazards and fire/explosion protection measures. Engineering analyses are systematic studies incorporating pertinent scientific principles and data. In this case the pertinent science stems from combustion and fire science, as well as the relevant physical and chemical principles governing the design and operation of the fire protection systems. Relevant data includes generic fire/explosion incident data, test data, and plant-specific operating experience. This information is to be evaluated along with the applicable codes and standards, expert opinion, and a general understanding of the relevant facility, products, and/or processes.

The three steps involved in an engineering determination of suitable protection measures for an industrial fire or explosion hazard are: (1) scenario identification; (2) consequence analysis; and (3) alternative protection evaluation. Scenario identification for an actual or hypothesized fire or explosion entails a description of the pre-fire/explosion situation, ignition sources, combustible materials ignited, and surrounding materials vulnerable to fire spread and fire or smoke or blast damage. Consequence analysis entails estimates of the extent of fire spread and damage with and without the installed or proposed fire or explosion protection. If these consequences are acceptable to the facility stakeholders, then another scenario can be considered, as indicated in Figure 1.1. If the consequences are not acceptable, then alternative protection should be evaluated. Acceptance criteria depend upon the stakeholder specific objectives and the designated scope of the analysis, as well as cost constraints. Alternative protection evaluation entails estimates of the extent to which the ignition and fire/smoke/blast damage potential can be reduced by alternative installed protection measures. Figure 1.1 is a flowchart of the three steps in the analysis, and the input to the first two steps. Explanations and examples of the three steps follow.

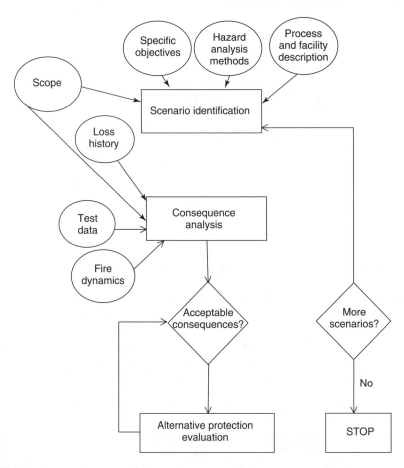

Figure 1.1. Flowchart for scenario identification, consequence analysis, and alternative protection evaluation

1.1.1 FIRE/EXPLOSION SCENARIO IDENTIFICATION

Guidelines for developing and analyzing fire scenarios are described in *Fire Safety Aspects of Polymeric Materials, Volume 4: Fire Dynamics and Scenarios* (National Academy of Sciences, 1978), and in Chapter 8 of the *SFPE Engineering Guide to Performance-Based Fire Protection* (2000). In accord with those guidelines, the following items should be identified to characterize the fire scenario through the time of self-sustained burning.

Pre-fire situation – The industrial process or operation, including start-up, maintenance, and shutdown, is the starting point in pre-fire situation considerations. A thorough review would consider potential mishaps as well as normal operations. Three techniques for systematic reviews of mishaps are the Failure Modes and Effects Analysis (FMEA), the What If Analysis, and the Hazards and Operability Study (HAZOP), described in various hazard evaluation references such as the Center for Chemical Process Safety's *Guidelines for Hazard Evaluation Procedures* (1992). Other elements to be evaluated in the pre-fire situation are: facility services (HVAC, electrical, gas, water, telecommunications, waste removal), building geometry and construction features, installed fire protection systems, and the fire protection interest and training of plant management, operating, and emergency response personnel. This background information is used in assessing the potential for fire initiation, fire spread, damage, and extinguishment.

Ignition sources – Equipment and personnel representing potential ignition sources should be identified. Quantitative descriptions to characterize the strength of the ignition source are helpful in establishing time-to-ignition and the potential effectiveness of various prevention measures. Quantification in the case of exposure fires would entail a description of flame size, heat release rate, and proximity to target material. In the case of electrical ignition sources, strength characterizations would include arc/spark/short energy or energy release rate.

Ignited material – The first material to be ignited depends on the flammability properties of target materials, their size, configuration, and proximity to the hypothesized ignition source. Quantitative ignitability criteria for a variety of combustible materials can be found, for example, in Chapter 6 of Drysdale (1985). An attempt should be made to determine if these criteria are satisfied in the hypothesized ignition scenario. Examples are presented in several chapters of this text.

Flaming or smoldering combustion – Some combustible materials may burn with or without a flame. Sometimes a material may smolder for a period of time before suddenly bursting into flame. Solids with low thermal conductivities, such as expanded plastics and rubbers, certain cable insulations, and combustible dust layers are often prone to smoldering combustion. The occurrence of smoldering versus flaming combustion is an important aspect of the fire scenario because heat release rates and flame spread rates are much lower during smoldering. Furthermore, smoke composition, particle size, and detectability differ considerably in flaming and smoldering combustion. Descriptions of fuel, ignition, and ventilation conditions conducive to smoldering combustion are available in Chapter 8 of Drysdale (1985) and in Chapter 2-11 in the *SFPE Handbook* (1995).

Fire spread and heat release rates for first ignited material – The rate of fire spread over the ignited material is a key factor determining the severity of the hypothesized fire. Data and analyses for surface flame spread rates are presented in Chapter 7 of Drysdale (1985), and in Quintiere's review in the *SFPE Handbook* (Chapter 2-14, 1995). Fire heat release rates depend both upon flame spread rates and mass burning rates as well as the effective heat of combustion (defined as the theoretical heat of combustion multiplied by the burning efficiency) of the ignited material. They are highly configuration dependent. Heat release rates for an assortment of materials and configurations have been compiled by Babrauskas in the *SFPE Handbook* (Chapter 3-1, 1995). Values for representative materials are listed in Tables A.3 and A.4 in Appendix A. Additional compilations are presented in Chapters 5 and 6 for warehouse commodities, and Chapters 7 and 8 for flammable liquids.

Fire spread to second material – Fire spread across a gap between the first ignited material and some adjacent combustible can occur via radiant or convective heating or via direct contact with burning fuel that may have dripped (for a thermoplastic polymer for example), flowed out of its container (flammable liquid), collapsed, generated windblown firebrands, or splattered as a result of ineffective extinguishment attempts. Flame radiation heat fluxes can be calculated by methods summarized in Appendix A. Fire plume and ceiling layer heat fluxes can be evaluated with engineering correlations presented in Chapters 3 and 5 and in several chapters of the *SFPE Handbook*. Target material ignition potential can be evaluated as described in Chapter 6 of Drysdale (1985), and in Chapter 7 of this text for the case of flammable liquids.

Example – A chronic fire hazard in semiconductor and electronics manufacturing plants has been the use of plastic and plastic-lined wet benches or tubs with electric immersion heaters

Figure 1.2. Clean room wet benches. © 2002 Factory Mutual Insurance Company, with permission

for heating various process liquids. These tubs, which are situated in clean rooms as shown in Figure 1.2, are used for cleaning, etching, and plating circuit boards and semiconductors in the wet bench. They often contain nonflammable liquids such as acids or precious metal solutions. Many fires have occurred when the liquid level fell below the heater while the heater remained energized in close proximity to the tub wall. The resulting rapid increase in heater surface temperature ignited the tub wall, and flame often spread to plastic exhaust hoods, ducting, and adjacent tubs. Although most of these fires have not been very extensive, the surrounding area is often susceptible to smoke damage and contamination, which usually occur before ceiling or duct sprinklers actuate. Several incidents of this nature are described in Factory Mutual Data Sheet 7-7 (1997) and in Semiconductor Safety Association papers such as Lotti (1986).

A key aspect of the pre-fire situation for this hazard is the heater and liquid-level controls. A typical arrangement would involve a liquid temperature sensor, often in the form of a high temperature limit switch, for heater control. When the liquid level falls below this sensor, it is measuring air/vapor temperatures which do not increase much above ambient room temperature, even though the heater surface temperature may be sufficiently high to ignite the tub wall. Liquid level cutoff switches are recommended in FM Data Sheet 7-7, but they are not always installed or used. More sophisticated control systems have been designed for newer heater designs. For example, Lotti (1986) describes a highly reliable special control system for immersion heaters in which the heater power supply is dependent on two series wired, normally open solid state relays attached to a controller that monitors the signals from five independent sensors: a liquid temperature sensor, a liquid level sensor, a heating element temperature sensor, an automatic timer, and a smoke or flame detector. Although this much complexity and expense may not be warranted in many facilities, fault tree analyses similar to those described in Henley and Kumamoto (1981) can be conducted to determine the incremental reliability with each additional sensor and interlock.

The ignition source in this scenario is obviously the energized heater. Quantification of this source involves specifying the heater power level and its physical size, shape, and location in the

INTRODUCTION AND PERSPECTIVE

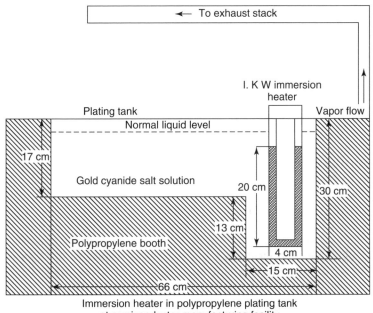

Immersion heater in polypropylene plating tank
at semiconductor manufacturing facility

Figure 1.3. Immersion heater near wall of wet bench

tub. In one representative incident, a 1 kW (0.95 Btu/sec) heater was immersed in a polypropylene plating tub shown in Figure 1.3. Vapors from the tub, which is one of several in a polypropylene booth used to process silicon wafers, are collected and exhausted through a wall duct and overhead exhaust duct. The actual heater size and location were not specified in the loss report, but we will assume the quartz heating element is 4 cm (1.6 in) in diameter, 20 cm (7.9 in) long, and situated near the end wall of the 30 cm (11.8 in) deep section of the tub.

The ignited material in this scenario is the polypropylene tub wall (or a polypropylene heater housing in some cases). According to the data in Table 5.11F of the *NFPA Fire Protection Handbook* (1986), the critical radiant heat flux for ignition of polypropylene is about 20 kW/m^2 (1.76 Btu/sec-ft^2). Since the lateral surface area of the quartz heating element is $\pi(4)(20) = 251$ cm^2 (39 in^2), the 1 kW heater can generate a radiant surface heat flux of 40 kW/m^2 (3.5 Btu/sec-ft^2). We assume that most of the heat transfer occurs via radiation to the tub walls when the liquid level falls below the bottom of the heater. The radiant heat flux on the tub wall is $\phi(40)$ kW/m^2, where ϕ is the radiation view factor between an element on the tub wall and heater. The value of ϕ corresponding to the critical heat flux for ignition is $20/40 = 0.50$. We can use the configuration factor tables in Appendix A to find the wall-to-heater separation distance D such that a wall element at the mid-elevation of the heater will receive a radiant heat flux of 20 kW/m^2. Using Figure A.2, the result is $D = 2$ cm (0.79 in). Therefore, we would expect the heater to ignite the tub wall when it is within 2 cm of the wall and remains energized for an indefinitely long time without being immersed in liquid.

The polypropylene tub fire in this scenario is most likely to produce flames with a high soot content. According to Table 5.11F of the *NFPA Handbook* (1986), the soot yield, Y_{smoke}, for a well ventilated polypropylene fire is about 0.08 g-soot/g-fuel. The heat release rate for a small polypropylene wall fire can be estimated from extrapolation of the data in Table A.3. The result is about 76 kW per m width (22 Btu/sec) per ft width of wall for a 30 cm (1 ft) high wall burning in the open with unrestricted air access. The smoke release rate, which can be considered to be the source term for smoke damage analyses, can be estimated by multiplying the mass burning rate (heat release rate divided by the product of heat of combustion and combustion efficiency) by the soot yield, i.e.

$$\dot{m}_{smoke} = \left(\frac{\dot{Q}}{\Delta H_c \chi}\right) Y_{smoke}$$

Using the values for ΔH_c and χ for polypropylene in Appendix A (43.4 kJ/g [18,700 Btu/lb] and 0.89, respectively) and the previous values for heat release rate and soot yield. The estimated smoke release rate is about 0.16 g/sec per m (1.07×10^{-4} lb/sec-ft) of tub wall burning. However, since this is based on laboratory tests with well ventilated samples burning, whereas the tub fire is probably not well ventilated, the estimate should only be considered accurate only to within an order-of-magnitude. Large-scale wet bench test data would be far preferable. Some tests of this nature are described in Chapter 4, but smoke generation rate data were apparently not measured.

As for the second material ignited, in this example it might well have been the plastic exhaust duct at the top of the tub, since the duct was probably directly exposed to substantial convective heating from the fire plume. If there is a strong downward ventilation flow over the wet bench, and exhaust ducting in the bench interior, the second material ignited can be the bench interior. Tests conducted with this wet bench configuration are described in Chapter 4.

Explosion scenarios – Combustion explosions (i.e. gas and dust explosions) are amenable to a scenario identification analysis directly analogous to that for fires. The pre-explosion situation should include a description of the various factors that influence the likelihood of ignition and the subsequent course of the explosion. Besides identifying the gas or dust to be ignited, some rough estimates of possible gas or dust concentration relative to the lower explosive limit (as given in Chapter 3-16 of the *SFPE Handbook*, for example) and relative to the stoichiometric or worst-case concentration (concentration producing the maximum rate of pressure rise) should be determined. As for the rate of flame spread and the heat release rate, these parameters are every bit as important for explosions as they are for fires. One complicating factor unique to combustion explosions is the effect of turbulence on flame speeds and associated energy release rates. Unfortunately, it is difficult to quantify the turbulence effect without access to large-scale test data for a relevant explosion configuration. A thorough discussion of these effects is beyond the scope of this text. The reader is referred to Chapter 3-16 of the *SFPE Handbook* (1995), and to references therein for a quantitative introduction to explosion protection evaluations.

1.1.2 CONSEQUENCE ANALYSIS

The consequences of a particular fire scenario can in principle be determined from knowledge of fire heat and smoke production rates and estimates of the extent of flame spread. Manual and automatic suppression system effectiveness in the given scenario should be assessed to estimate the extent of flame spread. When this assessment is conducted assuming installed suppression systems are functioning as designed, the resulting damage estimate (measured in dollars) at an industrial facility is called the *Loss Expectancy*. When the equivalent consequence analysis is conducted assuming installed suppression systems are not operational, the resulting financial damage estimate is called the *Maximum Foreseeable Loss*. Estimates of the loss expectancy and the maximum foreseeable loss are critical parameters needed by corporate risk managers and insurers.

Damages for industrial fires and explosions generally include property damage, injuries and fatalities, business interruption losses, and possible environmental damages. Traditionally, expected damages have been determined from loss history, i.e. from compilations of previous incidents in similar occupancies. However, loss expectancies will vary with the development

INTRODUCTION AND PERSPECTIVE 7

of new products, processes, and fire protection technology. Furthermore, maximum foreseeable losses are rare occurrences.

Two alternatives to historical loss data are full-scale testing and theoretical analyses. Since full-scale testing is often expensive and logistically prohibitive (especially for personnel injury and environmental damage evaluations), there is considerable incentive to proceed with theoretical analyses, particularly for 'what-if' variations in the basic fire scenario. Theoretical analyses sometimes involve deterministic evaluations, and other times utilize risk analysis (probabilistic) methodology. A brief explanation of how these methods can be used for consequence analyses is provided here with application to the example fire scenario of the semiconductor plant wet bench fire.

Property damage – Property damage caused by thermal loads imposed during a fire can be estimated through an assortment of analytical methods. Thermal loads generally consist of flame radiation heat fluxes (Appendix A) and convective heat fluxes associated with the fire plume and the hot ceiling jet (Chapters 3 and 5). Structural member fire resistance under these thermal loads (prior to or without sprinkler activation) can be estimated via methods described in Chapter 3. Thermal damage to equipment and products can easily be calculated with similar analyses if the equipment/product is a simple configuration incorporating materials with known thermal properties and damage threshold temperatures. For example, plastic equipment/products may satisfy this criterion where the damage threshold temperature is associated with softening (thermoplastics), decomposition (thermosets) or ignition. Since these damage temperatures depend upon specific polymer composition, small-scale test data are needed for pertinent polymers and blends. Fortunately, standardized testing is being conducted for several of the more commonly used industrial equipment components, such as grouped electrical cables and plastic exhaust ducting.

Nonthermal damage due to smoke, water, and corrosive chemicals released during a fire cannot currently be calculated with any deterministic analytical methodology. As with thermal damage, laboratory scale tests of nonthermal damage to various industrial equipment and components are beginning and data may be available in a few years to incorporate into damage estimates. In the meantime, historical loss data is the primary source of information for nonthermal damage.

Example – In the previous example of the polypropylene tub fire in a semiconductor plant, there was substantial thermal and nonthermal damage. Thermal damage occurred to electrical equipment, the exhaust ducting, and the plating booth in which the tub was situated. Approximately 200 silicon wafers in and around the booth were damaged and had to be discarded. Nearby electronic equipment was subjected to water damage from the two sprinkler heads that opened and controlled the fire. Widespread smoke damage occurred throughout the clean room. The total property damage was on the order of $1 million in this incident. Another incident, described in FM Data Sheet 7-6 (1986), involved an immersion heater igniting an empty hydrochloric acid tank and producing thermal to a structural steel roof to the extent of $800,000. A 1995 wet bench fire summarized in FM Data Sheet 7-7 (1997) contaminated 2300 in-process silicon wafers, caused approximately $12 million in damages, and required 74 delays to achieve pre-fire production levels. However, most incidents result in much less damage, with negligible thermal or nonthermal damage beyond the room since the exhaust system is usually equipped with multiple filters. According to FM Data Sheet 7-6 (1986), the average property damage in this type of plastic tank fire is about $63,000.

Injuries and fatalities – Fire injuries and fatalities are due to toxic combustion products as well as burns. Deterministic methods are available for calculating the extent of the danger zone associated with toxicity and thermal effects. These methods are discussed in Section 2.1.1. However, casualty estimates are not usually amenable to an entirely deterministic methodology because of uncertainties associated with transient exposure toxicity effects and human response to a developing fire. Engineering risk analyses incorporating relevant statistical and probabilistic data can be utilized as shown for example by Boykin *et al.* (1986) for transformer fires in occupied buildings. A similar approach, but with the addition of more complete fire and toxicity models, has been used in a National Fire Protection Research Foundation study to analyze casualties due to hypothesized fires in homes, offices, hotels, and restaurants (Gann *et al.*, 1991).

Example – Since the immersion heater fire scenario usually occurs in an unattended facility and does not involve a very rapid flame spread or extremely toxic combustion products, there are very rarely any serious injuries or fatalities. However, there have been several incidents in which fear of exposure to combustion or decomposition products from unknown chemicals in the tanks/tubs did significantly delay manual suppression of the fire. These fears can be alleviated (or accommodated when warranted) through emergency response planning, drills, etc., with the local fire department.

Business interruption – Business interruption losses refer to lost income incurred due to fire damage. Estimates of business interruption losses are based on the anticipated down time of the plant or process. It is inevitably a plant-specific determination best made by or in consultation with a professional adjuster. In the case of modern semiconductor fabrication facilities with a large backlog of orders to fill, business interruption losses can exceed property damage.

Environmental damages – Fire damage to the environment can occur either through airborne smoke and toxic combustion product deposition on vulnerable plants and animals, or through contaminated water runoff into a nearby river or lake. Although it is very difficult to quantify this type of loss, it warrants serious qualitative consideration. Its importance in pesticide and other toxic chemical fire scenarios should be obvious, but it can also be a major factor in other scenarios such as the August 1984 fire in a London warehouse complex (Kort, 1985) containing 318,000 kg (350 tons) of cocoa butter, among various other commodities. The cocoa butter melted and burned during the fire and flowed out of the warehouse and into the city storm sewers which empty into the London canal network. According to Kort (1985), the cocoa butter quickly killed most of the fish in the canals. Another unanticipated environmental effect from this fire was the widespread asbestos contamination originating from a layer of asbestos cement in the warehouse roof.

Perhaps the most environmentally notorious fire was the November 1986 Basel, Switzerland chemical warehouse fire in which toxic chemicals were washed into the Rhine River and destroyed most downstream aquatic life in France, Germany, and the Netherlands. In the aftermath of the fire, a huge holding tank for firefighting water runoff has been constructed in Basel to avoid future environmental catastrophes of this nature.

Explosion damage – Explosion damage mechanisms include pressure (and pressure-time impulse) damage and projectile/shrapnel damage. Explosion pressures within the enclosure containing the combustible gas or dust can be calculated using the methods described in Chapter 3-16 of the *SFPE Handbook* (1995). Blast wave pressures associated with commercial explosives and pressure vessel failures can be estimated from the methods described in Appendix C. These methods are also commonly used to estimate vapor cloud explosion pressures outside the vapor cloud itself, providing there is a credible basis for estimating the quantity of flammable vapor in the cloud and the effective blast yield (fraction of combustion energy released in the form of blast waves). Projectile-induced damage estimates are more questionable, but techniques described in Baker *et al.* (1983) and Army Technical Manual TM5-1300 (1991) are available.

1.1.3 ALTERNATIVE PROTECTION EVALUATION

When the consequence analysis suggests the fire or explosion can cause an unacceptably large loss, it is necessary to evaluate alternative protection. The potentially more cost-effective alternatives fall into the two broad categories of preventive measures and damage control measures. Preventive measure alternatives consist of less flammable materials and modifications to eliminate or reduce the frequency of ignition sources. Alternative damage control measures include improved detection, suppression, and smoke control systems, more fire resistant materials and structures, and improved drainage of water runoff.

A systematic comparison of these alternative protection measures sometimes entails estimating failure/success probabilities and using fault tree and event tree analyses such as those described in Henley and Kumamoto (1981) and the AIChE Guidelines for Chemical Process Quantitative Risk Analysis (1989). In other cases, such as the following example, it may be more appropriate to conduct deterministic analyses and/or fire testing.

Example – Returning to the wet bench electric heater fire scenario, several alternative preventive measures are possible. One modification could be the use of a less easily ignitable plastic in place of polypropylene. Data in Table 5.11F of the *NFPA Handbook* (1986) indicate that polytetrafluoroethylene (Teflon) and some rigid phenolics (probably with fire retardant additives) have critical radiant heat fluxes for ignition that are approximately twice as large as that for polypropylene. Several new halogen free polymers that are beginning to be produced commercially are also very resistant to ignition. Factory Mutual Research Corporation has recently developed a Flammability Test Protocol for clean room materials (1997) that includes time-to-ignition measurements at surface heat fluxes greater than the critical heat flux.

Alternatively, a less powerful heater or more reliable heater controls could be used as described previously. Another preventive measure suggested in the FM Data Sheet and from Lotti (1986) is the replacement of the electric heater with steam immersion coils. Factory Mutual Data Sheet 7-7 recommends that electric immersion heaters be replaced with other heating methods employing aqueous, or other noncombustible, heat transfer fluids.

Several polymers have significantly lower heat release rates and smoke release rates than polypropylene. For example, the data in Appendix A indicate that the heat release rate for polymethyl methacrylate (PMMA) is at least 30% lower than that of polypropylene for similar configurations. Using Equation [1.1] and Appendix A values for PMMA, the estimated smoke release rate for a PMMA tub wall fire would only be about 0.026 g-soot/sec per m (1.74×10^{-5} Btu/sec-ft) wall width, i.e. about one sixth of the corresponding value for polypropylene.

The FMRC Clean Room Materials Flammability Test Protocol includes heat release rate tests with 10 cm (4 inch) wide test samples exposed to an oxygen enriched atmosphere and an imposed heat flux at the sample bottom. A Fire Propagation Index is calculated from the heat release rate data and time-to-ignition test data. A Smoke Damage Index is calculated from the soot yield and Fire Propagation Index, and a Corrosion Damage Index is calculated from the yield of corrosive combustion products. Materials with sufficiently low values of these indices are listed as FMRC Clean Room Materials that do not, of themselves, require fixed fire detection or suppression equipment. Two materials, a fire retardant polypropylene and a fire retardant polyvinyl chloride, were listed in the 1998 FMRC Approval Guide as having satisfied the Clean Room Materials Flammability Test Protocol.

If low flammability bench materials are not utilized, a damage control alternative is the use of a smoke detector and suppression system for the bench itself. The detector signal could de-energize the heater and actuate an alarm as well as initiate the discharge of suppression agent. Sophisticated fiber optic laser beam smoke detectors, and air sampling systems combined with light scattering detectors are also being utilized in some clean rooms. Detector and gaseous suppression system design considerations for this type of application are discussed in Chapter 4 as well as in FM Data Sheet 7-7. Portable carbon dioxide or other gaseous agent extinguishers placed near the tanks would also be useful for constantly attended operations.

1.2 Statistical overview of industrial fires and explosions

The National Fire Protection Association (NFPA) compiles statistics on fires reported to public fire departments in the United States, and provides a statistical overview each year in the NFPA Journal. Occupancy categories correspond to the categories and groupings in the NFPA 901 reporting form. The groupings that might be considered 'industrial' are basic industry, utilities, manufacturing, and many of the storage properties. NFPA statistics for 1990 (Karter, 1991) indicate that there were 22,000 fires in the combined categories of basic industry, utilities and manufacturing, and an additional 39,500 fires in storage properties. These figures neglect the numerous fires that are not reported to the public fire department because they are extinguished by plant personnel or fixed suppression systems.

The US Occupational Safety and Health Administration (OSHA) maintains a database of workplace injuries. Their data for 1998 (Bureau of Labor Statistics, 2000) indicate that there were 4152 non-fatal workplace injuries due to fires, and another 1670 occupational injuries due to explosions. In addition, there were 206 fatalities in 1998 due to fires and explosions in US workplaces. These represent slightly less than 4% of all the workplace fatalities, most of which are due to vehicle accidents.

1.2.1 INDUSTRIAL OCCUPANCIES IN LARGE LOSS FIRES

Table 1.1 is a listing of US industrial fires and explosions with property damage of at least $30 million in 1990 dollars. The exact dollar amounts are not necessarily accurate because they are based on published data often from preliminary, unofficial estimates. Nevertheless, the tabulation of 64 incidents provides an overview of the types of facilities and events that have been responsible for the most costly US industrial fires and explosions.

Seventeen of these largest losses (27%) occurred in warehouses, with about half of these being used primarily for paper, plastic, or general commodity storage. There are also several flammable liquid warehouses, and several other cold storage warehouses. One common aspect of these warehouse fires was the failure of the installed sprinkler system to adequately control the fire. Warehouse storage sprinkler protection is discussed in Chapter 5.

Twelve of the largest losses (19%) occurred in petroleum refineries. The large quantities and concentrations of flammable liquids and vapors at these refineries provide both the potential and the realization of some very large fires and explosions. Property damage is often exacerbated by the high replacement costs of the sophisticated process equipment.

Other industrial occupancies with multiple occurrences in Table 1.1 are power plants (5 incidents), chemical plants (5), grain elevators (3), textile plants (2), telephone exchanges (2), ink manufacturing facility (2), and aluminum plants (2).

The British Fire Protection Association publishes annual compilations of losses in the United Kingdom causing at least £1 million. Data for the period 1985–86 indicate that the types of industrial facilities with the largest numbers of large-loss fires are general warehouses (13 incidents), textile mills (12), and the combined category of wood, furniture, paper, and printing plants (12). The relatively large number of textile mill fires is probably a manifestation of the large number of textile mills still in operation in the UK. The three largest industrial fire losses in this period were a £10 million ($15 million) fire at a general warehouse, a £9 million fire at a carpet warehouse, and a £8 million fire at a small tool manufacturing plant.

Perhaps the most costly industrial loss in recent history was the 1988 gas explosion and fire on the Piper Alpha oil production platform in the North Sea. According to a recent compilation of disasters by Swiss Re (1998), the insured loss due to the Piper Alpha catastrophe is estimated to be $2759 million. It also was responsible for the death of 167 workers. Extensive research into improved fire and explosion protection for offshore platforms was initiated soon after the Piper Alpha catastrophe.

Table 1.1. Largest US industrial fire and explosion losses sorted by property damage

Year	Location	Facility	Type of fire or explosion	Property damage[a] ($ million)	(1990 $ million)[a]
1989	Pasadena, TX	Chemical Plant	Vapor Cloud Explosion	750	758
1947	Texas City, TX	Ship & Chem Plant	Ammonium Nitrate Explosion	67	558
1999	Dearborn, MI	Powerhouse	Gas and Dust Explosions	650	540
1975	Browns Ferry, AL	Nuclear Power Plant	Electrical Cable Fire	227	507
1995	Lawrence, MA	Textile Plant	Flock Dust Explosion and Fire	500	427
1988	Norco, LA	Petroleum Refinery	Vapor Cloud Explosion	330	347
1953	Livonia, MI	Auto Transmission Plant	Flammable Liquid Fire	50	319
1999	The Grammercy Works, LA	Aluminum Plant	Pressure Vessel Bursts due to Steam Overpressurization	300	249
1996	LA	General Storage Warehouse	Incendiary Fire	280	233
1999	Richmond, CA	Oil Refinery	Gas/Vapor Explosion	247	205
1970	Linden, NJ	Oil Refinery	Pressure Vessel BLEVEs	50	175
1987	Pampa, TX	Chemical Plant	Vapor Cloud Explosion	160	173
1995	GA	Carpet Manufacturer	Oil Pool Fire	200	171
1969	Rocky Flats, CO	Nuclear Weapons Plant	Hydrogen Fire & Explosion	45	171
1999	Missouri	Power Plant	Gas Explosion	196	163
1984	Tinker AFB, OK	Jet Engine Repair	Roofing Fire	138	160
1985	Elizabeth, NJ	Warehouse	Aerosol Storage Fire	123	139
1975	New York, NY	Telephone Exchange	Electrical Cable Fire	60	134
1982	Falls Township, PA	General Warehouse	Aerosol Storage Fire	100	126
1984	Romeoville, IL	Oil Refinery	Vapor Cloud Explosion	100	116
1988	Henderson, NV	Rocket Propellant Plant	Ammonium Perchlorate Explosion	103	108
1973	Chicago, IL	Ink Manufacturer	Warehouse Storage Fire	37	104
1994	TX	Methanol Production Plant	Explosion?	116	103
1991	Madison, WI	Cold Warehouse	Food Storage Fire	100	100
1992	Missouri	Cold Storage Warehouse	Polystyrene Foam Insulation Fire	100	96.6

(continued overleaf)

12 INDUSTRIAL FIRE PROTECTION ENGINEERING

Table 1.1. (*continued*)

Year	Location	Facility	Type of fire or explosion	Property damage[a] ($ million)	(1990 $ million)[a]
1976	Galena Park, TX	Grain Elevator	Grain Dust Explosion	42	89
1995	TX	Industrial Chemical Storage	Under Investigation	100	85
1988	Los Angeles, CA	General Warehouse	Warehouse Storage Fire	80	84
1977	Fairbanks, AK	Oil/Gas Pumping Station	Gas Explosion	40	79
1980	Borger, TX	Oil Refinery	Vapor Cloud Explosion	50	75
1996	NJ	General Storage Warehouse	Under Investigation	84.5	70
1989	Baton Rouge, LA	Petroleum Refinery	Vapor Cloud Explosion	68.9	70
1998	IN	Plastic Warehouse	Warehouse Storage Fire	85	68
1980	New Castle, DE	Chemical Plant	Vapor Cloud explosion	45	67
1994	WV	Steel Mill	Structure Fire Due To Hydraulic Oil Spray	75	66
1993	LA	Petroleum Refinery	Refinery Coker Unit Fire	65.2	62
1998	KS	Grain Elevator Facility	Grain Dust Explosion	75	60
1992	Washington	Cold Storage Warehouse	Roof Insulation Fire	62	60
1978	Texas City, TX	Refinery	Pressure Vessel BLEVEs	32	60
1977	Westwego, LA	Grain Elevator	Grain Dust Explosion	30	59
1999	Martinez, CA	Oil Refinery	Naphtha Explosion and Fire	71	63
1983	Avon, CA	Oil Refinery	Flammable Slurry Fire	49	58
1986	Bismark, ND	Power Plant	?	50	55
1981	Louisville, KY	Soybean Processing Plant and Municipal Sewer	Flammable Vapor Explosion	40	55
1978	Bensonville, IL	General Warehouse	Warehouse Storage Fire	30	55
1995	NY	Paper Products Manufacturer	Paper storage Fire Resulting From Electrical Malfunction	62	53

Year	Location	Facility	Event	Loss
1988	Hinsdale, IL	Telephone Exchange	Electric Cable Fire	50
1987	Dayton, OH	Paint Warehouse	Warehouse Storage Fire	49
1979	Edison, NJ	General Warehouse	Aerosol Storage Fire	30
1993	LA	Petroleum Refinery	Refinery Coker Unit Fire	50
1996	WA	Plywood Manufacturing Plant	?	50
1996	IL	Paper Records Storage	Warehouse Storage Fire	50
1993	TX	Paper/plastic Warehouse	Warehouse Storage Fire	45
1989	Louisiana	Oil and Gas Drilling Platform	Natural Gas Fire	40
1995	VA	Steam Power Plant	Fuel Line Fire Resulting From Faulty Turbine	45
1996	NE	Sugar Refinery	Sugar Dust Explosion	44
1995	Rouseville, PA	Refinery	HC Fire	40
1994	IN	Aluminum Fabrication Plant	?	40
1990	Louisiana	Wharf Warehouse	Baled Rubber Storage Fire	34
1992	Virginia	Nylon Fiber Manufacturing Plant	Nylon fluff and finishing oil fire	34
1992	Texas	Petrochemical	Flammable Liquid Fire	32.3
1989	Maryland	Cold Storage Warehouse	Food Storage Fire	30
1990	Colorado	Combustible Liquid Tank Farm	Aircraft Fuel Fire	30
1990	Ohio	Ink and Dye Manufacturing Plant	Combustible Vapor Explosion and Fire	30

				53
				53
				52
				45
				42
				42
				41
				40.7
				38
				37
				36.2
				35
				34
				33
				31
				30.5
				30
				30

[a] Property damage based on data in *NFPA Fire Journal* annual compilations of large loss fires and on data in the J&H Marsh & McLennan thirty-year review (1998), with conversion of reported property damage to 1990 $ using FM Data Sheet 9-3. Supplemental data was also obtained from Hoyle *et al.* (2000)

Data on industrial fire losses in most other countries are more difficult to obtain, at least in English. One good source is the Marsh & McLennan 1998 thirty-year review of chemical and petroleum plant losses. It lists many refinery and gas processing plant explosions around the world with losses in excess of $100 million. Two other noteworthy losses in this category are the 1997 fire at a semiconductor fabrication facility in Taiwan (loss estimated to be $420 million), and the 1994 fire at a particularly large Russian vehicle assembly plant (loss estimated to be approximately $1000 million).

1.2.2 TYPES OF FIRES/EXPLOSIONS IN THE LARGEST LOSSES

The types of fires and explosions listed in Table 1.1 provide an overview of the types of challenges faced by fire protection engineers at industrial facilities. Eleven of the incidents (17%) involved flammable liquid fires. Flammable liquid ignitability, heat release rates in typical fire scenarios, and extinguishability are discussed in Chapter 7. Two of the historic fires described in Appendix B (the GM Livonia fire and the Sandoz Basel fire) were inherently flammable liquid fires.

Eight of the fires (12.5%) listed in Table 1.1 were plastic storage fires, which are described in Chapter 5. Appendix B includes a specific account of the Ford Cologne, Germany warehouse fire.

Six incidents (9%) in Table 1.1 involved dust explosions. Dust explosions are often followed by extensive fires, such as the Lawrence, Massachusetts textile plant fire shown in Plate 1. Five incidents (8%) were vapor cloud explosions. The only aspects of explosion protection discussed in this book are explosion prevention and blast wave calculations. An overview of explosion protection is provided in one chapter of the *SFPE Handbook* (Zalosh, 2001). A more comprehensive discussion of dust explosions can be found in Eckhoff (1997), and a more complete discussion of vapor cloud explosions is provided in the 1994 Guidelines published by the Center for Chemical Process Safety.

Other multiple incidents listed in Table 1.1 are gas explosions (3 incidents), electrical cable fires (3), aerosol product storage fires (3), gas fires (3), flammable vapor explosions (3), liquid container fires (2), Boiling Liquid Expanding Vapor Explosions (2 BLEVEs), and thermal insulation fires in cold warehouses (2). Some of these special types of fire scenarios are described in this text, whereas the explosion scenarios are covered in the previously cited explosion references.

1.2.3 FACILITIES INVOLVED IN MULTIPLE FATALITY FIRES AND EXPLOSIONS

Table 1.2 is a tabulation of 36 reported worldwide industrial fires and explosions with at least 20 fatalities during the period 1981–2000. In culling through the references cited at the end of the table, transportation accidents, mining disasters, and military ammunition incidents were omitted. Other omissions involved incomplete reports in which the site of the incident may or may not have been an industrial facility. When different references reported differing fatality numbers, both numbers are listed in the table.

Ten of the 32 catastrophes listed in Table 1.2 occurred in petrochemical plants such as refineries. These were explosions, flash fires, BLEVEs, or flowing liquid fires. The 726–967 total fatalities in these incidents were due to inadequate warning and no time for employees and other victims to escape. In at least one of these catastrophes (the 1984 disaster in a suburb of Mexico City), the majority of the victims were residents in a densely populated poor community immediately adjacent to a large petrochemical facility.

Nine of the catastrophes listed in Table 1.2 occurred in various types of factories. The majority of the 549–606 fatalities in these cases were primarily due to inaccessible or inadequate exits and egress paths. In the Bangladesh fire and the Chinese rainwear factory fire, workers were trapped on the upper floor and roof, while in the North Carolina chicken plant fire many workers were

Table 1.2. Industrial fires and explosions with more than twenty fatalities (1981–2000)

Date	Location	Plant	Type of fire/explosion	Fatalities (reference)
December 18–21, 1982	Caracas, Venezuela	Fuel Tank Farm	Fuel Oil Tank Boilover Fire	128[a] 153[d]
February 25, 1984	Sao Paulo, Brazil	Gasoline Pipeline	Gasoline Fire	89[d] 508[b]
August 16, 1984	Rio de Janeiro, Brazil	Gas Production Platform	Gas Fire	36[d]
November 19, 1984	Mexico City, Mexico	PEMEX LP-Gas Plant	BLEVEs	334[a] 550[b]
May 19, 1985	Priola, Italy	Petrochemical Plant	Ethylene Explosion	23[b]
June 25, 1985	Jennings, Oklahoma	Fireworks Factory	Fireworks Explosion	21[a]
April 26, 1986	Chernobyl, Ukraine	Nuclear Power Plant	Reactor Explosion/Fire	31[c]
1987	Grangemouth, UK	Petrochemical Plant	Hydrocarbon Explosion/Fire	67[b]
July 6, 1988	North Sea Platform	Piper Alpha Oil/Gas	Gas Explosions & Fires	167[a,b]
October 22, 1988	Shanghai, China	Oil Refinery	Petrochemical Explosion	25[d]
November 9, 1988	Maharashtra, India	Petrochemical Plant	Naphtha Fire	25[b]
October 23, 1989	Pasadema, Texas, USA	Petrochemical Plant	Ethylene Vapor Cloud Explo	23[a,b]
1989	Antwerp, Belgium	Chemical Plant	Aldehyde Explosion	32[b]
November 5, 1990	Maharashtra, India	Petrochemical Plant	Ethane/Propane Fire/Explo	35[b]
December 27, 1990	Dhaka, Bangladesh	Garment Factory	Garment Fire	23[a]
January 5, 1991	Guatemala	Geothermal Power Plant	?	21[a]
May 30, 1991	Dongguang, China	Rainwear Factory	Textile Fire	66[a] 71[d]
June 6, 1991	Brazil	Fireworks Factory	Fireworks Explosion	24[a]
July 12, 1991	Meenampatti, India	Fireworks Factory	Fireworks Explosion	38[a]
September 3, 1991	Hamlet, N Carolina, USA	Food Processing Plant	Hydraulic Oil Spray Fire	25[d]
December 10, 1991	Albania	Food Warehouse	Warehouse Fire	≥60[a]

(continued overleaf)

Table 1.2. (continued)

Date	Location	Plant	Type of fire/explosion	Fatalities (reference)
December 15, 1991	India	Cotton Mill	Textile Fire?	≥ 22[e]
April 22, 1992	Guadalajara, Mexico	Unspecified Factory	Hexane Vapor Explosion in Sewers	170[a]
April 29, 1992	New Delhi, India	Warehouse	Unspecified Chemical Explosion	43[d]
May 10, 1993	Bangkok, Thailand	Toy Factory	Plastics Fire	88[a] 240[d]
November 19, 1993	Kuiyong, China	Doll Factory	Unspecified Fire	81[d]
June 17, 1994	Zhuhai, China	Textile Mill	Unspecified Fire	76[d]
June 29, 1996	Piya, China	Unspecified Factory	Unspecified Explosion	36[d]
September 14, 1997	Visag, India	Petroleum Refinery	LPG Tank Fire & Explosion	34[d] 60[b]
September 20, 1997	Jin Jiang, China	Shoe Factory	Unspecified Fire	32[d]
December 11, 1998	Brazil	Fireworks Factory	Fireworks Explosion	41[a]
May 13, 2000	Enschede, Netherlands	Fireworks Factory	Fireworks Explosion	21[a]

[a]From individual reports on these fires
[b]From Table 9 of Khan and Abbasi, *J. Loss Prevention*, 1999
[c]There were 28 fatalities due to acute radiation sickness, and three others from non-radiation causes within the three months following the Chernobyl accident (IAEA, 1996)
[d]From UN Environmental Program Selected Accidents Involving Hazardous Substances http://www.unepie.org/apell/accident.html
[e]From Loss Prevention Association of India compilation, 1992

trapped because of an exit door locked from the outside. There were inevitably many other fire protection deficiencies in these nine factory fires, but details were not available for most of these incidents.

Two of the Table 1.2 catastrophes occurred on oil/gas production platforms, including Piper Alpha. The 203 reported fatalities in these cases were due to a combination of the factors described above, i.e. inadequate warning and time to escape, plus being isolated on the burning platform. Defects in installed fire and explosion protection were also important factors.

Five of the catastrophes were explosions in fireworks factories. In addition to a failure of preventive measures, these explosions involved failures to adequately isolate the inherently hazardous fireworks manufacturing operations and limit the quantity of explosive material. At least two of these five fireworks factories were situated in or immediately adjacent to densely populated towns.

Perhaps the best-known and most widely reported catastrophe cited in Table 1.2 is the Chernobyl nuclear reactor explosion and subsequent fire. Of the 31 reported near-term fatalities at Chernobyl, 28 were due to acute radiation sickness, mostly incurred during attempts to control the fire and install a cocoon around the reactor. Among other things, the Chernobyl disaster illustrates the need for adequate protection for plant fire brigades and other emergency response personnel dealing with special hazards.

There were two incidents in which the majority of the fatalities occurred far from the industrial plant site. In one case, a flammable liquid spill from a factory into the sewer system caused a subsequent series of explosions in a large section of the sewer system. Another case involved a gasoline pipeline in an industrial area in Brazil. Both these incidents are representative of many others that may not have originated in industrial facilities, but illustrate the need for awareness of the various fire and explosion scenarios associated with flammable liquid pipelines and discharges from industrial facilities.

1.2.4 IGNITION SOURCES

Figure 1.4 shows the most prevalent ignition source categories for large loss fires in manufacturing facilities and storage facilities as reported in the NFPA Journal annual compilations of large-loss fires. Electrical ignition sources are responsible for the greatest number (25%) of large loss manufacturing facility fires, whereas deliberately set open flame ignition sources are responsible for the greatest number (24%) of large loss storage facility fires. This is consistent with the preponderance of energized electrical equipment in manufacturing facilities, and with the large quantities of combustibles in storage facilities being targets for arsonists. Similar statistics compiled for insured industrial properties indicate that electrical ignition sources are responsible overall for about twice as many fires and explosions as incendiary/arson fires. Arcing and shorting of fixed wiring are the major ignition mechanisms in fires started by faulty electrical equipment/wiring. Chapter 9 discusses the consequences of these electrical ignition sources that initiate fires in electrical cables, transformers, and other electrical equipment.

Other ignition sources playing significant roles in industrial fires as represented by the percentages shown in Figure 1.4 include cutting/welding operations, hot objects, fuel fired equipment, and spontaneous ignition. Cigarettes, which have responsible for a large percentage of residential fires, are a relatively infrequent ignition source (3–4%) in industrial fires.

These ignition source statistics indicate that an engineering approach to seek better preventive measures for industrial fires might best focus on reducing electrically induced ignitions. This could entail implementing more reliable electrical controls and interlocks, installing electrical cables with higher ignition temperatures (or critical heat fluxes), and/or providing greater physical separation between electrical equipment and combustible materials.

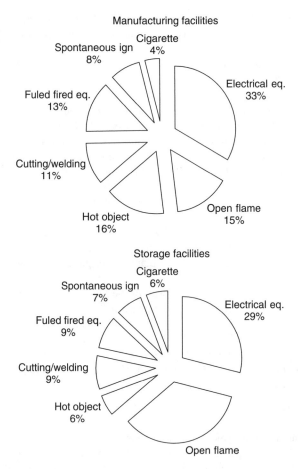

Figure 1.4. Ignition sources in industrial fires (based on NFPA large loss data excluding fires with unknown ignition sources)

1.2.5 NEED FOR AUTOMATIC DETECTION AND SUPPRESSION

Many of the loss statistics make a strong case for installing automatic detection and suppression systems to reduce losses. One indication of this is the distribution of facility operating status categories for large loss fires. As shown in the pie chart in Figure 1.5, more than half of the NFPA 1985 large loss fires were initiated with nobody on the premises. Only about 36% of these losses occurred while the plant was in full operation with a full staff available for manual detection and suppression. According to the time distribution of large loss fires given in Figure 1.6, many more fires are initiated in the interval 12:00 A.M. to 6:00 A.M. than in any other six hour period. UK large loss fires follow similar patterns with about 64% of the fires occurring at night, i.e. between 6:00 P.M. and 6:00 A.M. (Ward, 1988). During these peak periods for ignition there is a large probability that plant personnel will not be available for a prompt response to the fire. Hence, there is a clear need for automatic detection and suppression.

A more direct measure of the benefits of automatic suppression systems is the comparison of losses with and without automatic sprinkler systems available. In the NFPA loss statistics for the period 1980–1983 (Cote and Linville, 1986), the overall average losses with and without sprinklers are $8500 and $20,700, respectively. Thus, the absence of automatic sprinklers

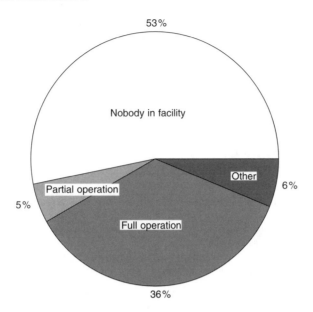

Figure 1.5. Facility operational status at time of fire (based on NFPA data for 238 fires in 1985)

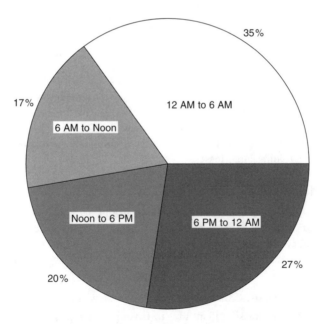

Figure 1.6. Time distribution of fires in industrial facilities (from NFPA data on 338 fires in 1985)

increased the average loss by 140%. In some occupancy categories, such as machinery manufacturing, the unsprinklered-to-sprinklered fire loss ratio is as high as 5.4. This type of data is important for risk managers to utilize to justify the cost of sprinkler installation.

The status of automatic suppression systems and detection systems in NFPA 1987 large loss industrial fires is summarized in Table 1.3. Suppression systems were not installed in half the

Table 1.3. Status of automatic suppression systems and detection systems in 1987 large loss fires (data from NFPA Journal, November/December 1988)

Automatic sprinkler system status	Number of fires
Not Installed	14
Overpowered by Fire	7
System Shut Off Before Fire	2
Not in Area of Fire Origin	3
Damaged by Explosion	1
Unknown	1
Total	28
Automatic detection system status	Number of fires
Not Installed	19
Functioned as Designed	4
Not in Area of Fire Origin	1
Installation Incomplete	1
Unknown	3
Total	28

large losses, and were either shut off, not in the area of fire origin, or damaged by an explosion in another 25% of the losses. The suppression systems were overpowered by the fires in the other 25% of the losses. Thus, 75% of the large losses occurred without any active automatic suppression, while the other 25% involved inadequate automatic suppression. Design criteria and testing needed for effective sprinkler protection are discussed in the context of warehouse fires in Chapters 5 and 8, and for electric cable trays in Chapter 9.

In the case of automatic detection systems, data in Table 1.3 indicate 75% of the losses occurred where there were no installed detectors at all or in the area of fire origin. On the other hand, automatic detection did function as designed in at least 14% of the large loss fires. Therefore, the provision of automatic detection and suppression systems is a favorable, but not necessarily a sufficient factor in avoiding large losses.

1.3 Historic industrial fires and explosions

It is unfortunate but inevitable that most advances in fire protection occur in response to catastrophic losses. Statistics are often not nearly as convincing as personalized accounts and pictures of large destructive fires and explosions. Appendix B contains descriptions of eight historic industrial losses that have had a large and lasting influence on contemporary fire protection engineering. These eight fires are:

- The 1911 Triangle Shirtwaist Garment Factory Fire in New York City
- The 1953 GM Auto Transmission Plant Fire in Livonia, Michigan
- The 1967 McCormick Place Convention Hall Fire in Chicago
- The 1975 Bell Telephone Exchange Fire in New York City
- The 1977 Ford Automobile Parts Warehouse Fire in Cologne, Germany
- The 1982 K Mart Warehouse Fire in Falls Township, Pennsylvania
- The 1989 Ameritech Telephone Exchange Fire in Hinsdale, Illinois
- The 1986 Sandoz Flammable Liquids Warehouse Fire in Basel, Switzerland Fire.

INTRODUCTION AND PERSPECTIVE 21

Table 1.4 lists several key factors involved in each fire. Plastics, flammable liquids, and electric cables were each the primary fuel involved in two fires, and combustible liquids were also involved in three other fires. Flammability aspects of plastics, combustible liquids, and electric cables are discussed in Chapters 5, 7, and 9, respectively.

Sprinkler systems were installed in the fire areas of only two of the eight fires, and those two systems were inadequate for the type of storage covered. Some of the lessons learned (albeit not necessarily implemented) about sprinkler protection and other aspects of these historical industrial losses are summarized in Section 1.3.1.

Besides motivating important developments in industrial fire protection (as indicated in the next to last row of Table 1.4), several of these fires also triggered broader societal changes. The most historic of these changes has been the impact of labor unions and labor laws engendered in response to the 145 fatalities in the 1911 Triangle Shirtwaist fire. From a business practice standpoint, an appreciation of the risk associated with relying on one facility for critical production/operations evolved first from the 1953 G.M. Livonia fire, and then again from the 1989 Hinsdale telephone exchange fire. From an international perspective, perhaps the most important societal impact is the Sandoz Basel fire because it is partially responsible for international agreements on the prevention and cleanup of toxic material releases affecting neighboring countries.

1.3.1 FIRE PROTECTION LESSONS LEARNED

The following are some of the major lessons learned and adopted throughout the industrial fire protection community as a result of these historic fires.

Need for fire walls and other passive barriers in large plants – The General Motors Livonia fire provided dramatic evidence of the need for fire walls, parapets, etc. to limit Maximum Foreseeable Losses in large, single structure industrial facilities. Prior to the G.M. Livonia fire, the trend in manufacturing facilities was to have large undivided spaces. Evidence of the effectiveness of a fire wall occurred in the aftermath of the Ford Cologne fire which destroyed the section of the warehouse on one side of the fire wall, while leaving the other side undamaged. The 1982 K Mart warehouse fire involved four failed fire walls because they were not supported by structural members of comparable fire resistance. Chapter 3 includes a brief description of structural and reliability considerations involved in designing effective fire walls and fire doors.

Need for roof deck fire spread tests – Fire spread along the asphalt and tar on the G.M. Livonia steel roof deck was a major factor in the destruction of the building. It stimulated the development of the roof deck fire tests described in Chapter 3 and the corresponding assessments of asphalt and tar loadings on the potential for self sustained fire spread along the roof deck. It also stimulated considerations of the factors involved in scaling up and generalizing the results of small-scale fire spread tests. These considerations continue to this day without one definitive, consensus scale-up method.

Need to regularly test sprinkler water flow rates and to be informed of water supply impairments – The shutdown of several city water pumps needed to provide McCormick Place with its design sprinkler system water supply was an important factor emphasized in the report of the fire. Municipal and plant officials are now sensitive to the need to communicate information on impairments and to implement contingency plans. Public water supply reliability is discussed in Chapter 2. In the case of industrial facilities with installed fire pumps, these pumps are tested regularly, with water flow rates measured as specified in NFPA 20. Insurance companies require that they be notified of water supply impairments, so that they can help arrange either alternative water supplies or fire watches.

Need to upgrade warehouse sprinkler protection to accommodate storage of more combustible commodities – The Ford Cologne, Germany warehouse fire is a classic example of the problems

Table 1.4. Key factors in eight historic fires

Fire factor	GM Livonia 1953	McCormick place Chicago 1967	K Mart Falls Township 1982	N.Y. Telephone 1975	Ford Cologne 1977	Triangle Shirtwaist N.Y.C. 1911	Hinsdale Telephone Exchange 1989	Sandoz Basel 1986
Occupancy	Automobile Transmission Manufacture	Exhibit Hall	General Warehouse	Telephone Exchange	Auto Parts Warehouse	Garment Factory	Telephone Exchange	Flammable Liquid Warehouse
Ignition source	Hot Work	Electrical	Lift Truck Wiring	Electric Motor	Cigarette	Cigarette	Arcing	Heat during Shrink Wrap
Primary fuel	Flammable Liquid	Plastic	Aerosol Cans	Electric Cable	Plastic	Cotton Fabric	Electric Cable	Flammable Liquid
Other combustibles	Asphalt Roof	Wood, Fabric	Flammable Liquid	–	Motor Oil	Oil Residue, Rags & Lint	–	Pallets
Sprinkler system	None in Fire Area	None in Fire Area	Inadequate	None	Inadequate	None	None	None
Fire wall	None in Manufacturing Area	None	Walls Breached	Cable Penetrations Breached	Effective	None	Effective	Partially Breached
Other factors	Exposed Steel	Aluminum, Inoperable Pumps & Hydrants	No Fire Doors	Smoke Damage, HCl Corrosion	Aisle Storage	Inadequate and Locked Exits	Delayed Response to Alarm	Rocketing Steel Drums, Environmental Disaster
Fatalities	6	0	0	0	0	145	0	0
Fire protection ramifications	Roof Tests, New Style Sprinklers, Many More Fire Walls	Heat & Smoke Vents, Water Supply Reliability, FPE Consultants	Aerosol Product Sprinkler Protection, Better Fire Walls	Smoke Control, Cable Penetration Seals	Large Drop & ESFR Sprinklers	Life Safety Code,	Fire Resistant Cables, Sensitive Smoke Detectors	Containment of Water Runoff, Improved Protection of Flammable Liquid Storage
Other ramifications	Multiple Manufacturing Sites	Electrical Code Enforcement	–	–	Sprinkler Protection in Europe	Labor Unions, Labor Laws	Reliability of Telephone Systems	International Environmental Regulation

associated with the introduction of higher challenge plastic commodities into a warehouse with fire protection designed for less flammable materials. The constraints involved in tailoring protection to different commodity classes, and the relaxing of many of these constraints when using Early Suppression Fast Response sprinklers are discussed in Chapter 5. Similarly, there was a significantly increased challenge to sprinkler systems in K Mart and other warehouses when liquefied petroleum gas propellants replaced nonflammable propellants in aerosol products. This was one of the motivations for doing several series of large-scale fire tests to establish different classes of aerosol product flammability and their associated protection requirements. Sprinkler protection requirements for flammable liquids in small containers are discussed in Chapter 8. Sprinkler protection requirements for aerosol products are described in NFPA 30B.

Need for smoke control in facilities with equipment vulnerable to damage from smoke and corrosive combustion products – Both the 1975 New York Telephone Exchange fire and the 1989 Hinsdale telephone central office fire demonstrated the need for smoke isolation and control to prevent widespread damage to electrical equipment far from the fire. The engineering principles of smoke venting and smoke control are discussed in Chapter 4.

Need for fire resistant electrical cables – The two historic telephone exchange fires also demonstrated the need to develop more fire resistant electric cables. Indeed, a large number of fire resistant cable insulations and jackets are now commercially available, and a variety of cable fire flammability tests are available to certify the increased levels of fire resistance. Cable flammability is discussed in Chapter 9.

Need for adequate emergency egress provisions for large numbers of workers – The Triangle Shirtwaist fire, which resulted in 145 deaths, tragically demonstrated the need for accessible emergency exits to accommodate escape for rapidly developing fires. The apparent locking of exit doors to eliminate pilfering and unauthorized early departure led to the development of the NFPA Life Safety Code as well as a number of workplace safety regulations.

Need for improved protection of flammable liquid warehouse – International attention to the plight of the Sandoz Basel flammable liquid warehouse in the 1986 fire helped arose renewed concern for the protection of flammable liquids in steel drums and other liquid containers. Descriptions of the resulting test programs and current sprinkler protection guidelines are discussed in Chapter 8.

Need for containment of contaminated water runoff – The widespread pollution of the Rhine River during the Sandoz Basel fire caused new requirements to be invoked on the containment of water runoff from fire protection systems at flammable liquid facilities. Some of these requirements are discussed in Chapter 8.

1.3.2 LESSONS NOT LEARNED

The following lessons are still going unheeded at many industrial facilities.

Need for compartmentation via reliable fire walls and doors in large manufacturing facilities – Although this lesson has been implemented in many facilities, it has been neglected in favor of more convenient material handling and economic considerations. Examples include large open areas for modern Flexible Manufacturing Systems, and many plants in which lack of maintenance and testing of fire doors has rendered them ineffective. Chapter 2 describes the engineering considerations involved in compartmentalizing facilities for hazard segregation. The reliability of fire walls and fire doors are discussed in Chapter 3.

Need to restrict storage of special hazard commodities in general purpose warehouses – Despite the demonstration in the Kmart fire that petroleum based aerosols and flammable liquids in small containers can rapidly overtax sprinkler systems designed for ordinary combustible commodities, many other warehouses still do not restrict storage of these special hazard commodities. Aerosol storage is discussed in Chapter 6, and the storage, of flammable liquids in small metal and plastic containers, is covered in Chapter 8.

Problems caused by residue of flammable liquids on building walls, ceiling, and floors – Even though the effect of flammable liquid residues on the ceiling and walls in the GM Livonia fire was well documented, many contemporary plants have similar widespread deposits that can be the crucial factor in the plant structure surviving a major fire. This problem is prevalent in many renovated plants in which flammable liquids were previously used in large quantities over long periods of time. The ignitability of thin liquid layers is discussed in Chapter 7. There is an analogous explosion hazard Chapter 3-16 of the *SFPE Handbook* (1995).

Need for automatic detection AND suppression systems in areas containing large quantities of electrical equipment and cables – Despite the vivid demonstration in the 1975 New York Telephone Exchange fire of the need for automatic detection and automatic suppression over grouped cables that can propagate fire and produce copious amounts of smoke and corrosive combustion products, there are still many unprotected computer facilities and telephone exchanges. Fire detection and suppression alternatives for electronic equipment in computer rooms and telephone exchanges are discussed in Chapters 4 and 8.

Need for adequate emergency egress provisions for large numbers of workers – Despite the Life Safety Code and government regulations, it is evident that many industrial facilities ignore the lesson first learned in 1911 in the Triangle Shirtwaist fire. For example, the 1991 Hamlet, North Carolina chicken plant fire resulted in 25 deaths, many of them due to workers trapped in the plant because of a door locked on the outside for 'security' purposes. The apparent dilemma of providing secure yet accessible emergency exits is discussed in Chapter 2.

Effective sprinkler protection for flammable liquids in plastic containers – The Sandoz Basel fire involved flammable liquids in various containers. Since then, a number of research programs have been conducted to determine effective sprinkler protection for warehouse storage of flammable liquid in containers. These programs have led to effective guidelines for flammable liquids in metal containers, but storage in plastic containers still remains an unanswered challenge, particularly in palletized storage applications. A large portion of Chapter 8 is devoted to warehouse storage of flammable liquids.

References

Baker, W. E., Cox, P. A., Westine, P. S., Kulesz, J. J. and Strehlow, R. A., *Explosion Hazards and Evaluation*, Elsevier, 1983.
Boykin, R. F., Kazarians, M. and Freeman, R. A., Comparative fire risk study of PCB transformers, Risk Analysis, December 1986.
Bureau of Labor Statistics, Number of Nonfatal Occupational Injuries and Illnesses involving Days Away from Work by Event or Exposure, OS TB 4/20/2000 Table R.31, http://stats.bls.gov/oshhome.html, 2000.
Center for Chemical Process Safety, *Guidelines for Evaluating the Characteristics of Vapor Cloud Explosions, Flash Fires, and BLEVEs*, CCPS, AIChE, 1994.
Cote, A. and Linville, J., eds, *Fire Protection Handbook*, Sixteenth Edition, National Fire Protection Association, Quincy, MA, 1986.
Drysdale, D., *An Introduction to Fire Dynamics*, 2nd Edition, John Wiley & Sons, 1998.
Eckhoff, R., *Dust Explosions in the Process Industries*, Butterworth Heinemann, 2nd Edition, 1996.
Fire Safety Aspects of Polymeric Materials, Volume 4: Fire Dynamics and Scenarios, National Academy of Sciences Publication NMAB 318-4, Technomic Publishing Co., Westport, CT, 1978.
FPA Large Fire Analysis for 1986, *Fire Prevention*, **212**, September 1988.
FMRC Clean Room Materials Flammability Test Protocol, Factory Mutual Research Corporation Test Standard Class Number 4910, September 1997.

Gann, R. et al., *Risk Assessment: Final Report*, National Fire Protection Research Foundation, 1991.

Guidelines for Hazard Evaluation Procedures, 2nd Edition, Center for Chemical Process Safety, AIChE, 1992.

Guidelines for Chemical Process Quantitative Risk Analysis, American Institute of Chemical Engineers Center for Chemical Process Safety, 1989.

Henley, E. and Kumamoto, H., *Reliability Engineering and Risk Assessment*, Prentice Hall, 1981.

Hoyle, W., McCleary, S. and Rosenthal, I., The Chemical Safety and Hazard Investigation Board's Process for Selecting Incident Investigations, pp. 69–100, *International Conference and Workshop on Process Industry Incidents*, Center for Chemical Process Safety, Orlando, Florida, October 2000.

J&H Marsh & McLennan, *Large Property Damage Losses in the Hydrocarbon-Chemical Industries A Thirty-year Review*, J&H Marsh & McLennan Consulting Services, 18th Edition, 1998.

Karter, M.J., Fire Loss in the United States During 1990, *NFPA Journal*, **85(5)**, September/October 1991.

Kort, J., On the Job: London, *Firehouse*, pp 48–52, January 1985.

Lotti, P.J., *Electric Immersion Heater Redundant Control System*, Semiconductor Safety Association paper, May 1986.

Major Fires: 1991-Worldwide, *Loss Prevention*, **14(4)**, Loss Prevention Association of India, October–December 1992.

NFPA Fire Analysis Division, Automatic Sprinkler Systems Do Have an Impact in Industry, *Fire Journal*, January 1987.

Natural Catastrophes and Man-made Disasters 1998: Storms, Hail and Ice Cause Billion-Dollar Losses, *Sigma*, **1/1999**, Swiss Re, 1999.

Plastic and Plastic-Lined Tanks with Electric Immersion Heaters, Factory Mutual Loss Prevention Data Sheet 7-6, June 1986.

Redding, D. and Pauley, Jr., P., Large-Loss Fires in the United States During 1985, *Fire Journal*, November 1986.

SFPE Engineering Guide to Performance-Based Fire Protection Analysis and Design of Buildings, Society of Fire Protection Engineers, National Fire Protection Association, 2000.

SFPE Handbook of Fire Protection Engineering, Society of Fire Protection Engineers, National Fire Protection Association, 1995.

Semiconductor Fabrication Facilities, Factory Mutual Loss Prevention Data Sheet 7-7, January 1997.

Structures to Resist the Effects of Accidental Explosions, Department of the Army Technical Manual TM5-1300, 1991.

Taylor, K.T., The Large-Loss Fires of 1986, *Fire Journal*, November/December 1987.

Taylor, K.T. and Norton, A. L., Large-Loss Fires in the United States During 1987, *Fire Journal*, November/December 1988.

Ward, R., FPA Large Fire Analysis for 1985, *Fire Prevention*, **207**, March 1988.

2 PLANT SITING AND LAYOUT

Fire protection can play an important role in plant siting and layout decisions, particularly when hazardous materials and public safety issues are involved. This chapter is intended to provide an engineering framework for evaluating plant sites and layouts so as to promote effective fire and explosion protection. These considerations and analyses can also help provide answers to officials responsible for public safety and land-use planning, such as is required by the European Commission's Seveso II Directive (Christou et al., 1999).

2.1 Fire protection siting considerations

The major fire protection considerations involved in a plant site review are: (1) safe separation distances from exposing or exposed properties; (2) water supply accessibility and reliability; (3) capabilities and locations of local firefighting facilities; (4) applicability of local fire and building codes; and (5) local environmental effects.

2.1.1 SAFE SEPARATION DISTANCES

Flame radiation

Although fires can spread via a variety of propagation modes, when there is spatial separation between the exposed and exposing properties/structures, the dominant threat to be evaluated is usually flame radiation. A generalized engineering approach for assessing safe separation distances for flame radiation exposures would be as follows:

1. Postulate a worst-case exposure fire scenario in terms of the extent of the materials or structures that are burning.
2. Determine the heat release rate and/or the effective flame radiation temperature and emissivity.
3. Calculate the flame emissive power using the methods described in Appendix A.1.
4. Calculate the flame height using equations given in Appendix A.2.
5. Determine the flame-target configuration factor using the information in Appendix A.3 or a good thermal radiation reference.
6. Calculate the radiant heat flux, q'', impinging on the target from:

$$q'' = \phi \mathbf{E} \tau \qquad [2.1.1]$$

where ϕ is the configuration factor, **E** is the flame emissive power, and τ is the atmospheric transmissivity.

7. Compare the calculated q'' to the critical heat flux for ignition or for structural damage of the exposed structure/property.
8. Repeat the calculations this time accounting for wind effects on flame height and configuration factor for a wind tilted flame and a downwind target.
9. If the calculated radiant heat flux is greater than the ignition/damage threshold heat flux, and the separation distance cannot be increased, evaluate the feasibility of shielding the exposed structure with a more fire resistant wall material and/or with outside sprinklers.

Example – Consider a situation in which two wood frame buildings face each other with a separation of 6.1 m (20 ft). Imagine a scenario in which a 3.7 m (12 ft) high by 7.6 m (25 ft) wide wall facing the adjacent building, is ignited by vandals as occurred in the fire shown in Figure 2.1. Is the 6.1 m separation adequate to prevent flame propagation to the adjacent building?

The heat release rate for a 3.7 m high wood wall fire is about 1040 kw/m (300 Btu/sec) per foot of width, according to the data in Table A.3 of Appendix A. The corresponding flame height calculated using equation (A.7) is 5.1 m (17 ft). The emissive power is the radiant heat release rate per unit flame surface, which for a radiant heat release fraction of 0.26 (Table A.2) is

$$0.26(1040)/5.1 = 53\,\text{kW/m}^2 (4.67\,\text{Btu/sec-ft}^2)$$

This value is significantly lower than would be calculated using the flame temperatures, cited in Appendix A, and an emissivity of unity. The lowest expected flame temperature of 1100 K would produce an emissive power of 83 kW/m². Since this is higher than the 53 kW/m² calculated using the heat release rate, it will be used in the interest of estimating a conservative safe separation distance.

The configuration factor needed for equation [2.1.1] can be found from Figure A.1 of Appendix A. It is about 0.25 for a target at mid-flame height and the given combination of flame height, wall width, and separation distance. Therefore,

$$q'' = (0.25)(83) = 20.75\,\text{kW/m}^2 (1.83\,\text{Btu/sec-ft}^2)$$

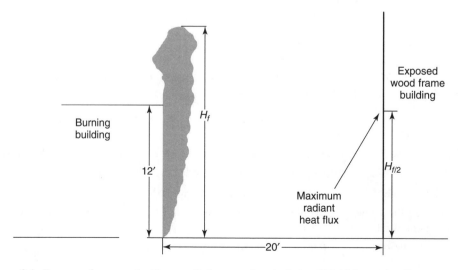

Figure 2.1. Exposure fire example. Flame radiation sample calculation 12 ft high wood wall exposure fire

assuming an atmospheric transmissivity of unity. Since this value is significantly higher than the threshold for pilot ignition of many woods (about $12\,kW/m^2$, or $1.06\,Btu/sec\text{-}ft^2$), flame spread would be expected. Wind effects would not be expected to affect this conclusion; the wind would be expected to produce some flame leaning above the wall and slightly increase the already high heat flux on the exposed building. It would be wise to install either a fire resistive outer wall layer or outside sprinklers.

The preceding example was relatively simple because only the outside wall of the exposing building was assumed to be burning. If the exposure fire had started within the exposing building, it would be much more difficult to calculate heat release rates and flame heights, even if the internal materials burning were specified. In lieu of an applicable calculation procedure and data for this situation, a more empirical approach is needed.

The approach utilized in both the NFPA and Factory Mutual standards on exposure fire separation distances (NFPA 80A, 1980; FMDS 1-20, 1979) is to introduce generic classes of exposure fire severity. The categories in NFPA 80A were established from considerations of the factors affecting the severity of the various test fires in the St. Lawrence Burns series of building fires (Shorter et al., 1960; McGuire, 1965).

Radiant fluxes near the flames in the first twenty minutes of the St. Lawrence Burns were conservatively (and probably unrealistically) estimated to be as high as $360\,kW/m^2$ ($31.7\,Btu/sec\text{-}ft^2$). The design-basis flame emissive powers and corresponding maximum acceptable configuration factors for the three categories of fire severity given in NFPA 80A are as follows:

NFPA category for exposure fire	Emissive power (kW/sq-m)	Configuration factor
Severe	358	0.035
Moderate	179	0.07
Light	90	0.14

The product of emissive power and configuration factor for all three categories is $12.6\,kW/m^2$ ($1.19\,Btu/sec\text{-}ft^2$), the assumed critical heat flux for wood and many other materials (Shorter et al., 1960). The maximum separation guide numbers in NFPA 80A are based on these configuration factors plus an assumed flame extension to leeward of $1.5\,m$ ($5\,ft$). The exposure fire category criteria involve fire loading per unit area and flame spread rating.

The four Factory Mutual exposure fire categories for determining safe separation distances are based on the flammability classification of the commodity. Tables of safe separation distances are given for a variety of exposed wall configurations.

If outside sprinklers are needed to shield the exposed wall from the impinging flame radiation, the effect of spray parameters on the radiation transmissivity, τ, can be inferred from the data correlation in Appendix A.4.

Reductions in required safe separation distance attributable to the water sprays can be estimated from equation [2.1.1], as well as from the guidelines in the FM and NFPA standards. The preceding analysis does not account for other combustible materials located between the exposure fire and the exposed building. Yard storage, trash, brush, or other combustibles located near the exposed building could be ignited by the exposure fire and produce sufficiently high heat fluxes to cause the fire to spread to the exposed building if it has a combustible outer surface. If this scenario is foreseeable, external sprinklers or a fire resistant covering should be considered.

Toxic fire plumes

Fire plume dispersal can be an important, and possible dominant, factor in determining safe separation distances. Toxic vapor and particulate concentrations in the downwind plume have on numerous occasions caused emergency evacuations of thousands of people. Some well documented examples in the United States include a metal processing plant fire (Duclos *et al.*, 1989), several pesticide plant fires (Diefenbach, 1982), and several fires involving swimming pool chemicals (oxidizers) (Custer, 1988). In some of these fires, the chemical itself is toxic or emits toxic vapors. In other cases, the combustible material is innocuous prior to combustion, but produces toxic combustion products.

The fires involving water treatment chemicals for swimming pools are particularly interesting because the toxic plumes are produced from the addition of water to the burning chemicals. For example, the Springfield, MA 1988 fire (Custer, 1988) involved trichloroisocyanuric acid which is an oxidizer that decomposes exothermally when exposed to water, and produces chlorine vapor. Chlorine has an OSHA specified maximum permissible exposure level (PEL) of 0.5 ppm for long term exposures and an Immediately Dangerous to Life and Health (IDLH) value of 10 ppm for short duration exposures. One fire occurred at the building in which the isocyanuric acid was stored when rainwater entered through open windows and soaked the fiberboard drums containing the oxidizer. A dry pipe sprinkler system also actuated and exacerbated the oxidizer decomposition. Springfield public safety officials ordered three separate evacuations which at one point extended to a distance of about 10 km (6 miles) in the downwind direction, as illustrated in Figure 2.2. Firefighters eventually suppressed the fire with copious water applications.

Calculation procedures exist to estimate safe separation or emergency evacuation requirements. The concepts involved are briefly reviewed here, with the actual equations described in the references.

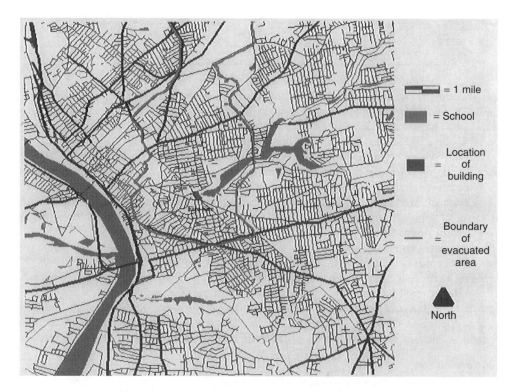

Figure 2.2. Evacuation area surrounding Springfield fire site

Besides spreading radially by entraining ambient air, the buoyant smoke plume emitted from a burning structure or outdoor fire drifts downwind as it rises. When the plume half-width (associated with a prescribed smoke concentration) equals or exceeds the plume rise height, the lower edge of the plume reaches ground level. Vapor/particulate ground level concentrations gradually decay by dilution as additional air is entrained into the plume.

In general, two values of the ground level concentration distribution dictate safe separation or emergency evacuation distances. The lower value is the irritation threshold for sensory or pulmonary irritant substances. The upper level is the tenability limit for incapacitation and possibly death after prolonged exposure. In the case of narcotic gases, such as CO and HCN, there is no irritation threshold, and the tenability limit for incapacitation is due to loss of consciousness. In the case of irritant gases such as HCl, the irritation threshold "represents unpleasant and quite severely disturbing eye and upper respiratory tract irritation" (Purser, 1995). The tenability limit for incapacitation due to a irritant gas is associated with severe pain in the eyes and/or upper respiratory tract, with copious lacrymation, mucus secretion, and pulmonary edema (Purser, 1995). The tenability limit for respirable particulate (diameters less than about 5 μm) is the concentration at which smoke deposits physically clog the airways.

Threshold concentrations for sensory irritation and for incapacitation vary with individuals according to their age, health, and in the case of narcotic gases, with the time of exposure. Concentration ranges to account for these variations are shown in Table 2.1 for three commonly encountered toxic gases. The tenability limit for particulate also depends on the material and subject, but even with inert particles it is less than $5\,g/m^3$ (Purser, 1995). Evacuation zones are selected to preclude concentrations somewhere between the Immediately Dangerous to Life and Health (IDLH) concentration (typically an order-of-magnitude less than the tenability limit) and the Threshold Limit Value (TLV) for eight-hour interval repeated exposures. US Department of Labor limits on eight-hour exposure to diesel particulate is in the range $160\,\mu g/m^3$ to $400\,\mu g/m^3$ (Mine Safety and Health Administration, 1998).

Safe separation or emergency evacuation distances should be estimated to preclude anyone from being exposed to these concentrations without self contained breathing apparatus. A suggested procedure to estimate these distances is as follows:

1. Estimate the heat release rate for the worst-case fire.
2. Estimate the yields or generation rates of toxic gases and particulates in the worst-case fire.
3. From the heat release rate and assumed wind speeds, calculate the buoyancy flux and plume rise height (Briggs, 1969; Wackter and Foster, 1986; Mills, 1987; Crowl and Louvar, 1990). Alternatively, choose some other starting point for smoke plume dispersal calculations.
4. Decide whether to use either empirical Gaussian plume correlations of plume dispersal (Wackter and Foster, 1986; Mills, 1987; Crowl and Louvar, 1990), or some type of fluid dynamic model, such as the NIST ALOFT model (Walton and McGratton, 1998, 1996). In either case, the atmospheric stability category or some measure of atmospheric turbulence needs to be

Table 2.1. Tenability limit concentrations of common toxic combustion products (Purser, 1988)

Gas	Irritation limit	Incapacitation 5-min exposure	Incapacitation 10-min exposure
Carbon Monoxide	–	6000–8000 ppm	1400–1700 ppm
Hydrogen Cyanide	–	150–200 ppm	90–120 ppm
Hydrogen Chloride	75–300 ppm	300–16,000 ppm	300–4000 ppm

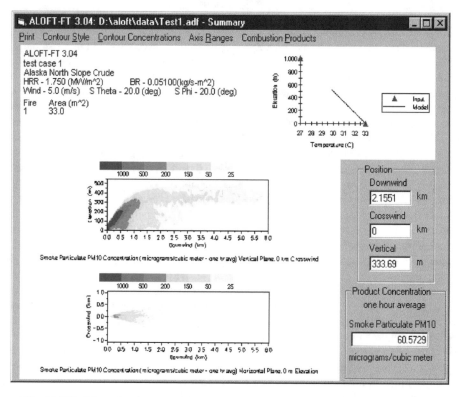

Figure 2.3. ALOFT-PC smoke plume model: sample particulate concentration output (from NIST BFRL)

assumed. Using the selected model, calculate plume widths and concentration distributions. Figure 2.3 shows sample output obtained with the NIST ALOFT-PC model.

5. Determine the distance at which the concentrations are equal to some fraction (depending on the desired level of conservatism) of the irritation threshold or the tenability limit for incapacitation.

6. If necessary, revise the calculations to account for any adsorption of vapors onto surfaces/structures (Galloway and Hirschler, 1989; Crowl and Louvar, 1990).

Mills (1987) has shown how this procedure can be used to estimate safe separation distances for an ethyl chloride diked spill fire. Assuming HCl is the design basis toxic combustion product, Mills calculated downwind concentrations for a range of assumed burning rates and wind speeds. His results showed that the highest assumed wind speed (5 m/s or 11 mph) produced the largest downwind ground level concentrations because the plume did not rise as high as with the lower wind speeds. Effects of adjacent buildings are discussed in Mills' paper in terms of dispersion rather than adsorption. The adsorption of HCl onto building surfaces, and the subsequent reaction with or diffusion into these surfaces, has been modeled by Galloway and Hirschler (1989) using empirical estimates of mass transfer coefficients based on experiments with various building interior surfaces.

Yamada (1998) has shown how smoke plume dispersal modeling can be used to plan large-scale crude oil burn experiments, and subsequently how the field data compared with the model predictions.

PLANT SITING AND LAYOUT

Flammable vapor clouds

The release of a large quantity of flammable gas or vapor produces a cloud that is diluted with air as it drifts downwind. If there is no ignition source in the immediate vicinity of the release site, the downwind extent of the cloud in which flammable concentrations of vapor exist represents the downwind danger zone. Any person or object within the flammable cloud would be immersed in flame when/if ignition occurs. Thus, the safe separation distance for this hazard is the downwind distance to the lower flammable limit concentration. The downwind extent of the flammable cloud depends upon the following parameters:

1. The vapor release rate or, in the case of a virtually instantaneous release, the total quantity of vapor.
2. The vapor temperature and molecular weight.
3. Wind velocity.
4. The atmospheric stability category, or some other measure of atmospheric turbulence level.
5. Size and elevation of the release site.
6. Terrain and size, location, and geometry of buildings and structures in the vicinity of the release site.

Once these parameters have been determined, the downwind extent of the flammable cloud can be determined either by calculation or by wind tunnel testing. Calculations need to account for the gravitational spreading effects due to the fact that most flammable gases are heavier than air. The combination of gravitational spreading and atmospheric turbulent diffusion almost invariably requires computer based numerical solutions. Descriptions of many of the commercially or publicly available computer models can be found in the *Guidelines for Use of Vapor Cloud Dispersion Models* (AIChE, 1996). Up to date information on the availability of codes such as ALOHA developed under government sponsorship for emergency response preparedness can be found on the World Wide Web site of the Environmental Protection Agency Chemical Emergency Preparedness and Prevention Office.

Blast wave exposure

Safe separation distances for blast wave exposure can be estimated using the blast wave correlations and structural damage pressure criteria given in Appendix C and the references therein. The procedure entails identifying the site and strength (energy for an ideal blast wave) of the explosion, and then determining whether blast wave pressures on the exposed property would exceed the relevant structural damage thresholds. If the blast wave pressures are indeed capable of causing significant damage, the exposed plant management has the options of increasing separation distance, hardening the exposed structure, or erecting some type of barricade.

Example – Imagine a new plant building being located about 200 m (656 ft) away from the explosives storage building at a fireworks plant. Approximately 500 kg (1100 lb) of a Class A explosive (readily detonable) with a specific energy of 5000 kJ/kg (2154 Btu/lb) are stored in the explosives warehouse. What type of construction will allow the exposed building to withstand a worst case explosion in the fireworks plant without incurring significant structural damage?

The blast energy associated with the 500 kg of explosive is $500(5000) = 25 \times 10^5$ kJ $= 25 \times 10^8$ N-m (2.37×10^6 Btu). Using equation (C.1), the nondimensional distance, $\overline{R}$, to the target structure is

$$\overline{R} = (200\,\text{m}) \left[\frac{(10^5\,\text{N/m}^2)}{(25 \times 10^8\,\text{Nm})} \right]^{1/3} = (200\,\text{m})\,(0.0342\,\text{m}^{-1}) = 6.84$$

According to Figure C.2, the peak pressure in the incident blast wave at this value of $\overline{R}$ is

$$(P_s - P_0)/P_0 = 0.031$$

This would be the peak over pressure experienced by a wall situated parallel to the direction of blast wave propagation, 200 m from the center of the explosion. A wall at the same distance but situated normal to the direction of blast wave propagation would experience the reflected blast wave peak pressure given by

$$(P_r - P_0)/P_0 = 0.062$$

Thus, the side of the exposed building facing the fireworks plant should be designed to withstand a peak over pressure of

$$P_r - P_0 = 101{,}000\,\text{Pa}\ (0.062) = 6262\,\text{Pa}\ (0.91\,\text{psig})$$

According to the damage threshold pressure data in Table C.1, an over pressure of 6262 Pa (0.91 psig) can break large plate glass windows and can produce minor damage to brick walls if the positive phase impulse (area under the pressure versus time curve) is greater than about 110,000 Pa-msec (16 psig-msec). The positive phase impulse at a normalized distance of 6.84 from the center of the explosion can be obtained from Figure C.3 for the incident blast wave ($\overline{I_s} = 0.005$). Thus, from Equation C.3, the positive phase duration becomes:

$$I_s = \overline{I_s}\frac{(P_0^2 E)^{1/3}}{a_0} = 0.005(10^{10}\,\text{N}^2\text{m}^{-4}25\times 10^8\,\text{Nm})^{1/3}/(330\,\text{ms}^{-1})$$

$$= 44.3\,\text{Nm}^{-2}\text{sec} = 44.3\,\text{kPa-msec}\ (6.4\,\text{psig-msec})$$

The positive phase impulse associated with the reflected blast wave pressure loading is equal to about twice the value of I_s at this distance. Therefore, $I_r = 88$ kPa-msec (13 psig-msec), which is only about 20% less than the critical value for minor structural damage to a brick wall.

In view of the preceding calculation, the exposed plant should have a more blast resistant construction than brick wall, at least for the wall facing the fireworks plant. Structural steel walls or reinforced concrete walls would provide a significant margin of safety against blast damage according to the data in Table C.1.

It is often useful to present the results of plant siting blast wave analyses in the form of blast wave contours. Each contour represents the locations that experience the designated value of the peak pressure in the incident blast wave. These contours would be circles if the explosion originated from a single known point. If the origin of the explosion can be anywhere within a prescribed boundary (such as a building), each contour is the locus of circles with centers along the prescribed boundary. This is indicated in Figure 2.4, which shows the 0.3 bar (4.35 psig) contour and the 0.1 bar (1.45 psig) contour for an explosion in the explosives plant with the rectangular plan view shown in the figure.

In the example shown in Figure 2.4, the 0.1 bar contour extends beyond the plant property line. Let us assume this is unacceptable to the Authority Having Jurisdiction for plant siting approval. The primary way to reduce the extent of the blast wave contours is to reduce the blast wave energy by reducing the amount of explosive material. In the case of the explosives plant this can be achieved by dividing the building into three smaller buildings, each with one third the inventory of explosive material. Blast walls are needed between the buildings so that an explosion in one will not trigger additional explosions in shock sensitive explosives in the adjoining building. This is illustrated in Figure 2.5. The blast wave scaling laws can be used to show that the distance to a given blast pressure will be $(1/3)^{1/3} = 0.693$ times the distance for the original contours shown in Figure 2.4. As indicated in Figure 2.5, this reduction is sufficient to bring the 0.1 bar contour within the property line.

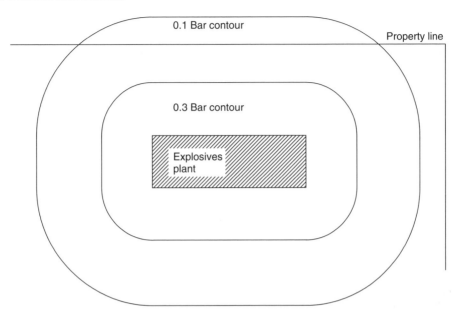

Figure 2.4. Blast wave contours for explosives plant explosion

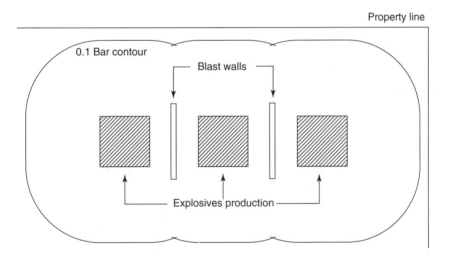

Figure 2.5. Calculated blast wave contours for redesigned explosives plant

Similar calculations can be performed for other types of explosions using the equivalent energy and blast wave pressure calculation methods described in Appendix C. Sometimes the results of several different explosion scenarios at one plant are combined in the form of risk contours. A blast wave risk contour is an enclosed curve representing the locations at which there is a certain probability of a blast wave of designated magnitude. For example, suppose the explosion indicated in Figure 2.4 had a probability of occurrence of 10^{-5} per year. Furthermore, suppose there was a similar probability of occurrence for an explosion involving the unloading of explosive material off a truck. Since the origin and energy involved in the second explosion would be different, the blast contours would also be different. However, the 0.10 bar contours from the two explosions could be combined to represent the 10^{-5} per year blast wave risk contour.

This type of analysis is often used in the siting of large petrochemical plants in Europe. The large quantities of flammable gases and liquids at these plants generate concerns for flammable vapor cloud hazards, pool fire thermal radiation hazards, and explosion hazards (often vapor cloud explosions). Each hazard can be represented in the form of risk contours so as to provide a consistent, visual representation of the overall risk of siting the plant at the proposed location.

2.1.2 WATER SUPPLIES

Water supply accessibility

Desirable plant sites should have access to a strong water supply for plant fire protection systems. Specifically, the water supply should provide the flow rates and pressures needed by automatic sprinkler systems and hose streams. Sprinkler system flow rates depend on the flow rate per specified head and the expected maximum number of heads opened in a worst-case design basis fire. An equivalent specification for water supply estimates, albeit not necessarily equivalent for sprinkler effectiveness, is the product of the water design density (flow rate per unit floor area) and the expected maximum area involved in the design basis fire. Actual water flow requirements are slightly higher because of hydraulic losses through piping, as calculated for each hydraulically designed sprinkler system.

Sprinkler system water capacity requirements cannot be determined from theoretical calculations; they are based on a combination of loss data and large-scale fire tests. The tests demonstrate the ability of a specific sprinkler discharge to control or suppress the fire, as well as the number of heads opened. Test results are scenario specific and not readily extended to other scenarios with variations in combustible configuration, ceiling height, ignition location, etc. Nevertheless, the test data are a necessary complement to loss data (primarily data on the number of heads opened in various incidents) with frequently incomplete information on occupancy and sprinkler system arrangement, and fire initiation scenario.

Code and standard specifications for sprinkler/hosestream water demands are usually presented in terms of occupancy categories. NFPA 13 (1996), for example, has five general occupancy categories: one light hazard category, two ordinary hazard categories, and two extra hazard categories. Generic definitions and representative occupancies for these categories are given in NFPA 13. Hydraulic design water demands for these five categories vary from a few hundred gpm (400–500 L/min) to about 2500 gpm (9500 L/min), and demand durations vary from 30 min to 120 min. NFPA 13 specified hose stream flow rates required to supplement the sprinkler system flow rates vary from 100 gpm (380 L/min) for the light hazard category to 1000 gpm (3800 L/min) for highest hazard category. Thus the total (sprinklers plus hose streams) water demand per NFPA 13 varies over a range from about 300 gpm (1100 L/min) to about 3500 gpm (13,000 L/min) depending on occupancy category and available water pressure. Storage occupancy sprinkler requirements, some with larger water demands, are also contained in the newer editions of NFPA 13. The experimental basis for warehouse storage occupancy sprinkler system demands is reviewed in Chapter 5; special storage protection is discussed in Chapter 6.

The British Loss Prevention Council (LPC) sprinkler requirements (LPC, 1973) are also specified in occupancy categories with water flow rates (water densities multiplied by areas) slightly lower than the NFPA 13 flow rates for most corresponding hazard categories. The water flow rate requirements for the LPC and CEN higher challenge warehouse storage categories are dependent on storage configuration and height, with some flow rate requirements being lower than the comparable NFPA standard and others being greater.

Factory Mutual Data Sheet 3-26 (1983) has two general occupancy categories for water demands for pipe schedule sprinkler systems with a 50 psig (345,000 Pa) supply pressure and

$\frac{1}{2}''$ orifice heads, and 16 generic industry/occupancy categories for other sprinkler configurations. Total water demands (sprinkler system plus hose streams) vary from about 550 gpm (2080 L/min) to about 3400 gpm (13,000 L/min). Required demand durations are either one or two hours. Thus, a water reservoir capacity in the range 125 to 1500 m^3 (33,000–408,000 gal) would be required for plant firefighting. In addition to these specifications, there are a myriad of other Factory Mutual water demand specifications for storage and special hazard occupancies, some of which require higher water flow rates.

Actual water demands in a particular incident are entirely dependent on the decisions of when to cutback or curtail flow, as well as the hosestream flows actually employed. These decisions are usually made by the fire brigade chief based on an assessment of the degree of fire control/suppression being achieved. Sardqvist (1999, 2000) has reviewed data on fire brigade water flow demands in Europe, and on available models and data correlations for estimating required flow rates based on incident data.

Once the plant water demand is determined from one of the aforementioned specification systems, available water supply sources can be evaluated in terms of the adequacy of flow rate, pressure, and duration. Water supply testing usually entails measuring the static (negligible flow) pressure and the residual pressure at a measured flow rate through hydrant connections.

In comparing water supply to water demand, the possibility of future changes at the proposed site location is also an important factor. Process or storage changes (either via expansion or changes in materials and equipment) usually entail increases in water demand. The available water supply can deteriorate as new facilities are built in surrounding areas. Although these changes are difficult to anticipate, there should be some significant margin of extra available supply to accommodate demand increase and supply deterioration. In the case of anticipated supply deterioration, this margin is sometimes expressed as a 10-psig cushion between the supply pressure (at the base of the riser or the pump suction flange) and the demand pressure (at the total demand flow rate).

Water supply example

Suppose a new warehouse is to be constructed about 700 ft away from the town water main connection. In order to allow for the most flexibility in storage commodities and storage heights, the warehouse sprinkler protection in accord with NFPA 13 will consist of 12 ESFR sprinklers flowing 121 gpm at the hydraulically most remote head, plus a 250 gpm hose stream demand. A (booster) fire pump, taking suction at 20 psig from a connection to the town water main, will provide the needed pressure for the sprinklers and hose stream. A town water main flow test is made near the anticipated connection to the plant inlet line, and the following results are obtained: Static Pressure, $P_{stat,} = 77$ psig; Residual Pressure $= 45$ psig at a flow rate of 1580 gpm.

The sprinkler system water demand is 12(121 gpm) = 1450 gpm plus the additional flow due to most sprinklers being at a higher pressure than the hydraulically most remote head. This additional flow is estimated to be about 10% of the flow associated with the nominal 1450 gpm flow rate, i.e. roughly 150 gpm. Thus, the total water demand = 1450 + 150 + 250 = 1850 gpm, which is to be supplied at 20 psig at the fire pump intake.

The available pressure at the town water main connection can be estimated from the Hazen-Williams correlation, i.e. $P = P_{stat} - kQ^{1.85}$, where for this water main, $k = (77 - 45)/(1580)^{1.85} = 3.87 \times 10^{-5}$ psi/(gpm)$^{1.85}$. At the 1850 gpm water demand, $P = 77 - 3.87 \times 10^{-5}(1850)^{1.85} = 34$ psig. The supply pressure at the fire pump inlet will be 34 psig minus the pressure drop due to the flow through 700 ft connecting line, and minus the pressure drop through the required backflow preventer. If a 16-inch concrete connecting line is used,

Hazen-Williams correlation calculations indicate that the pressure drop will be about 4 psig. The pressure drop through a 16-inch diameter backflow preventer depends on the particular design, but is probably about 10 psig. Thus, the supply pressure for a 1850 gpm flow at the pump inlet is $34 - 4 - 10 = 20$ psig. This is, coincidentally, the nominal demand pressure for the required flow at the pump inlet. However, it does not provide any cushion between supply and demand pressure to allow for pipe deterioration over the years. Therefore, the possibility of either putting in a larger, new water main, or providing an on-site water supply should be considered.

Potential on-site water sources include gravity tanks, and pump stations fed from either surface tanks or nearby surface bodies of water. The acceptability of these sources depends on their reliability as well as their contribution to satisfying the total plant water demand. Pumps, gravity tanks, and suction tanks are discussed in Section 2.2.5 in the context of sprinkler system layout. Public water supply reliability is discussed next.

Water supply reliability

There are several reports of large losses in which the impairment or unavailability of the public water supply was a major factor. One such loss is the Chicago McCormack Place fire (Appendix B) in which two of four 2500 gpm (9450 L/min) pumps at an unattended pumping station failed to operate. The reduced water supply impeded manual firefighting efforts, which would have been problematic anyway because of the absence of automatic sprinklers in the main fire area.

A more definitive demonstration of the effects of an unreliable water supply is the 1965 Cambridge, MA warehouse fire described in 'City Water Main Repairs Cause Severe Loss' (*NFPA Journal*, 1965). Water was supplied to the warehouse via an 8 in dead-ended line on the Cambridge, MA water distribution system. The warehouse supply line was fed from two mains; one on First Street and the other on Main Street. At the time of the fire, both mains were impaired. The 30 cm (12 in) diameter First Street main had been shut off to allow for cleaning and relining. (There was a 10 cm (4 in) temporary line in service, while the 30 cm (12 in) main was being cleaned.) The 60 cm (24 in) diameter Main Street main had sprung a leak and was shut off for repairs two and a half hours before a fire started in the warehouse. Thus, both mains were out of service and the water pressure on the fifth story of the warehouse (where the fire started) was less than 5 psig (34,500 Pa). The lack of a water supply for the installed sprinklers eventually caused the ware house to be gutted. It was a $1 million loss in 1965 dollars, equivalent to about $4 million in 1987 dollars.

How can the reliability of public water supplies be assessed prior to a large loss? Two methods used are: (1) reliability guidelines, and (2) probabilistic reliability theory.

Reliability guidelines specify minimum levels of redundancy for water supplies. For example, a single unlimited capacity water source (e.g. a large lake) might be considered acceptable for a high risk facility if there are two separate suction lines and fire pumps. Similarly, two separate ground water sources, each with its own pumping stations, would be acceptable unless the ground water sources are subject to drought or freezing problems. A single supply line at the end of public distribution network might be considered unreliable, as would a single pumping station and ground water source. Whether or not such an unreliable water supply would be tolerable depends on the inherent fire risk at the facility as specified for example by the expected fire frequency and the Maximum Foreseeable Loss in the event the water supply is unavailable.

The use of probabilistic reliability theory most likely would involve some type of fault tree analysis. An example is the fault tree analysis of the Lawrence Livermore National Laboratory water supply as described in Hasegawa and Lambert (1986). The unavailability (probability of it not being available when needed for a fire) of the existing water supply was calculated to be

PLANT SITING AND LAYOUT

3.6×10^{-4}. The most effective single measure to increase water supply reliability at Livermore would be to install a water-level sensor in each of the three storage tanks, rather than rely on one sensor for all three tanks. Installation of these three sensors was determined to produce a factor of 50 reduction in system unavailability.

If a public water main outage is the dominant failure mode, as occurred in the 1965 Cambridge, MA warehouse fire, published data on the frequencies of water main breaks and outages can be utilized to quantify the water supply reliability. For example, Table 2.2 provides a compilation of the frequencies, f_b, of water main breaks in various municipalities. Data on water main leaks is harder to obtain, but the compilation in Walski and Pelliccia (1982) indicates that the median ratio of main leaks requiring repair to main breaks is 0.29. Thus a good approximation would be to multiply the data in Table 2.2 by 1.29 to estimate the frequency of main breaks and leaks, f_{bl}. The corresponding water main probability of outage is

$$p = f_{bl} L t_r \qquad [2.1.2]$$

where p is the probability of water main being unavailable, f_{bl} is the frequency of main breaks and leaks per mile-yr, L is the length (mile) of main from water source to plant connection, and t_r is the average time to repair break/leak (yr).

Equation [2.1.2] assumes that f_{bl} is sufficiently small for p to be much smaller than one. Since the highest value of f_b in Table 2.2 is 1.8 breaks/leaks per mile-yr, this assumption should be valid for water main lengths on the order of a mile or less and for repair times no longer than a few days.

Table 2.2. Frequency data for public water main breaks

City	Main breaks per mile year	Reference
Binghamton, NY:		
Sand Spun Cast Iron Mains	0.088	Walski and Pelliccia, 1982
Pit Cast Iron Mains	0.055	Walski and Pelliccia, 1982
Boston, MA	0.036	O'Day, 1982
Chicago, IL	0.054	O'Day, 1982
Denver, CO	0.156	O'Day, 1982
Houston, TX	1.290	O'Day, 1982
Indianapolis, IN	0.083	O'Day, 1982
Los Angeles, CA	0.043	O'Day, 1982
Louisville, KY	0.123	O'Day, 1982
Milwaukee, WI	0.234	O'Day, 1982
New Orleans, LA	0.680	O'Day, 1982
New York, NY:		
Manhattan	0.170	O'Day, 1982
Other Four Boroughs	0.50	O'Day, 1982
Manhattan High Break Area	1.250	O'Day, 1982
Nottingham, England	0.262	O'Day, 1982
Philadelphia, PA	0.270	Goulter and Kazemi, 1989
San Francisco, CA	0.106	O'Day, 1982
St. Catharines, Ontario	0.430	Goulter and Kazemi, 1989
St. Louis, MO	0.077	O'Day, 1982
Troy, NY	0.167	O'Day, 1982
Washington, DC	0.116	O'Day, 1982
Winnipeg, Canada	1.760	Goulter and Kazemi, 1989

The following empirical relationship for t_r is presented by Walski and Pelliccia (1982):

$$t_r = 6.5(D)^{0.285} \text{ hr} = 7.4 \times 10^{-4}(D)^{0.285} \text{ yr} \qquad [2.1.3]$$

where D is the water main diameter in inches.

Based on equation [2.1.3], a 30 cm (12 in) diameter water main would take 13 hr to repair, at least in Binghamton N.Y. (Walski and Pelliccia, 1982).

Equations [2.1.2] and [2.1.3] should be used with site specific values for f_{bl} and t_r as well as L. Table 2.3 provides data on how f_{bl} can be adjusted by relative frequency factors such as the type of soil and pipe material, the pipe diameter, and the presence of previous breaks on or near the water main in question. One possibly important factor (particularly in applications where corrosion is the dominant failure mechanism) not listed in Table 2.3 is the pipe age. However, O'Day (1982) and Goulter and Kazemi (1989) indicate that pipe age is not as important as the other factors listed. In a complex municipal water distribution system, the effective water main length to be used in equation [2.1.2] is the length of main which, if unavailable, would reduce the water supply below the required level. (Hydraulic analysis computer programs can be used to determine how flow rates are altered by shutting off different pipe segments.) When more than one section of main is needed to provide adequate flow, L should be the sum of all critical section lengths.

Example – We can use the data in Table 2.2 to calculate the probability that the Main St water main in the 1965 Cambridge, MA fire would be unavailable. Although a site-specific value of f_{bl} is not available, a range of values can be estimated from the data in Table 2.2. We can use the value of f_b for Boston (0.036 breaks per mile-yr) as the low end of the range, and the median value of f_b (0.12 breaks per mile-yr), adjusted by the relative frequency for mains over 51 cm (20 inches) in diameter, as the high end of the range. Thus

$$f_{bl} > (0.036)(1.29) = 0.046 \text{ breaks/leaks per mile-yr},$$

and

$$f_{bl} < (0.12)(1.29)(4.7) = 0.73 \text{ break/leak per mile-yr}.$$

The calculated time to repair the 61 cm (24 in) diameter main, based on equation [2.1.3], is

$$t_r = 7.42 \times 10^{-4}(24)^{0.285} = 1.83 \times 10^{-3} \text{yr} = 16 \text{ hr}$$

The section length in question was about 1/4 mile, so equation [2.1.1] is

$$p > (0.046)(0.25)(0.00183) = 2.1 \times 10^{-5}$$

Table 2.3. Factors affecting water main break frequencies

Factor	Relative break frequency	Reference
No previous breaks	0.68	Walski and Pelliccia, 1982
One or more previous breaks	7.3–9.4	Walski and Pelliccia, 1982
Type of Soil:		
River gravel	0.65	O'Day, 1982
Sand and gravel	0.8	O'Day, 1982
Clay	1.5	O'Day, 1982
Pipe Diameter:		
15 cm (6 inch)	3.7	O'Day, 1982
30 cm (12 inch)	0.68	O'Day, 1982
51 cm (>20 inch)	4.7	Walski and Pelliccia, 1982

and

$$p < (0.73)(0.25)(0.00183) = 3.3 \times 10^{-4}$$

The calculated probabilities are low, but not sufficiently low to ignore the possibility of losing water access, particularly in view of the consequences that actually occurred. The overall probability of losing water access would also include the possible unavailability of the 20 cm (8 in) diameter, 400 m (1/4 mile) long supply line leading directly to the warehouse. Based on the calculated value of t_r for this line, the overall probability (both the 24 inch main and the 8 inch pipe) of water supply outage due to water main leaks/breaks is about 1.7 times the preceding calculated values of p, i.e. between 3.6×10^{-5} and 5.6×10^{-4}.

How large can the probability of the public water supply being unavailable be before it significantly degrades plant fire protection? One way to answer this question is to compare the water supply availability to the overall failure rates for automatic sprinkler systems. Chandler (1987) has reviewed published data on sprinkler system reliability from various sources. There is a wide variation in the data depending on the source and the criteria for system failure. NFPA data (Automatic Sprinkler Performance Tables, 1970) for the period 1925–1969 showed that sprinkler failure, in the sense of allowing excessive fire spread, occurred with a probability of 0.038. The lowest failure rates are reported for Australia and New Zealand, where Marryat's statistics for the 100 year period 1886–1986 (Marryat, 1988) indicate a system failure probability (defined as an uncontrolled fire in a sprinklered facility) of 5.4×10^{-3}. Sprinkler systems in Australia and New Zealand are apparently more reliable because water flow alarms are hardwired to local fire brigades, and required to undergo weekly inspection and testing.

Based on the NFPA data (Automatic Sprinkler Performance Tables, 1970), public water supply unavailabilities calculated for the Cambridge warehouse incident are two to three orders-of-magnitude smaller than the total probability of a sprinkler system failing to control a fire. Therefore, this failure mode would not have been an important factor in sight selection. However, the data in Tables 2.2 and 2.3 indicates that pipe break/leak frequencies can be one to two orders-of-magnitude higher in certain cities, particularly if the site is near an area where previous water main breaks occurred. In those cases, public water supply unavailability can significantly decrease the probability of successfully controlling a fire. Similar considerations apply to the probability of water capacity being effectively lost due to prolonged droughts. In other words, these considerations are ordinarily minor factors, but local conditions can increase their importance to the point of significantly decreasing the overall reliability of water based fire protection systems.

2.1.3 LOCAL FIREFIGHTING ORGANIZATIONS

The capabilities of local firefighting organizations can have an important impact on the overall level of plant fire protection. The Insurance Services Office Fire Suppression Rating Schedule (Fire Suppression Rating Schedule, 1980) provides some quantitative guidelines to measure public firefighting capabilities. Relevant factors are: (1) fire department communication facilities for reporting fires and for dispatching apparatus; (2) apparatus and equipment for engine and ladder-service companies; (3) personnel on duty and on call; and (4) training requirements. The availability of special equipment (such as a foam truck) and mutual aid from neighboring towns are other factors affecting local firefighting resources. Pre-fire planning, especially in conjunction with accessible drawings of plant layouts and installed fire protection, is a less tangible factor, but should also be considered.

In the case of an evaluation for a specific plant site, the estimated response time to the site would be another vital factor. Historical data on response times to nearby facilities might be used if the new plant is not expected to significantly increase traffic or alter accessibility.

Many large industrial facilities maintain plant fire brigades with special equipment and fire-fighting experience. The availability of these brigades in providing assistance at other industrial facilities can be an important attribute, particularly in rural areas that do not provide a paid on duty public fire department.

2.1.4 LOCAL CODES AND ATTITUDES

Local building and zoning codes vary substantially and may affect the desirability of a plant site in terms of acceptability of the plant to the local community, requirements for installed fire protection equipment, and the chances of future exposure problems on adjoining property. These local or even national code variations can cause a drastic change in fire protection strategy. One area of variation is the differing requirements for installed fire resistance versus installed fire suppression systems. Other differences involve requirements for smoke control, toxicity of combustion products from building materials, and the listing/approval specifications for fire fighting equipment such as sprinkler systems. Members of the European Community are eliminating these national differences in fire protection equipment requirements, so that site selection within the EC countries should no longer be a factor in this regard.

As indicated in Section 1.2.3, many industrial fires are ignited deliberately. The propensity for arson or suspicious fires does vary significantly from one community to another. Statistics on the numbers of suspicious fires and arson arrests/convictions in different towns and cities are available (for example in *Firehouse* magazine). Variations among nations are also striking and are indicative of the cultural differences and attitudes regarding deliberately set fires.

2.1.5 LOCAL ENVIRONMENTAL EFFECTS

Earthquake effects

Special provisions are needed for sprinkler installations intended to withstand earthquakes. Most of these provisions involve providing flexible connections between sprinkler piping and building structural supports. It is also important to provide bracing in order to prevent or dampen excessive vibrations associated with these flexible couplings. Guidelines on earthquake resistant sprinkler piping connections and bracing can be found in the NFPA Sprinkler Installation Standard (NFPA 13, 1994) and in a special FM Data Sheet on this subject (FM Data Sheet 2-8, 1982).

Another important industrial fire hazard in earthquake prone locations is the increased possibility of releases from flammable gas/liquid piping and storage vessels. Prevention measures are analogous to those for sprinkler piping. Guidelines are usually provided in regional/local building codes. These codes also prescribe building structure designs to accommodate seismic loads.

Ambient temperature extremes

Locations with extremely low or high temperatures can present special fire protection problems. Subfreezing ambient temperatures require precautions to prevent water freezing in sprinkler piping and water storage tanks. This includes heating the storage tank, placing the yard piping below the frost line, and either heating the building in which the sprinkler system is installed or using a dry pipe system. The latter option is less expensive but inherently results in a less effective and less reliable sprinkler system.

Extremely high ambient temperatures can present problems for materials susceptible to spontaneous heating and ignition as discussed in Chapter 6. Ambient temperatures exceeding the flash point of flammable liquids to be used at that location may generate the need for special fire and explosion protection measures. One other detrimental effect of high ambient temperatures is the possible need to either use higher temperature settings for thermal detectors and sprinkler

PLANT SITING AND LAYOUT 43

links, or to accept a greater chance of spurious actuations. Air conditioned buildings can virtually eliminate these problems, but air conditioning has the drawbacks of being expensive for large storage buildings and also possibly requiring use of a flammable liquefied gaseous refrigerant.

Water runoff capacity

Sprinkler system and hose stream water runoff can present problems at certain plant locations. This could occur if the sewer system cannot be utilized for runoff or is undersized for a design basis fire water supply. Another example is a location immediately adjacent to an environmentally sensitive river, lake, or aquifer. This problem seems to be escalating and has required the construction of special runoff containment and treatment facilities such as the one in Basel, Switzerland.

2.2 Plant layout for fire/explosion protection

2.2.1 GENERAL PRINCIPLES AND PROCEDURES

Locations of plant equipment and structures play an important role in determining plant fire and explosion hazards. The following general principles are offered for laying out plant equipment and structures to provide an inherent layer of fire and explosion protection:

1. Segregate hazards according to different levels of flammability and explosibility. Isolate the most hazardous materials and processes so that special precautions can be maintained in these hazardous areas. Limit the size and extent of any one hazardous area so that the entire plant is not jeopardized by any single incident.
2. Eliminate ignition sources from the vicinity of the most flammable and explosive materials and equipment.
3. Provide passive barriers for fire containment and explosion resistance. These barriers should effectively limit fire/explosion propagation and damage even in the absence of active detection and suppression.
4. Locate in-situ automatic fire detection and suppression systems components so that they will respond rapidly and effectively with minimal reliance on plant personnel.
5. Provide accessibility for manual fire fighting.
6. Provide adequate numbers and locations of emergency exits for rapid evacuation and rescue.

A somewhat longer set of similar principles has been suggested by Lees (1980, Chapter 10) for the layout of chemical process plants. The longer list includes facilitating plant operations, and minimizing cost and land usage. The latter often conflicts with the objective of maintaining safe separation distances.

Lees cites several layout techniques in which these principles can be implemented. These techniques generally start with a process flow sheet, a plot plan, and a set of equipment specifications. The Institution of Chemical Engineers (Mecklenburgh, 1982) provides systematic guidance on how these can be combined with British regulatory codes and standards and safe separation calculation procedures to obtain a suitable plant layout.

The emphasis in this chapter is on good fire protection engineering practice applicable to a wide assortment of industrial facilities. This is presented via elaboration and illustrative examples of the use and misuse of the six principles above.

2.2.2 HAZARD SEGREGATION AND ISOLATION

Hazard segregation may be based upon material properties or upon classifications in applicable codes and standards. One obvious criterion would be materials with special hazards. These would include highly reactive materials, toxic materials, unstable materials (prone to self-decomposition), oxidizers, and flammable gases, vapors, and dusts/powders. NFPA 49 (1988) and 325M (1988) provide listings and brief property descriptions of these types of hazardous chemicals. Segregation of these materials allows special protection measures to be implemented without burdening the entire plant or building.

An illustrative application of the need to isolate flammable gases and vapors exists in aerosol can filling operations with flammable liquefied gas propellants. Often, none of the other materials used in these filling operations warrant the explosion protection measures needed for butane-propane mixture propellants. Therefore, the location of the 'gassing room' for the charging of butane-propane propellants into aerosol cans is a critical factor in obtaining effective explosion protection. NFPA 30B (1990) specifies that the gassing room be separated from adjacent buildings by at least 1.5 m (5 ft), and from inside manufacturing areas by noncommunicating walls. It also specifies that the gas room be constructed with a combination of blast resistant and deflagration venting walls in accord with the NFPA 68 guidelines. The blow out wall should face away from the manufacturing building as shown, for example, in Figure 2.6.

In isolating the gassing room at an aerosol plant, there is a need to provide access for the conveyor lines carrying cans into and out of the gassing room. Figure 2.6 shows one scheme for providing conveyor access without running the conveyor lines outdoors. The enclosure between the gas house and the main building represents an intermediate hazard or transition between the gas house and the main production facility.

Isolation of flammable gas/vapor/powder equipment also includes piping and ducting. For example, if an aerosol filling main production facility is to remain free of explosion hazards, flammable propellant piping leading from storage tanks to the gassing room should not pass through or alongside the production facility. An analogous situation exists in plants with combustible powder or dust transported through pneumatic tubing or dust collector ducting.

After isolating special hazards, Lees (1980) suggests that layout proceed by dividing areas into hazard initiators, hazard transmitters/augmenters, and targets/victims. Hazard initiators usually involve the manufacturing operations or utilities. Hazard transmitters/augmenters usually involve the storage of large quantities of combustible materials. Potential targets or victims usually involve

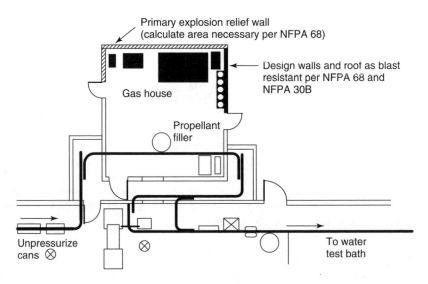

Figure 2.6. Propellant gas house layout at an aerosol production facility

areas where many employees are located or where there is some critical equipment or resource. Passive barriers, typically in the form of fire walls, are often needed to separate the initiators from the transmitters from the targets/victims. This is illustrated in the plant layout example shown in Figure 2.7.

Hazard segregation based on different fire protection requirements might be based on the classifications for automatic sprinkler protection as specified, for example, in the NFPA Standards. Table 2.4 is a list (not complete) of some NFPA Standards with different hazard categories and corresponding sprinkler density, water supply and sprinkler demand area requirements.

The categories and classification criteria in these standards are based on material/commodity flammability and extinguishability properties. Detailed descriptions are given in the appropriate chapters: Chapter 5 for Warehouse Commodities, Chapter 6 for Roll Paper and Aerosol Products, and Chapter 7 and 8 for Flammable Liquids. Unfortunately, the approaches/properties/tests used to establish classification criteria are not consistent for the various standards. A direct quantification of fire suppression requirements would be desirable. For example, early suppression sprinkler system requirements could logically be based on Required Delivered Densities (RDD) of water reaching the top of the commodity array at a particular time after ignition (Fleming, 1995). Effective utilization of these early suppression systems may ultimately require hazard segregation based on commodity RDD values.

Hazard segregation to establish suitable fire protection requirements should account for building fire resistance as well as commodity flammability. Fire resistance of building ceilings walls, and

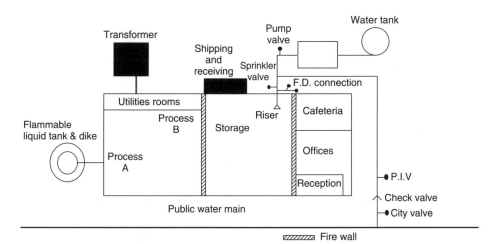

Figure 2.7. Plant layout example isolating initiators, transmitters, and targets

Table 2.4. NFPA standard sprinkler protection classifications

Material/Occupancy Classification	NFPA Standard
General occupancy	NFPA 13
Warehouse commodities	13, 230
Roll paper	13, 230
Flammable liquids	30
Aerosol products	30B
Liquid & solid oxidizers	49A
Transformers, cable trays, tanks	15

doors is discussed in Chapter 3. NFPA 220 and Factory Mutual Data Sheet 1-1 offer generic building fire resistance classifications and recommended flammability tests to establish classifications for specific building materials and products.

2.2.3 IGNITION SOURCE ISOLATION

Guidance and priorities for isolating ignition sources can be found in industrial fire loss statistics. One excellent source is the NFPA review of large loss fires (property damage of at least $500,000) in the period 1975 through 1984 (Redding and O'Brien, 1985). Table 2.5 summarizes the reported distribution of ignition sources responsible for large loss fires in mercantile, manufacturing, and storage properties.

The first two ignition source categories in Table 2.5, electrical origin and incendiary/suspicious ignitions, account for over half the large loss industrial fires with known ignition sources (34% have unknown ignition sources). Thus high priorities in ignition source isolation should be (1) remote locations of open fired equipment, (2) maintenance of tight security to prevent access of arsonists, and (3) elimination of ignition-prone electrical equipment from areas with high concentrations of combustible and easily ignitable materials.

Open fired equipment includes furnaces, heaters, dryers, and flares. It would be prudent in most cases to locate this equipment far from sources of flammable liquids and vapors. In the case of the aerosol manufacturing plant, cans are passed through dryers upon emerging from water baths to check for can leaks. There is a need to provide adequate separation between the water bath (source of flammable vapor) and the dryer (potential ignition source even if it isn't open fired equipment). In many plants, heaters are often used to shrink wrap the finished product. These are frequent ignition sources if not adequately isolated and maintained.

Security precautions to prevent entry of potential arsonists include supervision of exits and entrances. However, exit location supervision should not interfere with emergency evacuation and rescue capacity as described in Section 2.2.7. This may entail the use of alarmed exits and entrances.

The National Electrical Code (NFPA 70, 1989) is the most prominent and generally applicable Standard concerning ignition-prone electrical equipment. One particular aspect of the National Electrical Code that is especially relevant to ignition source isolation is the section on electrical equipment in hazardous locations (Articles 500 through 517). The intent is to eliminate electrical

Table 2.5. Large loss industrial fire ignition sources (based on data reported by Redding and O'Brien, 1985)[a]

Ignition source	Mercantile	Manufacturing	Storage
Electrical Equipment Arcing and Shorting	36%	25%	19%
Open Flame, or Incendiary Device	33	12	30
Hot Object in Processing, Electrical or Other Equipment	–	14	5
Cutting or Welding Torch	–	8	6
Fuel-fired Heating or Other Equipment	9	10	7
Spontaneous Ignition	–	6	5
Cigarette	4	3	4

[a]Based on 1227 fires with known ignition sources

ignition sources from areas possibly containing flammable gas-air mixtures, or combustible dust or fiber suspensions or accumulations. Classifications for flammable gas, combustible dust, and combustible fiber hazardous locations are designated Class I, II, and III locations, respectively.

An important and relevant subclassification for NEC hazardous locations pertains to the likelihood of flammable gas or dust concentrations being present under normal and accidental conditions. The subclassification scheme is as follows:

- *Division 1* locations are those in which flammable concentrations exist continuously, intermittently, or periodically under normal operating or maintenance conditions; or where equipment failure may simultaneously produce such concentrations and electrical ignition sources.
- *Division 2* locations are those containing normally confined flammable gases, vapors, or dusts; or those locations immediately adjacent to Division 1 locations.

Electrical equipment are rated for specific Class/Division classifications. In Class I Division 1 locations, electrical equipment are required to be either 'explosion proof', 'intrinsically safe,' or 'purged.' An explosion proof rating implies the equipment will contain an internal explosion (of a specific vapor-air mixture) and will have a surface temperature below the ignition temperature of the mixture. An intrinsically safe rating implies the equipment will not release sufficient electrical or thermal energy under normal conditions *and* under electrical faults during abnormal conditions to ignite the most readily ignitable concentration of vapor for that classification. In Class I Division 2 locations, equipment are required to be 'nonincendive,' which means it will not be an ignition source during normal operation (less restrictive than the 'intrinsically safe' rating). Tests and listings of electrical equipment rated for use in hazardous locations are provided by Underwriters Laboratories and Factory Mutual Research Corporation in the US and by their counter parts, such as UL-Canada and T.U.V., in other countries.

The British counterpart to the NFPA/NEC hazard classifications is British Standard 5345: Part 1, which defines the following three categories of hazardous areas:

- *Zone 0* in which a flammable gas-air mixture is continuously present, or present for long periods.
- *Zone 1* in which a flammable gas-air mixture is likely to occur sometime during normal operation.
- *Zone 2* in which a flammable gas-air mixture is not likely to occur during normal operation.

Thus, the BS Zone 0 and Zone 1 correspond to two subcategories of the NFPA/NEC Division 1 category, while the Zone 2 category is loosely equivalent to the NFPA/NEC Division 2 category.

Since the presence and extent of the NFPA and BS hazardous location areas depend on the potential formation of flammable vapor-air mixtures, it is important to consider both the size and likelihood of a flammable liquid/vapor release, *and* the effectiveness of ventilation in rapidly diluting the released vapor. It is sometimes necessary to make site-specific measurements with flammable vapor detectors or dust concentration probes. In most cases, generic estimates of these hazardous areas can be found in the standards. Examples of NFPA/NEC Division 1 and Division 2 locations for flammable vapors and gases are given in NFPA 497A (1986) for representative process equipment and potential release sites. A few example diagrams are reproduced here in Figures 2.8 and 2.9. The example in Figure 2.8 is for a ground level leakage source of flammable liquid in a building. If the building is adequately ventilated, it is unlikely that flammable mixtures will form anywhere except in a below grade sump or trench in which the leak will be contained. Therefore, the sump/trench is classified as Division 1, while a 0.91 m (3 ft) high area surrounding the leakage sight and a semicircular area of 1.5 m (5 ft) radius is classified as Division 2. The bulk of the building is considered to be a nonhazardous nonclassified region. On the other hand,

48   INDUSTRIAL FIRE PROTECTION ENGINEERING

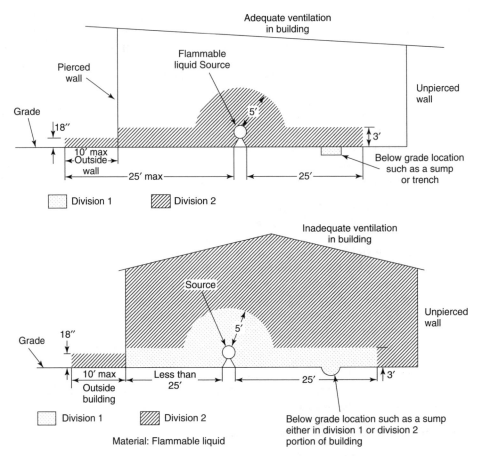

Figure 2.8. Division 1 and Division 2 areas in flammable liquid buildings. Reprinted with permission from NFPA 497®, *Classification of Flammable Liquids*. Copyright © 1997 National Fire Protection Association, Quincy, MA 02269. This reprinted material is not the complete and official position of the National Fire Protection Association, on the referenced subject which is represented only by the standard in its entirety

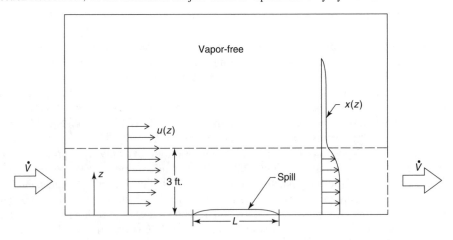

Figure 2.9. Air velocities and vapor concentrations in a well ventilated enclosure with a combustible liquid spill

PLANT SITING AND LAYOUT

if the building is not adequately ventilated, the lower 0.91 m is considered to be Division 1 and the bulk of the building is deemed to be Division 2.

How much ventilation is adequate to prevent flammable vapor-air mixture formation throughout a large section of the building? According to NFPA 497A, the mechanical ventilation should be equivalent to natural ventilation in an open area enclosed with at most one wall and a roof. Since this is too ambiguous to quantify, the approach suggested here is to evaluate the ventilation required to maintain the volume-average concentration below one fourth of the vapor lower flammable limit. This in turn requires an estimate of the vapor generation rate, which could either be obtained empirically or via the following theoretical analysis.

Consider a spill or leak of flammable liquid onto the floor such that the area of the spill is A. The actual value of A could be estimated from either the size of the sump/trench or, if there is no nearby sump, from the volume spilled and the equilibrium spill layer thickness as described in Chapter 7. The mass generation rate of vapor, E_v, is given by

$$E_v = M_v k A P_{sat}/(RT_L) \qquad [2.2.1]$$

where E_v is the vaporization rate (kg/min), M_v is the vapor molecular weight (kg/kg-mole), k is the mass transfer coefficient (m/min), P_{sat} is the liquid saturation vapor pressure (Pa) at T_L, R is the ideal gas constant (8314 J/(kmol-°K), and T_L is the liquid temperature (°K).

Several different empirical correlations for the mass transfer coefficient are available. The simplest correlation, which is intended for outdoor releases, is (EPA-OSWER-88-001, p. G-3)

$$k = 0.25 u^{0.78} \left(\frac{18}{M_v}\right)^{1/3} \text{cm/s} = 0.15 u^{0.78} \left(\frac{18}{M_v}\right)^{1/3} \text{m/min} \qquad [2.2.2]$$

where u is the wind speed (m/s), which is usually measured at an elevation of 10 m.

A more comprehensive correlation, which is based on laminar flow over a flat plate representing the spill surface (AIChE CCPS, 1996), is

$$k = 0.664 \left(\frac{D_v}{L_{sp}}\right) Sc^{1/3} Re^{1/2} \qquad [2.2.3]$$

where D_v is the vapor molecular diffusivity (m²/min), L_{sp} is the liquid pool dimension in direction of u (m), Sc is the vapor Schmidt number (kinematic viscosity/vapor diffusivity), and Re is the Reynolds number (velocity × L_{sp}/kinematic viscosity).

The vapor diffusivity is usually evaluated based on the diffusivity of water vapor, D_w, and the molecular weights of water and the vapor in question, as

$$D_v = D_w \left(\frac{18}{M_v}\right)^{1/2} = 1.435 \times 10^{-3} \left(\frac{18}{M_v}\right)^{1/2} \text{m}^2/\text{min} \qquad [2.2.4]$$

Combining equations [2.2.3] and [2.2.4] with the definitions of Schmidt and Reynolds number (and approximating the vapor kinematic viscosity with that of air, v_a) yields

$$k = \frac{0.664}{L} \left(\frac{18}{M_v}\right)^{1/3} D_w^{2/3} v_a^{1/3} \left(\frac{uL}{v_a}\right)^{1/2} \qquad [2.2.5]$$

The EPA Guidelines for Risk Management recommend that the spill area A in equation [2.2.1] be taken as the area corresponding to the full liquid volume with a depth of 1 cm, unless this calculated area is larger than the room floor area. They further recommend for their submittals that the air velocity u be input as about 0.1 m/s for a room/building ventilation rate of 0.5 volume changes per hour and typical sized ventilation fans. Of course, the actual effective value of u

depends on the local air flow velocity over the spill which depends on ventilation duct location as well as ventilation rate. EPA does not offer guidance on how the value of u should be adjusted for different ventilation rates and duct locations, so the options are to conduct in-situ measurements or flow field calculations. In the context of hazardous location determination ventilation rates for flammable liquid spills, we can consider u to be the average velocity between floor level and the 3 ft elevation level.

Once E_v is determined from equations [2.2.1] and [2.2.2] or [2.2.5], the volumetric ventilation rate needed to dilute the vapors to one-fourth of the LFL can be calculated as

$$V = 4E_v/(\chi_{lfl}\rho_v) \qquad [2.2.6]$$

where V is the required ventilation rate (m³/min), χ_{lfl} is the lower flammable limit volume fraction, and ρ_v is the vapor mass density (kg/m³).

Using the ideal gas law for ρ_v and equation [2.2.1], we obtain

$$V \geq \frac{4kAP_{sat}T_a}{\chi_{lfl}P_aT_L} \qquad [2.2.7]$$

where P_a and T_a denote ambient pressure and temperature, respectively.

If the ventilation inflow is below elevation, $h = 3$ ft, sweeps across the floor and exits at the opposite wall as sketched in Figure 2.9, the ventilation rate may be written as $V = uhw$, where w is the room width. In this case, u appears on both sides of equation [2.2.7], so that an explicit solution can be obtained for the minimum ventilation velocity to achieve the desired vapor dilution.

Similar approximations can be used to estimate vapor or gas leakage rates and required ventilation rates in other scenarios such as from vented storage tanks or leaky pipe fittings, or more catastrophic failures. Even when adequate ventilation is provided, there will inevitably be local regions of flammable vapor concentrations in the immediate vicinity of the release site. As an example, Figure 2.10 shows the Division 1 and Division 2 boundaries given in NFPA 497A for a well ventilated building containing a flammable liquid tank or vessel with a vent line emerging on the roof and an emergency dump tank adjacent to the building. In view of the multiple leakage

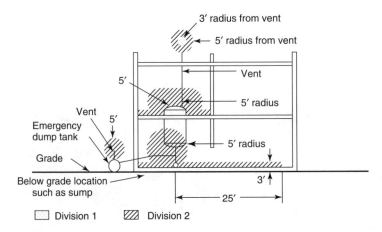

Figure 2.10. Division 1 and Division 2 classified areas in an adequately ventilated building containing a flammable liquid vessel or tank. Reprinted with permission from NFPA 497A®, *Classification of Flammable Liquids*. Copyright © 1992 National Fire Protection Association, Quincy, MA 02269. This reprinted material is not the complete and official position of the National Fire Protection Association, on the referenced subject which is represented only by the standard in its entirety

PLANT SITING AND LAYOUT 51

sites and Division 2 areas in the situation shown in Figure 2.10, it sometimes becomes more cost effective to classify the entire building as Division 2 rather than restrict unrated electrical equipment to the unclassified regions of the building.

In the case of combustible dusts processed in mixers, grinders, extruders, etc., guidance on classified Class II (dust) areas is offered in NFPA 497B (1991). Electrical equipment in Class II, Division 1 areas is required to be free of ignition sources, whereas equipment in Division 2 areas need only be dust tight. The relevant classification considerations are: (1) the potential for a large dust cloud; (2) accumulation of deep (>3 mm (1/8 in)) layers dust that are not readily discernible on equipment, floors, etc.; (3) the type of dust cloud or layer produced upon failure of dust collection equipment; and (4) the composition of the dust. In the case of unenclosed or only partially enclosed dust processing equipment producing a dust cloud, the Division 1 boundary either to a radius of 6.1 m (20 ft) or to the edge of the visible cloud. This often leads to an entire room being designated as Division 1, and a Division 2 area extending 3 m (10 ft) beyond it through a frequently opened door (NFPA 497B, 1991).

One other consideration in isolating ignition sources from combustible dusts and vapors is the potential for electrostatic discharge. This concern often leads to strict requirements for grounding equipment and for avoiding charge generations on operating personnel. In the case of at least one aerosol manufacturing facility, this entailed building the gas rooms with floors containing stainless steel grids extending up through the surface of the concrete.

2.2.4 PASSIVE BARRIERS

Passive barriers for fire resistance and containment include fire walls, doors, and roofing. Construction and test methods for these barriers are described in Chapter 3. Passive barriers for explosion isolation include damage limiting construction in the form of blast resistant walls and explosion venting walls and roofing. NFPA 68 and 69, and their counterparts in other countries, describe requirements for effective explosion resistance and explosion venting.

2.2.5 SPRINKLER SYSTEM LAYOUT

Sprinkler system layout considerations addressed here are:

1. Where should sprinkler heads be located?
2. Where should the water tank and pumphouse be located?
3. Where should sprinkler control valves be located?

Sprinkler heads are needed in buildings of combustible construction or combustible contents. Since building contents are often relocated, it is judicious to install sprinklers throughout most industrial structures. Interior ceiling sprinklers, which form the main line of defense, need to be supplemented with special sprinklers for the following situations (NFPA 13):

- concealed spaces, such as raised floors and suspended ceilings of combustible construction or containing electrical cables capable of self-sustained burning;
- on outside loading docks and similar structures used for temporary storage;
- in enclosed equipment processing combustible materials; and
- in enclosed structures such as the pumphouse if it is of combustible construction or contains diesel fuel.

In view of pump net positive suction head requirements, the pumphouse should be located close to the suction tank as illustrated in Figure 2.11. Pumphouse and water tank locations should

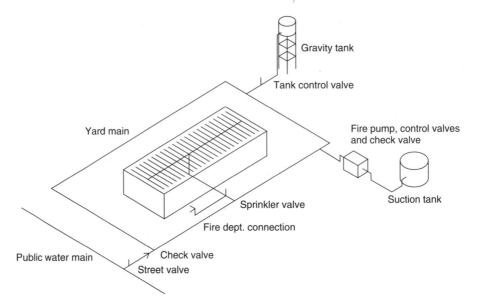

Figure 2.11. Sprinkler system layout

account for potential fire exposures as discussed in Section 2.1.1. If lack of yard room prevents adequate separation, fire resistant coatings or sprinkler heads may be needed for steel work. NFPA 22 (1984) specifies that structural steel for water tanks should be protected if the tank is within 6.1 m (20 ft) of combustible buildings or doors and windows from which flame can be emitted.

Accessibility is an important consideration in both pumphouse and sprinkler valve yard locations. Both the Plant Emergency Organization and the public fire department should be cognizant of these locations, which should be readily identifiable on a plan drawing of the plant fire protection services.

2.2.6 ACCESSIBILITY FOR MANUAL FIREFIGHTING

Accessibility here entails adequately wide roadways on the plant site for fire apparatus, and building/plant layout that allows hoselines to be run without obstruction. Lees (1980) suggests that adequate width for process plant roadways is 6.1 m (20 ft). Potential obstructions for hoseline runs between hydrants and fire sites include large process equipment and structures, trenches, waterways, and railroad spur tracks.

The author is aware of at least two examples of obstructed hoseline runs that significantly increased fire severity. One incident involved an effluent trench alongside a building with hydrants located on the roadway side of the trench. A trench fire ignited the building wall and prevented firefighters from laying hoselines across the trench to attack the building fire. The second example involved a railroad spur line between the main plant buildings and a storage structure. Hydrants were located on the main plant side of the railroad spur. When fire broke out in the storage shed hoselines were quickly laid across the tracks. Coincidentally, a small freight train traveled down the spur unaware of the fire. The train severed the hoselines and prevented any effective attack on the storage structure fire.

2.2.7 EMERGENCY EXITS

Current specifications for emergency egress have evolved from a series of tragic fires involving many lost lives. The first such incident in an industrial occupancy was the Triangle Shirtwaist

Company fire in New York City in 1911. One hundred forty five workers, mostly young girls, were killed when they were trapped in the Triangle fire. As described in Appendix B, insufficient numbers and sizes of exits, inward opening doors, and locked doors were factors contributing to the large number of fatalities. Public indignation led to the establishment of a NFPA committee that developed the first edition of the Life Safety Code in 1913 (NFPA 101, 1990).

The primary objective of emergency exits is to allow escape or rescue in a time less than the time to incapacitation due to the fire. Several different theoretical and empirical formulations to estimate escape time are described in NFPA 101 (1990) and by Pauls (1995). One simple formulation reviewed by Pauls (1995) is

$$t_e = N/(\text{Fw}) + L_e/v \qquad [2.2.8]$$

where t_e is the escape time for a group of occupants, N is the number of escapees, F is the flow rate of escapees through last exit (persons/m-s), w is the width of the last exit (m), L_e is the last exit to the first person in the escape group (m), and v is the walking velocity of the escapees (m/s).

In using equation [2.2.8] to evaluate escape times from tall office buildings, Pauls suggests a value of F of 1.1 persons/m-s and a value of v of 1.0 m/s down a stairway. The author is unaware of recommended values for industrial occupancies.

Equation [2.2.8] indicates that the two variables relevant to plant layout are the distance to the furthest exit, and the effective width of the smallest exit. NFPA 101 specifies values for these parameters for a variety of industrial occupancy categories. NFPA 101 also specifies the need for multiple exits and escape routes in hazardous occupancies.

Many computer codes, with varying levels of sophistication, are available for estimating egress times for specific fire scenarios. Some of these are deterministic, while others account for the randomness inherent in evacuee locations at fire initiation, door settings, etc. Although these codes are primarily used for analyzing escape times from public facilities, there is no inherent reason why they should not also be applicable to industrial fire scenarios.

2.2.8 COMPUTER AIDED PLANT LAYOUT

The advent of computer modeling is beginning to impact several aspects of plant layout fire and explosion protection considerations. In the case of special hazard isolation via spatial separation, models of flame radiation, blast waves from vapor cloud explosions, and flammable and toxic vapor dispersion are available with graphical output that can be used interactively with plot layout overlays. Most of the models are commercially available for use on personal computers, but the license fees can be expensive.

In some cases, the special hazard consequence models have been combined with accident probability calculations to produce risk contours plots, i.e. contours representing the area in which the probability or frequency of a blast wave pressure or a vapor concentration exceeds some specified value. These contours can be used as a computer aided design tool to evaluate alternative plant layouts and plant sites (for example, Ramsey et al., 1982; and the AIChE Guidelines fr Chemical Process Quantitative Risk Analysis, 1989). The Institution of Chemical Engineers has developed a simplified design procedure for plot layout based on contour plotting (Mecklenburgh, 1982).

Analogous computer models have been developed for interior compartment fires. HAZARD I, developed at the National Institute of Standards and Technology Center for Fire Research, is one such computer model that is now being used to explore alternative building and room layouts from the standpoint of fire resistance and successful escape and rescue. The applicability and accuracy of this and similar computer models is yet to be determined for most industrial fire scenarios.

References

Automatic Sprinkler Performance Tables. 1970 Edition, *Fire Journal*, NFPA, July 1970.
Briggs, G., Plume Rise, U.S. Atomic Energy Commission, 1969. (Available as TID-25075 from National Technical Information Service.)
Brighton, P.W.M., Evaporation from a Plane Liquid Surface into a Turbulent Boundary Layer, *J. Fluid Mechanics*, **159**, 323–345, 1985.
Chandler, S.E., A Review of Some of the Literature Relating to the Performance and Reliability of Sprinklers, Building Research Establishment Note N136/86, 1987.
Christoe, M., Amendola, A. and Smeder, M., The Control of Major Accident Hazards: The Land-Use Planning Issue, *Journal of Hazardous Materials*, **65**, 151–178, 1999.
City Water Main Repairs Cause Severe Loss, *NFPA Fire Journal*, 12–16, November 1965.
Crowl, D.A. and Louvar, J.F., *Chemical Process Safety Fundamentals with Applications*, PTR Prentice Hall, Englewood Cliffs, NJ, 1990.
Custer, R., Swimming Pool Chemical Plant Fire Springfield, Massachusetts (June 17, 1988), U.S. Fire Administration Technical Report Series, 1988.
Diefenbach, R.E., Pesticide Fires, in *Hazardous Materials Spills Handbook*, Bennett, G., Feates, F. and Wilder, I., eds., McGraw Hill, 1982.
Duclos, P., Binder, S. and Riester, R., Community Evacuation Following the Spencer Metal Processing Plant Fire, Nanticoke, Pennsylvania, *J. of Hazardous Materials*, **22**, 1–11, 1989.
Factory Mutual Loss Prevention Data Sheet 1-20, 'Protection Against Fire Exposures,' Factory MA, 1979.
Factory Mutual Loss Prevention Data Sheet 2-8, 'Earthquake Protection for Sprinkler Systems,' Factory Mutual Research Corporation, October 1982.
Factory Mutual Loss Prevention Data Sheet 3-26, 'Fire Protection Water Demand for Sprinklered Properties,' Factory Mutual Research Corporation, 1983.
'Fire Suppression Rating Schedule,' Insurance Services Office, New York City, 1980.
Fleming, R. 'Automatic Sprinkler Systems Calculations', *SFPE Handbook for Fire Protection Engineering*, 2nd Edition, SFPE, NFPA, 1995.
Galloway, F.M. and Hirschler, M.M., A Model for the Spontaneous Removal of Airborne Hydrogen Chloride by Common Surfaces, *Fire Safety Journal*, **14**, 251–268, 1989.
Goulter, I.C. and Kazemi, A., Analysis of Water Distribution Pipe Failure Types in Winnipeg, Canada, *Journal of Transportation Engineering*, **115**(2), March 1989.
Guidelines for Use of Vapor Cloud Dispersion Models, Second Edition, Center for Chemical Process Safety of the American Institute of Chemical Engineers, 1996.
Guidelines for Chemical Process Quantitative Risk Analysis, Center for Chemical Process Safety of the American Institute of Chemical Engineers, 1989.
Hasegawa, H.K. and Lambert, H.E., Reliability Study on the Lawrence Livermore National Laboratory Water Supply System, *Proceedings of the First Intl. Symposium on Fire Safety Science*, 1986.
Heseldon, A.J.M. and Hinkley, P.L, Measurement of the Transmission of Radiation through Water Sprays, *Fire Technology*, **1**, 130, 1965.
Kawamura, P. and Mackay, D., The Evaporation of Volatile Liquids, *Proceedings of the 2nd Annual Seminar on Chemical Spills*, Environment Canada, 1985.
Lees, F.P., *Loss Prevention in the Process Industries*, Butterworth, 1980.
Loss Prevention Council, 'Rules of the Fire Offices' Committee for Automatic Sprinkler Installations,' 29th Edition, London, 1973.
Marryat, H.W., 'Fire: A Century of Automatic Sprinkler Protection in Australia and New Zealand 1886–1986,' Australian Fire Protection Association, Melbourne, 1988.
Mecklenburgh, J.C., 'Hazard Assessment of Layout,' Inst. of Chemical Engineers Symposium Series No. 71, 1982.
McGuire, J.H., Fire and the Spatial Separation of Buildings, *Fire Technology*, **1**, 278–287, 1965.
Mills, M.T., Modeling the Release and Dispersion of Toxic Combustion Products from Chemical Plant Fires, *Proceedings Intl. Conf. on Vapor Cloud Modeling*, AIChE, 1987.
Nelson, H.E. and MacLennan, H.A., Chapter 3-14, "Emergency Movement," *The SFPE Handbook of Fire Protection Engineering*, National Fire Protection Association, 1995.
NFPA 13, 'Standard for the Installation of Sprinkler Systems,' National Fire Protection Association, 1996.
NFPA 30B, 'Code for the Manufacture and Storage of Aerosol Products,' National Fire Protection Association, 1990.
NFPA 49, 'Hazardous Chemicals Data,' National Fire Protection Association, 1988.
NFPA 70, National Electrical Code, Chapter 5, Special Occupancies, National Fire Protection Association, 1989.
NFPA 80A, 'Recommended Practice for Protection of Buildings from Exposure Fires,' National Fire Protection Association, Quincy, MA, 1980.
NFPA 101, 'Life Safety Code,' National Fire Protection Association, 1990.
NFPA 325M, 'Fire Hazard Properties of Flammable Liquids, Gases, and Volatile Solids,' National Fire Protection Association, 1988.

NFPA 497A, 'Recommended Practice for Classification of Class I Hazardous (Classified) Locations for Electrical Installations in Chemical Process Areas,' National Fire Protection Association, 1986.

NFPA 497B, 'Recommended Practice for Classification of Class II Hazardous (Classified) Locations for Electrical Installations in Chemical Process Areas,' National Fire Protection Association, 1991.

Pauls, J., 'Movement of People,' Chapter 3-13, *The SFPE handbook of Fire Protection Engineering*, National Fire Protection Association, 1995.

Purser, D.A., 'Toxicity Assessment of Combustion Products,' *The SFPE Handbook of Fire Protection Engineering*, Section 2/Chapter 8, p 2–141, National Fire Protection Association, 1995.

O'Day, K., Organizing and Analyzing Leak and Break Data for Making Main Replacement Decisions, *Journal American Water Works Association*, November 1982.

Ramsay, C.G., Sylvester-Evans, R. and English, M.A., 'Siting and Layout of Major Hazardous Installations,' Inst. of Chemical Engineers Symposium Series No. 71, 1982.

Redding, D. and Pauley, P., Jr., Large-Loss Fires in the United States During 1985, *Fire Journal*, November 1986.

Sardqvist, S., Fire Brigade Use of Water, Interflam, *Proceedings of the Eighth International Conference*, pp. 675–683, Interscience Communications, 1999.

Sardqvist, S., Correlation Between Fire Fighting Operation and Fire Area: Analysis of Statistics, *Fire Technology*, 2000.

Shorter, G.W., McGuire, J.H., Hutcheon, N.B. and Legget, R.F., The St. Lawrence Burns, *NFPA Quarterly*, **53**, 300–316, April 1960.

Wackter, D. and Foster, A., 'Industrial Source Complex (ISC) Dispersion Model User's Guide – Second Edition – Vol 1,' EPA Report EPA-450/4-86-005a, 1986.

Walski, T.M. and Pelliccia, A., Economic Analysis of Water Main Breaks, *Journal American Water Works Association*, March 1982.

Walton, W.D. and McGrattan, K.B., ALOFT-FT(trademark) A Large Outdoor Fire plume Trajectory model – Flat Terrain Version 3.04. National Institute of Standards and Technology, NIST SP 924, 1998.

Walton, W.D., McGrattan, K.B. and Mullin, J.V., 'ALOFT-PC A Smoke-Plume Trajectory Model For Personal Computers,' National Institute of Standards and Technology, *Arctic and Marine Oilspill Program Technical Seminar Proceedings*, pp. 987–997, 1996.

Yamada, T., Smoke Plume Trajectory From In-Situ Burning of Crude Oil in Tomakomai: Field Experiments and Prediction With ALOFT-PC., National Research Institute of Fire and Disaster, Tokyo, Japan. National Institute of Standards and Technology, Gaithersburg, MD NISTIR 6242; October 1998. National Institute of Standards and Technology. Annual Conference on Fire Research: Book of Abstracts. November 2–5, 1998, Gaithersburg, MD, Beall, K.A., Editor, 95–96 pp, 1998. Available from National Technical Information Service, PB99-102519.

3 FIRE RESISTANT CONSTRUCTION

Fire resistance is defined as the time during which a structural element can withstand fire exposure as imposed by a standard fire test. Fire resistant construction requires all structural members, including walls, roofs, beams, and columns, to have some specified fire resistance rating, which is often a minimum of two hours (FM Data Sheet 1-1, 1983) for industrial occupancies. During this time, fire spread via the paths shown in Figure 3.1 is supposed to be limited by the appropriate fire barriers. Spread through a wall should be limited by fire wall construction. Spread through wall openings should be limited by fire door closings, and by wall penetration seals with the same fire resistance as the wall. Spread over and within the roof is limited by fire resistant roofing construction. All these items are discussed in this chapter. Smoke propagation through ducting is discussed in Chapter 4.

Most modern fire resistant industrial construction consists of reinforced concrete or insulated steel frame. The properties of these and other construction materials are reviewed here before discussing fire resistance calculations and fire resistance testing. After this general discussion of fire resistance, the chapter focuses on fire walls, fire doors, roofing, and water spray protection of exposed structures.

3.1 Construction materials

3.1.1 STEEL

Steel is the major structural material in industrial buildings. Its strength, ductility, consistency, and availability render it uniquely desirable for structural framework and for concrete reinforcing. However it is significantly weakened at fire temperatures such that very lightweight unprotected members can fail after only 5–10 minutes direct exposure to an intense fire (Fitzgerald, 1986).

The actual fire resistance of a steel member is determined by its structural load, size and shape, constraints, fire heat flux history, and material properties. Critical structural properties include yield strength and ultimate strength. Relevant thermal properties include the coefficient of thermal expansion, density, specific heat, and thermal conductivity.

The yield strength is generally defined (Fitzgerald, 1986, p. 126) as the stress that produces a permanent deformation of 0.2% of the original length of the test sample. It is also approximately equal to the stress at the plateau region of a stress-strain curve for steel at temperatures of 200 °C (392 °F) and lower. The ultimate strength is the maximum stress in a stress-strain curve. Beyond the ultimate stress, increasing deformation occurs with decreasing stress until the steel ruptures. The room temperature yield stress for A-36 structural steel is approximately 36,000 psi (250 MPa),

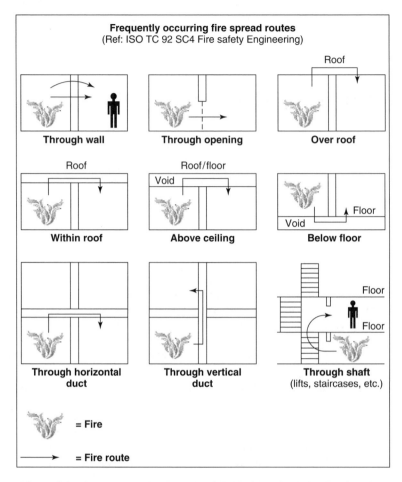

Figure 3.1. Fire spread paths (from ISO TC 92, SC4 Fire Safety Engineering)

and its room temperature ultimate stress is about 54,000 psi (370 MPa), although somewhat higher values are sometimes measured for both values (Lie, 1972).

The curves in Figure 3.2 show the ultimate strength and the yield strength variation with temperature, normalized by their respective values at room temperature, for St 37 mild structural steel. The yield strength decreases monotonically with temperature, while the ultimate stress peaks at about 250 °C (482 °F) and then decreases rapidly with increasing temperature. There is considerable spread in the data for both yield strength and ultimate strength. For example, the temperature at which the strength of St 37 is reduced by 40% is in the range 320 to 500 °C (610 to 930 °F) depending on the data scatter and whether we refer to yield strength or ultimate strength. The 40% reduction temperature is significant because the American Institute for Steel Construction specifies a maximum permissible design stress of approximately 60% of the yield strength for structural steel buildings (Milke, 1995).

The critical temperature for steel fire resistance is the steel temperature at which its strength is reduced to the point that it cannot support its applied load. This temperature depends on the precise structural failure criterion as well as the structure configuration, design load, and steel composition. Various examples listed in Table 3.1 span the range from 730 to 1220 °F

FIRE RESISTANT CONSTRUCTION

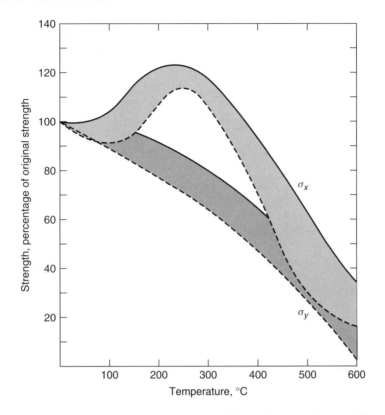

Figure 3.2. Temperature effect on steel strength (modified with permission from Lie, 1972)

Table 3.1. Critical temperatures for steel structures

Steel	Configuration	Load ratio[a]	Critical temperature [F (C)]	Reference
ASTM A-36	Statically Determinate Beam	0.6	880–1110(470–600)	Lie, p 162
ASTM A-36	Statically Indeterminate Beam	0.6	1080–1220(580–660)	Lie, p 162
ASTM A-36	Statically Determinate Beam	0.6	890(475)	Milke[b]
ASTM A-36	Statically Determinate Beam	0.4	1050(565)	Milke[b]
ASTM A-36	Statically Indeterminate Beam	0.6	890–1040(475–560)	Milke[b]
ASTM A-36	Statically Indeterminate Beam	0.4	1050–1140(565–615)	Milke[b]
ST 37	Statically Determinate Beam	0.6	730–890(390–475)	Lie, p 162
ST 37	Statically Indeterminate Beam	0.6	890–1020(475–550)	Lie, p 162
ST 37	Long Column ($L/r > 100$)	0.3	970(520)	Lie, p 168
ST 37	Short Column ($L/r < 100$)	0.3–0.5	790(420)	Lie, p 168
ASTM A-36	Long Column ($L/r > 100$)	0.52	940(505)	Milke
Various	Short Columns ($L/r = 23 - 87$)	–	507–753 (945–1387)	Talamona et al. (1996)[c]

[a]Load ratio values are the ratio of the applied (design) load to the load that would generate a stress equal to the room temperature yield stress
[b]Critical temperatures calculated by Milke (1995) were based on the analysis recommended by the European Convention for Constructional Steelwork.
[c]Critical temperatures reported by Talamona et al. (1996) are based on measured temperatures at observed column failure times.

(388 to 660 °C) for beams and from 790 to 1020 °F (421 to 549 °C) for columns. Values for statically indeterminate structures are higher than those for statically determinate structures. Lie (1972) recommends using a representative value of 790 °F (421 °C) for a statically determinate beam and 970 °F (521 °C) for a statically indeterminate beam.

The coefficient of thermal expansion for steel increases appreciably as the steel temperature is increased. The relationship quoted by Milke (1995) is

$$\alpha = (6.1 + 0.0019\Delta T) \times 10^{-6} \qquad [3.1.1]$$

where α is the coefficient of thermal expansion (in/in-°F), and ΔT is the steel temperature rise above 100 °F.

The linear expansion corresponding to equation [3.1.1] can be substantial at temperatures approaching 538 °C (1000 °F). For example, a 15 m (50 ft) long steel beam would be elongated by about 10 cm (4 in), which could be enough to collapse constrained walls or ceilings. The forces on fire walls due to heated beam expansions are discussed in Section 3.4.

Thermal property data for steel and other construction and insulation materials are listed in Table 3.2. The thermal conductivity and the mass density of steel are significantly higher than for most other materials. On this basis, steel temperatures would be expected to be spatially uniform and to lag significantly behind the local gas temperatures. The actual time lag is analyzed in Section 3.2.

Table 3.2. Thermal properties of construction and insulation materials (data from Appendix A of *SFPE Handbook* (1995) for most materials)

Material	Temperature (C)	k (W/m-C)	ρ (kg/m^3)	c (kJ/kg-C)	α (m^2/sec × 10^7)
Aluminum	20	204	2707	0.896	8.42
Asbestos Cement	20	0.175	750	–	–
Ceramic Fiber Blanket	260	0.055			2.18
	538	0.115	96.12	0.263	4.55
	816	0.202			8.00
	1093	0.208			11.40
Concrete					
Light Weight	20	0.61	1200	0.84	60.
Normal Weight	20	1.64	2300	0.84	8.5
	600	1.1	2300	1.25	3.8
Fiber Board	20	0.048	240	–	–
Magnesia-85%	38	0.067	270	–	–
Mineral Wool (Sprayed)		0.17	250	–	–
Plaster					
Cementitious	20	0.21	750	–	–
Metal Lath	20	0.47	1440	0.84	4.0
Steel					
1% Carbon	20	43	7800	0.473	1.17
	300	40	7800	0.47	1.1
	600	33	7800	0.47	0.90
Wood					
Maple-Oak	30	0.17	540	2.4	1.3
Gypsum Board	20	0.24	678	0.90	3.9
	100	0.24	649	3.0	1.2
	300	0.12	675	0.80	2.2

3.1.2 STEEL INSULATION

Protective insulation is often applied to steel structures in order to achieve a desired level of fire resistance. Some of the protective materials employed include magnesia, vermiculite, concrete, sprayed mineral wool, and intumescent/ablative coatings. Thermal property data for several of these materials are listed in Table 3.2. Some of the materials have thermal conductivities four orders-of-magnitude lower than that of steel. Fire resistance calculations to account for the low thermal conductivities of insulating material effects are described in Section 3.2. The effect of steel insulation on fire resistance test results is discussed in Section 3.3.

There are several practical considerations associated with the selection and evaluation of fire resistant steel insulation. For example, many of the commercial insulations require careful application and curing procedures to keep the insulation intact and properly attached to the steel structure. Furthermore, weathering, aging, and hose stream resistance tendencies of steel insulation are also important for certain applications. When warranted, special tests of these characteristics are sometimes conducted to supplement the fire resistance tests. Commercial insulations that have been certified to achieve a specified level of fire resistance and to have passed certain of these supplemental tests are listed in the Factory Mutual Approval Guide and/or the Underwriters Laboratories Fire Resistance Directory or in certification listings of other testing organizations.

3.1.3 CONCRETE

The inherent compressive strength of concrete (a typical room temperature ultimate compressive strength of 60,000 psi = 410 MPa) makes it an attractive material for columns and load bearing walls. Use of steel reinforcing bars to support tensile loads allows concrete to also be used extensively for beams and floor slabs. The ratio of reinforcing bar cross-sectional area to concrete area and the tensile strength and compressive strength of the steel and concrete, respectively, determine whether a beam will fail in tension or compression (Fitzgerald, 1986).

The variation of concrete compressive strength with temperature depends on the type of aggregate in the concrete. Siliceous aggregate concrete with cement-to-aggregate ratios of about 1:6 start weakening at temperatures of about 482 °C (900 °F) as indicated by the graphs in Figure 3.3. Higher proportions of cement start weakening at lower temperatures. Carbonate and lightweight aggregates remain relatively unaffected by temperature until about 649 °C (1200 °F). According to Fleischmann (1995), the critical temperature at which concrete is rendered structurally ineffective (as measured by strength reductions of about 50% of the room temperature values) is about 1200 °F for siliceous concrete and about 760 °C (1400 °F) for carbonate and lightweight aggregates.

The thermal expansion of concrete is similar to that of steel for temperatures up to about 538 °C (1000 °F). Thermal expansion of a concrete floor slab heated from below can cause large thermal thrust forces to be exerted on surrounding structures restraining the concrete expansion. These thermal thrust forces can be the limiting factor determining the fire resistance of concrete slabs. Calculation procedures to account for this effect are described by Fleischmann (1995).

3.2 Fire resistance calculations

Several different computational approaches are available for estimating the fire resistance of structural members. The three approaches reviewed by Milke (1995) are:

1. Empirical correlations of fire resistance test data.
2. Heat transfer analyses to calculate the fire induced heat fluxes and temperatures in structures.
3. Structural response computer analyses.

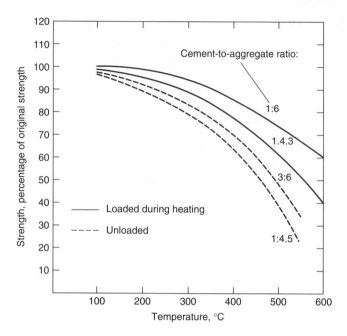

Figure 3.3. Temperature effect on concrete strength. (Reproduced with permission from Lie, 1972)

Empirical correlations are discussed briefly in Section 3.3.2. Simple heat transfer analyses to estimate the time required to reach a critical temperature are described in this section. More complicated heat transfer and structural response analyses requiring computer codes are reviewed by Milke (1995) and Barnett (1989). The more complicated heat transfer analyses are needed when it is important to account for either nonuniform temperature distributions in the structure or the nonlinearities associated with temperature dependent material properties or surface re-radiation, or for porous materials in which water evaporation and migration is a significant factor in determining fire resistance. A computer model that deals with the latter factor in concrete slabs has recently been reported by Ahmed and Hurst (1995).

The heat transfer equation applicable to the case of a uniform temperature steel structure exposed to a fire is

$$M_s c_s \frac{dT_s}{dt} = A(q_c'' + q_r'') \qquad [3.2.1]$$

where M_s is the mass of steel structure (kg), c_s is the specific heat of steel = 0.47 kJ/kg-°C, T_s is the steel temperature (°C) at time t (sec), A is the area of steel structure exposed to fire (m²), q_c'' is the convective heat flux to steel structure (kW/m²), and q_r'' is the radiative heat flux to steel structure (kW/m²).

If the fire induced heat fluxes are constant (a reasonable approximation *only* when $q_r'' \gg q_c''$ and $T_s < 200(q_r'')^{0.25}$, where q_r'' is in kW/m² and T_s is in K), the solution to equation [3.2.1] is

$$T_s T_o = \frac{(q_c'' + q_r'')At}{M_s c_s} = \frac{(q_c'' + q_r'')Dt}{(M_s/L)c_s} = \frac{(q_c'' + q_r'')t}{\rho_s(V_s/A)c_s} \qquad [3.2.2]$$

where D is the heated perimeter of steel structure (m), (M_s/L) is the steel structure mass per unit length (kg/m), (V_s/A) is the steel thickness (m), and ρ_s is the mass density of steel (7800 kg/m³).

The convective and radiative heat fluxes depend upon the steel temperature and its location relative to the fire. The simplest case might be for steel located within the flame, but is still

FIRE RESISTANT CONSTRUCTION

sufficiently cool for re-radiation to be negligible, and is sufficiently small to avoid local flame quenching. In this simplified case, the radiant heat flux is in the range 100–220 kW/m² (8.8 to 19.4 Btu/sec-ft²) (Appendix A). The convective heat flux will be at most 100 kW/m² (Alpert and Ward, 1984), with the upper limit corresponding to a cold surface located near the flame tip. Experimental data cited in Chapter 8 (Table 8.4) for steel objects immersed in pool fires suggest that the surface-average total (radiant + convective) heat flux is about 110 kW/m² (9.7 Btu/sec-ft²) for a large (at least 1 m (3.3 ft) in length) cool object and about 160 kW/m² (14 Btu/sec-ft²) for a small cool object. Substituting the former value into equation [3.2.2], the time required for a steel structure, completely engulfed in the fire, to reach its critical temperature, T_c, is

$$t_c = 33(V_s/A)(T_c - T_0) \text{ sec} \quad [3.2.3]$$

where t_c is the time (lower bound estimate) for a flame immersed steel structure to be heated to its critical temperature, and $T_c - T_0$ is in °C.

The value of t_c given by equation [3.2.3] is considered a lower bound because q_c'' and q_r'' will decrease as the steel is heated to T_c.

In the case of a steel column, with a characteristic thickness of 1 cm (0.4 in) and $T_c - T_0 = 400\,°C$ (752 °F), the lower bound $t_c = 33(0.01)(400) = 132$ sec. However, the caveat $T_s < 200(q_r'')^{0.25}$ for this heat flux would limit the applicability of equation [3.2.3] to steel temperatures less than 375 °C, which is about 45 °C less than the critical temperature of about 420 °C. More realistic estimates of t_c can be obtained from solutions to equation [3.2.1] for time varying heat fluxes as discussed below.

If the steel structure is located on the ceiling directly above the flame, the cold ceiling convective heat flux in the fire plume impingement region is given by (You & Faeth, 1979)

$$q_c'' H^2 Ra^{1/6}/Q_c^2 = < \begin{array}{l} 38.7 \text{ for } r/H < 0.16 \\ 1.88 r/H^{1.65} \text{ for } r/H > 0.16 \end{array} \quad [3.2.4]$$

where H is the ceiling (steel) height above top of burning fuel, or more precisely, above the plume virtual origin, Q_c is the fire convective heat release rate, and

$$Ra = \frac{g Q_c H^2}{\rho_a c_p T_a v_a^3} = \text{Rayleigh Number}$$

ρ_a is the ambient air density (1.16 kg/m³), c_p is the ambient air specific heat (1.01 kJ/kg-°C), v_a is the ambient air kinematic viscosity (1.6 × 10⁻⁵ m²/s), and T_a is the ambient air temperature (usually 293 °K).

As the steel structure is heated, the convective heat flux decreases since it is directly proportional to the local temperature difference between the steel and the hot gases. The correlation developed by Alpert (1987) is

$$0.24 H q_c''/(k_g \Delta T) = 0.36 Re^{0.61} \quad [3.2.5]$$

where ΔT equals $T_g - T_s$, which is the plume-steel temperature difference (°C), k_g is the thermal conductivity for plume gas which is $0.026(T_g/T_a)^{0.78}$ W/m-K, u_{max} is the maximum gas velocity at ceiling elevation of plume (m/s), b is the plume half-width at ceiling elevation (m), v is the plume gas kinematic viscosity which is $v_a(T_g/T_a)^{1.78}$, and

$$Re = u_{max} 2b/v = \text{plume Reynolds number} \quad [3.2.6]$$

According to plume theory (Alpert, 1987),

$$Re = 0.47 H^{2/3} (Q_c g/p)^{1/3}/v \quad [3.2.7]$$

where p is atmospheric pressure (100 kPa).

If q_c'' from equation [3.2.5] is used in equation [3.2.1], the solution is

$$\vartheta = 1 - e^{-t/\tau} \qquad [3.2.8]$$

where

$$\theta = \frac{T_s - T_a}{T_g - T_a}$$

is the nondimensional temperature rise of steel, and

$$\tau = \frac{(M_s/L)c_s H}{1.5 k_g D Re^{0.61}}$$

is the characteristic time constant.

The plume gas temperature rise needed to calculate T_s from θ is

$$T_g - T_a = 0.068 T_a Q_c^{2/3} H^{-5/3} \qquad [3.2.9]$$

where Q_c is in kW, H is in m, T_a is in °K, and $T_g - T_a$ is <1000°C (maximum temperature in plume).

For example, if a steel beam with a mass per unit length of 15 kg/m (10 lb/ft) and a heated perimeter of 30 cm (1 ft) is located on a 5 m (16 ft) ceiling directly above a fire with a convective heat release rate of 10 MW (9.5 Btu/sec-ft²) (and if we are willing to make the questionable/conservative assumption that the beam is sufficiently short for the entire beam to be effectively engulfed the plume), the unprotected beam fire resistance can be estimated as follows:

$$T_g - T_a = 0.068(293)(10000)^{2/3}(5)^{-5/3} = 632\,°C$$

$$Re = 0.47(5)^{2/3}[(10^4)(9.82)/(100)]^{1/3}/[(1.6 \times 10^{-5})(925/293)^{1.78}] = 1.1 \times 10^5$$

$$k_g = 0.026[(925)/(293)]^{0.78} = 0.064\,W/m\text{-}K.$$

$$\tau = \frac{(15)(0.47)(5)}{1.5(0.064 \times 10^3)(0.30)(1.1 \times 10^5)^{0.61}} = 1030\,\text{sec}.$$

The time required to heat the beam to a critical temperature of 500°C (932°F) corresponding to the critical value of θ denoted by θ_c is

$$t_c = -\tau \ln(1 - \theta_c)$$
$$= -1030 \ln[1 - (480)/(632)] = 1470\,s \qquad [3.2.10]$$

Thus, the beam would reach its critical temperature after 26 minutes of exposure to the convective heat flux from this fire.

Plume radiant heating can be added to the preceding solution by modifying the equation for τ as follows:

$$\tau' = \frac{(M_s/L)\dfrac{c_s}{D}}{1.5\dfrac{k_g}{H}Re^{0.61} + \sigma\varepsilon_s(T_g^2 + T_c^2)(T_g + T_c)} \qquad [3.2.11]$$

where σ is the Stefan-Boltzmann constant ($5.67 \times 10-11\,kW/m^2\text{-}K^4$), and ε_s is the steel surface emissivity (0.7–0.9).

If equation [3.2.11] is evaluated for the preceding example and the resulting $\tau' = 195\,\text{sec}$ substituted into equation [3.2.10], the revised solution is $t_c = 278\,\text{sec}$ (i.e. 4.6 min). The plume

FIRE RESISTANT CONSTRUCTION

radiant heat flux corresponding to $T_g = 652\,°C$ (1206 °F) and an emissivity of unity is 41.5 kW/m² (3.6 Btu/sec-ft²). The combined convective plus radiative cold wall heat flux is 56 kW/m² (4.9 Btu/sec-ft²). If this total heat flux is treated as constant and substituted into equation [3.2.2], the calculated lower bound value of t_c would be 201 seconds, i.e. 77 seconds less than the value calculated using equations [3.2.10] and [3.2.11].

Equations [3.2.5–3.2.9] do not account for the decreasing ceiling gas temperature outside of the plume intersection with the ceiling. In applying these equations to a steel beam, the calculations will underestimate the fire resistance of the beam because they overestimate the average gas temperature and heat flux along the entire length of the beam. It is possible to calculate the average gas temperature by using the ceiling jet temperature correlations (described by Alpert and Ward, for example) and integrating along the length of the beam. Thus,

$$\overline{(T_g - T_a)} = \frac{(T_g - T_a)_p}{L/(2H)} \left[0.18 + \int_{0.18}^{L/(2H)} 0.318 r^{-2/3} dr \right]$$

where $\overline{(T_g - T_a)}$ is the average gas temperature, and $(T_g - T_a)_p$ is the temperature rise of the plume at the ceiling, as given by equation [3.2.9].

In the preceding example, integration of this correlation over an 8 m (26 ft) beam length (such that the maximum value of r/H is 0.80), would yield an average ceiling jet temperature of 437 °C (819 °F). Since this temperature is less than the 500 °C (932 °F) critical temperature for the steel beam, one might be tempted to conclude that the beam does not require protection. However, it would probably be prudent to provide steel protection because the calculated value is sufficiently close to the critical temperature and the unconfined ceiling jet temperatures do not account for any accumulation of hot combustion gases under the ceiling jet. A more precise analysis in this case should also account for the virtual origin of the plume (as given in Chapter 5 for warehouse fires) and possible nonuniform steel temperatures over the beam length.

The most commonly used form of steel protection is a layer of thermal insulation. Solutions to the heat conduction equation for the case of an insulated steel structure have been presented in graphical form by Lie (1972) and Milke (1995). The steel temperature is assumed to be uniform and the one-dimensional transient heat conduction equation is solved for the layer of insulation applied to the steel. The relevant parameters are θ, defined as above, and

$$\text{Fo} = \alpha_i t / h^2$$

and

$$N = \frac{\rho_i c_i h}{c_s (M_s / LD)}$$

where α_i is the thermal diffusivity of the insulation (m²/s), h is the insulation thickness (m), ρ_i is the density of insulation (kg/m³), c_i is the specific heat of insulation (kJ/kg-K), and the other parameters are as previously defined.

Graphs of θ as a function of the Fourier number, Fo, with N as a parameter are shown in Figure 3.4.

As an example, suppose 1 cm (0.4 in) of mineral wool insulation is applied to the steel beam and exposure fire described in the previous example for unprotected steel directly above the fire:

$$N = (250)(0.8)(0.01)/[(0.47)(15)(0.30)^{-1}] = 0.085.$$

The value of Fo for $\theta = 0.78$ can be read by interpolation in Figure 3.4 to be Fo = 18. Therefore,

$$t = 19(0.01)^2/(1 \times 10^{-6}) = 1800\,\text{s} = 30\,\text{min}.$$

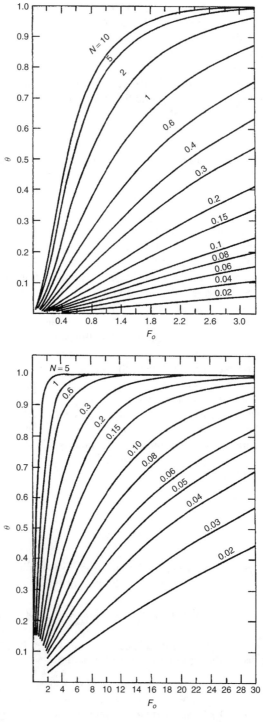

Figure 3.4. Dimensionless temperature of insulated steel versus insulation Fourier number. (Reproduced with permission from Lie, 1972)

FIRE RESISTANT CONSTRUCTION

Thus the 1 cm thick insulation renders the fire resistance 6.5 times (30/4.6) as great as the resistance of the exposed steel beam heated by the fire plume. Doubling the insulation thickness would approximately double the fire resistance of the beam in this case.

3.3 Fire resistance tests

Although the calculations described in Section 3.2 are useful for structural elements such as beams and columns, they become unwieldy in the case of structural assemblies such as fire doors, walls, and ceilings. Standardized fire resistance tests have traditionally been employed for both structural assemblies and structural components. These tests, which include furnace exposure tests and direct fire/burner exposure tests, are described in this section along with some empirical correlations of test results.

3.3.1 FURNACE EXPOSURE TESTS

Furnace exposure tests entail subjecting the structural assembly to a prescribed furnace gas temperature-time exposure achieved by regulating the furnace fuel and air flow rates. The temperature-time histories are based primarily on historical tests (circa 1916) with wood fires in poorly ventilated compartments. Figure 3.5 shows the temperature-time curves adopted by the International Standards Organizations (ISO 834, 1975), the American Society for Testing and Materials (ASTM E-119, 1983), and the British Standards Institution (BS 476-1972). The three curves have a rapid temperature rise during the first thirty minutes of exposure, and a much slower temperature rise over the next $6\frac{1}{2}$ hr. The ISO and BSI curves indicate slightly higher temperatures (30–40 °C (86–104 °F) higher) than the ASTM curve for exposure times beyond 60 minutes. Similar curves are used in most other countries (see Figure 4.8.11 of Lie, 1988).

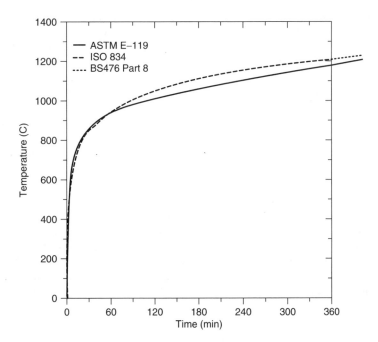

Figure 3.5. Comparison of temperatures in fire resistance tests

Two important complications inherent in the use of standard temperature-time exposures are: (1) differences between the anticipated enclosure fire temperature development and the furnace temperature-time curve; and (2) furnace heat fluxes on the test assembly depend on other factors besides furnace gas temperature. The first complication is usually addressed by specifying a furnace fire resistance that produces an area under the temperature-time curve that is equal to the area under the anticipated temperature-time curve. The historical rationale for the so-called 'equal area hypothesis' is discussed by Drysdale (1985). Lie (1988) and Drysdale (1995) describe relatively simple methods for calculating temperature-time curves for ventilation limited and for fuel limited enclosure fires.

An inherent limitation of the 'equal area hypothesis' is that it is better suited to convective heating (which is linearly proportional to temperature increase of the hot gas relative to the structure) than to radiant heating (proportional to the fourth power of gas temperature) which is the dominant heat transfer mode in many fire exposure scenarios. An alternative way to relate standard furnace exposure test data to anticipated fire scenarios is through the use of empirical correlations. For example, Latham et al. (1987) have measured steel structure temperatures in both furnace exposure tests and actual enclosure fires with varying degrees of fuel loading, ventilation, and enclosure thermal inertia. They compared their results to the following two correlations relating furnace exposure time to the parameters of an anticipated enclosure fire.

The Swedish correlation (Pettersson et al., 1976) is:

$$t_e = c'q \left(\frac{A_s}{A_v \sqrt{h}} \right)^{1/2} \qquad [3.3.1]$$

while the German correlation approved by the CIB exposure fire working group (CIB, 1986) is

$$t_e = c''qw \text{ (min)} \qquad [3.3.2]$$

where t_e is the furnace exposure time needed to produce the same maximum steel temperature measured in an enclosure fire, c' and c'' are coefficients which depend on the enclosure surface thermal properties, q is the enclosure fuel load density in MJ/m^2 of enclosure surface area, A_s is the enclosure surface area (m^2), A_v is the enclosure vent area (m^2), h is the height of enclosure vent (m), and w is the dimensionless ventilation factor accounting for enclosure vent geometry.

According to Latham et al. (1987), both correlations produced good agreement with data in most cases. They recommend equation [3.3.1] because Pettersson et al. (1976) allows for a continuous variation of c' with thermal properties, whereas the CIB (1986) currently allows only three tabulated values of c''. Coincidently, the value of both c' and c'' for a 'standard' enclosure surface is about 0.07, even though the units of c'' (min-m^2/MJ) are different than those of c'.

The second complication associated with furnace exposure tests implies that different furnaces can produce different fire resistances for the same structural assembly. The radiant heat flux on the tested assembly depends on the emissivity and characteristic dimension of the furnace gases, and the thermal inertia of the furnace walls. Thus, oil-fired furnaces are likely to produce gases with higher emissivities (because of their higher soot content) than gas-fired furnaces, which are used predominantly in North America. Furnace wall linings with low thermal inertia tend to heat up more rapidly and therefore produce higher radiant heat fluxes than those with high thermal inertia. ISO 834 Amendment 2 (1980) recommends new furnaces and furnaces being relined use lining materials with a minimum thickness of 5 cm (2 inch) and a thermal inertia at 500 °C (932 °F) such that

$$(k\rho c)^{1/2} \leq 600 \, W - s^{1/2}/m^2 - K.$$

Several refractory lining materials can satisfy the preceding relationship.

FIRE RESISTANT CONSTRUCTION

Table 3.3. ASTM E-119 temperature endpoint criteria (from Milke, 1995)

Structural member	Location	Maximum temperature F(C)
Walls/Partitions	Unexposed Side	250(139)[a]
Steel Columns	Average	1000(530)
	Single Point	1200(649)
Floor/Roof Assemblies	Unexposed Side	250(139)[a]
and Loaded Beams	Steel Beam (Average)	1100(593)
	Steel Beam (Single Point)	1300(704)
	Pre-Stressing Steel	800(426)
	Reinforcing Steel	1100(593)
	Open-Web Steel Joists	1100(593)
Steel Beams/Girders	Average	1000(530)
(Not Loaded)	Single Point	1200(649)

[a] Maximum temperature cited refers to the maximum temperature rise above initial conditions

Endpoint criteria for determining the duration of successful fire resistance are based on either excessive heat transmission or structural failure. Excessive heat transmission is measured in terms of the temperature of the unexposed side of the assembly, or the ignition of a cotton target placed near any opening on the unexposed side. Structural failure is specified either in terms of an observed loss of load-bearing capacity or in terms of a structural element reaching a specified maximum acceptable temperature. According to Milke (1995), floor and roof assemblies and load-bearing walls are always tested under load (in the US), whereas steel beams and columns may be tested with or without load. Table 3.3 shows the maximum acceptable temperatures specified in ASTM E-119 for various structural members. The specified maximum temperature for implied structural failure varies from 800 °F to 1300 °F (426–704 °C).

There are some special failure criteria for certain structures. For example, fire resistive walls are often subjected to a hose stream immediately after the test or to another sample that has been subjected to an exposure of one-half the desired resistance. If water passes through the wall upon hose stream application, the test is deemed a failure. In the case of assemblies with wood components, ignition of the wood constitutes failure independent of the other failure criteria.

Since flame transmission is one failure criterion, it is important to specify or at least measure the furnace pressure. ISO 834 specifies a furnace over-pressure of 10 ± 2 Pa (0.04 ± 0.008 inches of water) for assemblies intended to provide fire confinement. This over-pressure is supposed to exist over at least the upper two-thirds of wall height. Pressures generated during enclosure fires and their effect on smoke control are discussed in Chapter 4.

Compilations of fire resistance ratings for common construction assemblies are given in Fitzgerald (1986) and Latham (1987). In the case of simple walls constructed of 38–51 mm ($1\frac{1}{2}$–2 in) thick plaster on metal lath, the fire resistance varies from 1 to 2 hours, depending on the type of plaster and its thickness. One-half inch (13 mm) thick gypsum wallboard on wood studs will have a fire resistance of 15 minutes. Plaster protected steel columns have resistance ratings of from 1–4 hours, depending on the type of plaster and thickness. Resistance correlations for some other structures are discussed in the following Section.

3.3.2 EMPIRICAL CORRELATIONS

The following correlations are intended to provide estimates of the expected fire resistance measured in furnace exposure tests with different types of exposed structures. A more complete set of correlations and standard calculation methods for fire barrier assemblies made of either structural steel, various forms of concrete, masonry, or timber and wood, has been compiled by ASCE and SFPE (1995).

Steel columns – Unprotected steel columns have been found to have furnace endurance times proportional to $(W/D)^n$, where W is column weight per unit length (M_s/L), D is the heated perimeter of a column section, and n is an exponent equal to 0.7 for $W/D < 10$ lb/ft-in, and to 0.8 for $W/D > 10$ lb/ft-in. If the furnace gas temperature was invariant with time and re-radiation from the steel surface is neglected, equations [3.2.2] and [3.2.3] would suggest an exponent equal to unity. The empirical coefficients of proportionality are listed in Table 3.4.

Some of the various types of column insulation protection include concrete encasement, gypsum wallboard, and sprayed-on lightweight insulation. Fire resistance correlations for these types of insulated columns have been compiled by Milke (1995) and are reproduced here as Table 3.4. Several of the correlations indicate that the resistance is proportional to $(W/D)h$, where h is the insulation thickness. Gandhi (1988) has recently noted that this type of correlation should be modified for large columns because the column web has a greater influence on resistance than the column flange. Gandhi attributed this modification to the preferential heating of the flange whereas the column endurance is limited by the temperature increase of the web.

Steel beams – The fire resistance of steel beams has been found to vary with the parameter

$$h(W/D + 0.6)$$

where h is the thickness of a particular spray-applied insulation (in), W is the weight of the beam per lineal foot (lb/ft), and D is the exposed perimeter of the beam cross-section (in).

According to Milke (1995), this parameter is used as the basis for evaluating tradeoffs between beam weight and insulation thickness needed to achieve a given furnace endurance rating. This scaling parameter is applicable for $W/D \geq 0.37$ lb/ft-in, and $h \geq 0.375$ in.

Concrete slabs – The fire endurance of concrete walls and floor slabs depends upon the wall/slab thickness, type of aggregate, steel reinforcing properties, the applied load, and the type of restraints. Fleischmann (1995) has reviewed the analytical and empirical methods for evaluating these effects.

One noteworthy problem discussed by Fleischmann (1995) is the 'thermal thrust force' caused by local heating and expansion of a large concrete slab. This local thermal expansion is resisted by the cooler surrounding portion of the slab. The resisting force is called the thermal thrust, T_f, which for a given load and fire exposure varies with the ratio T_f/DE, where D is the heated perimeter of a slab cross-section normal to the thrust direction, and E is the unheated concrete modulus of elasticity. The value of this parameter can be calculated from an empirical relationship for the slab midspan deflection as a function of desired fire endurance (Fleischmann, 1995). Using this methodology, it is possible to determine whether the slab restraints can withstand the thermal thrust without deflecting more than the slab itself can deflect. The step-by-step calculation procedure is explained by Fleischmann and in various concrete design handbooks referenced therein.

Wood structures – The fire resistance of an exposed wood structure is limited by the rate of wood charring since the char has essentially no load bearing capacity. According to White (1995), charring occurs when the heated wood temperature reaches about 290 °C (554 °F), and there is a steep temperature gradient in the uncharred wood. Therefore, most of the uncharred wood remains cool enough to retain its original strength.

In the case of a wood beam, the fire endurance can be calculated as the time required for the beam section modulus to decrease sufficiently for applied stress to reach the ultimate stress. The

FIRE RESISTANT CONSTRUCTION

Table 3.4. Empirical equations for steel columns (from Milke, 1995)

Member/Protection	Solution	Symbols
Column/Unprotected	$R = 10.3(W/D)^{0.7}$, for $W/D < 10$ $R = 83(W/D)^{0.8}$, for $W/D \geq 10$ (for critical temperature of 1000 °F)	R = fire endurance time, min. W = weight of steel section per linear foot, lb/ft. D = heated perimeter, in.
Column/Gypsum Wallboard	$R = 130(hW'/D)/2075$ where $W' = W + (50hD/144)$	h = thickness of protection, in. W = weight of steel section and gypsum wallboard, lb/ft.
Column/Spray-Applied Materials and Board Products	$R = [C_1(W/D) + C_2]h$ Spray-Applied C_1 C_2 Cementitious Material 69 31 Mineral Fiber Material 63 42 Board Products Fiber-reinforced calcium silicate 63 26 Vermiculite-Sodium silicate 44 30	$C_1 \& C_2$ = material constants for specific protection
Column/Concrete Cover	$R = R_o(1 + 0.03 m)$ where $R_o = 10 \left(\dfrac{W}{D} \right)^{0.7}$ $+ 17 \left(\dfrac{h^{1.6}}{k_c^{0.2}} \right) \left\{ 1 + 26 \left[\dfrac{H}{\rho_c c_c h(L + h)} \right]^{0.8} \right\}$ $D = 2(b_f + d)$	R_o = fire endurance at zero moisture content of concrete, min. m = equilibrium moisture content of concrete, % by volume b_f = width of flange, in d = depth of section, in. k_c = thermal conductivity of concrete at ambient temperature Btu/hr ft °F
Column/Concrete Encased	for concrete-encased columns use: $H = 11 W + (\rho_c c_c/144)(b_f d - A_s)$ $D = 2(b_f + d)$ $D = (b_f + d)/2$	H = thermal capacity of steel section at ambient temp., $= 0.11W$ Btu/ft-F. c_c = specific heat of concrete at ambient temp., Btu/lb. °F. L = inside dimension of one side of square concrete box protection, in. A_s = cross-sectional area of steel column, in^2

relevant equations given in White (1995) are:

$$M/S(t) = \alpha \sigma_0 \qquad [3.3.3a]$$

$$S(t) = [(W - 2C_w t)(D - jC_D t)^2]/6 \qquad [3.3.3b]$$

where M is the design load applied moment (in-lb), $S(t)$ is the beam section modulus at time t (in^3), t is the time from ignition to beam failure (min), α is the fraction of room temperature stress at failure of remaining uncharred portion of beam, σ_0 is the room temperature ultimate stress for bending rupture (psi), D is the original depth of beam (in), W is the original beam width (in), j is 1 for 3-sided exposure, and 2 for four-sided exposure, C_D is the charring rate in depth direction of beam, and C_W is the charring rate in direction of beam width.

A typical wood charring rate is approximately 0.6 mm/min (1.4 in/hour) for conditions encountered in standard furnace fire resistance tests (Odeen, 1985). According to White (1995), the charring rate parallel to the wood grain is about twice that value.

The wood time-to-ignition should be added to the time calculated from equations [3.3.3a,b]. Odeen (1985) has reported ignition time data as a function of the impinging radiant heat flux for different wood surface treatments. At a radiant heat flux of 20 kW/m^2 (1.76 Btu/sec-ft^2), the time-to-piloted-ignition is in the range 1–10 minutes.

Odeen (1985) also presents empirical results relating the char layer thickness at failure to the beam dimensions and the ratio of applied load to beam strength at room temperature. His results indicate that the char layer thickness can be as large as 25% of the beam width, i.e. approximately half the wood is charred, before failure occurs in a beam exposed to fire on all four sides.

Gypsum wallboard can significantly increase the fire endurance of wood structures. White (1995) provides data on the fire resistance of both gypsum wallboard and plywood of varying thickness. He also reviews qualitative rules-of-thumb relating the endurance of multiple layers of wood/wallboard to the endurance of the individual wood layers and beams.

3.3.3 HIGH INTENSITY FIRE RESISTANCE TESTS

Many industrial fire scenarios involve more rapid and more intense fire development than is represented by the standard furnace exposures described in Section 3.3.1. Flammable liquid fires and warehouse rack storage fires are two examples of rapidly growing fires for which the standard fire resistance tests would overestimate fire resistance of exposed structures. The tests described here are intended to provide a more applicable measure of fire resistance for these scenarios.

Mobil Research and Development Corporation (Warren and Coronna, 1978) developed a propane burner firebox test to measure the fire resistance of protective coatings for structures and equipment exposed to hydrocarbon liquid/gas fires. The firebox has an open top, and sidewall openings to allow air access. Gas temperatures reach 1093 °C (2000 °F) within 10 minutes compared to about three hours in ISO 834 (1975). The total (radiant + convective) heat flux in the Mobil firebox is reported to be 230 kW/m^2 (20 Btu/sec-ft^2), which is significantly larger than the average heat fluxes measured for large objects immersed in hydrocarbon pool fires (see Table 8.4). Under these conditions, two inches of concrete protection, maintained steel temperatures below 538 °C (1000 °F) for only one hour, compared to three hours under the ASTM E-119 furnace exposure test.

ASTM P-191 (1986) describes a high intensity fire test for structures (columns, beams, and walls) exposed to hydrocarbon liquid pool fires. The actual test does not necessarily have to be a pool fire if the specified heat flux to exposed surfaces can be achieved in a furnace or other test facility. A total heat flux of 173 ± 8 kW/m^2 (15.2 ± 0.7 Btu/sec-ft^2) must be reached in five minutes and maintained for the duration of the test. This heat flux is supposed to be generated with a gas temperature of 927 to 1260 °C (1700 to 2300 °F) and a gas velocity 10 m/s (33 ft/s). Structural failure temperature criteria are similar to those in ASTM E-119. Optional specifications for accelerated weathering and aging tests are also provided.

A variety of other high intensity fire resistance tests are used for special equipment such as flammable liquid piping components. Some of these tests involve burner flames, others involve pool fires or jet flames. These ad-hoc test methods may be appropriate for scenarios in which there is a well defined design basis fire that is more severe those described here.

FIRE RESISTANT CONSTRUCTION

3.4 Fire walls

3.4.1 GENERAL CRITERIA FOR FIRE WALLS

Fire walls are intended to prevent fire spread from one side to the other for a specified period of time. The following three criteria must be satisfied to achieve this function:

1. The wall should have a fire resistance rating at least equal to the specified period of fire confinement.
2. The wall should prevent fire spread around or over it as well as through it.
3. The wall should remain standing despite the collapse of the building roof or framing as a result of fire exposure on one side.

The fire resistance criterion can be satisfied by a standard fire exposure test as described in Section 3.3.1. A 4-hour fire resistance is called for in the Factory Mutual Data Sheet for Maximum Foreseeable Loss Fire Walls (FM Data Sheet 1-22, 1985), which is intended to provide fire confinement despite the loss of automatic sprinkler protection. Since the entire wall cannot be tested in the furnace, it is important that a representative section be tested with representative structural loading.

The second criterion may entail the use of wall extensions such as parapets and the use of fire resistant construction on the roof and exterior wall sections adjacent to the fire wall. Factory Mutual recommendations in Data Sheet 1-22 (1985) call for a parapet height of at least 76 cm (30 in) above the top surface of the roof, while British Building Regulations cited in Cooke (1985) specify one-half this value (15 inches or 38 cm). FM Data Sheet 1-22 (1985) also recommends that at least 7.6 m (25 ft) of the roof surface on either side of the fire wall be covered with gravel or slag or an outdoor fire resistant coating. Other recommendations in the FM Data Sheet 1-22 (1985) cover the location of combustible structures and heat and smoke vents on the roof near a fire wall, and the maximum elevation (3 ft or 0.94 m) of wall penetrations such as pipes, conduits, cables, and ducts. Duct penetrations should incorporate breakaway connections and fusible link actuated dampers. Fire door designs and performance are discussed in Section 3.5.

The stability criterion for an effective fire wall requires special construction techniques to achieve stability despite the deformation and eventual failure of building structures on the fire side of the wall. Typical designs and design basis loads are described in Section 3.4.2.

3.4.2 FIRE WALL DESIGN

Figure 3.6 shows three types of fire walls constructed between columns on a double column line. The walls differ only in the connection between the wall and the adjacent structural steel sitting on the columns. If the wall is fastened to the building framework on both sides of the wall, it is called a tied fire wall. If there are no ties to the adjacent framework, and the wall is self-supporting, it is called a cantilevered fire wall. If the wall is fastened to the framework on one side and independent of the framework on the other side, it is called a one-way fire wall.

Tied fire walls

In order to maintain support during a fire, the columns and steel framework adjacent to tied fire walls should have a fire resistance rating at least equal to that of the wall. The forces exerted on a wall tied to a purlin heated by a fire are illustrated in Figure 3.7. The top sketch in Figure 3.7 shows the heated beam pushing the fire wall that is restraining it from thermal expansion. It can be shown (Cooke, 1985) that the horizontal force, P_1 (lb$_f$), is

$$P_1 = \alpha_s A_b E_s (T_b - T_0) \qquad [3.4.1]$$

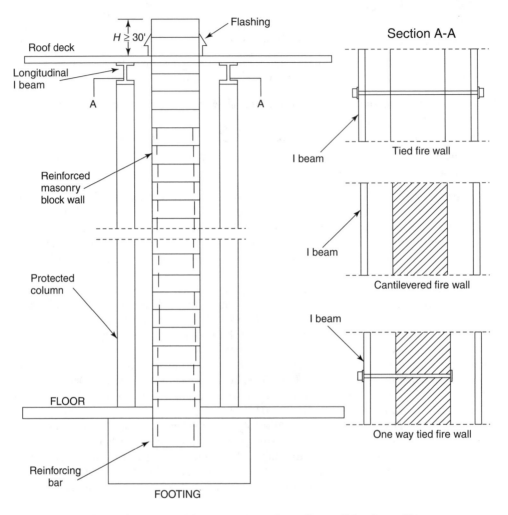

Figure 3.6. Types of fire walls (shown for purlin parallel to fire wall)

where α_s is the coefficient of thermal expansion of steel (°F^{-1}), A_b is the beam cross-sectional area (in^2), E_s is the Young's modulus of elasticity for steel (psi), and $T_b - T_0$ is the temperature rise of fire heated beam (°F).

As the steel weakens from heating, the roof loading causes the purlin to sag and pull the fire wall toward the fire as illustrated in the middle sketch in Figure 3.7. FM Data Sheet 1-22 (1985) suggests that the pulling force can be estimated by treating the sagging beam as a cable subjected to a vertical force per unit length, W. The parabolic approximation to the catenary curve of a sagging cable is

$$P_2 = \frac{W L_b 2}{8 \delta_c} \quad [3.4.2]$$

where P_2 is the tension in collapsed beam an pull on fire wall (lb$_f$), L_b is the beam span perpendicular to fire wall (in), δ_c is the sag of midpoint of collapsed beam (in), and W is the roof weight per unit beam length (lb$_f$/in).

FIRE RESISTANT CONSTRUCTION

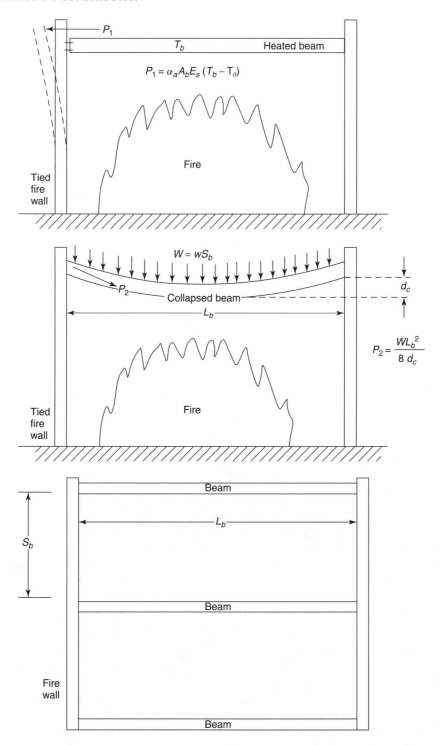

Figure 3.7. Forces on tied walls due to heated/collapsed beams perpendicular to wall

The roof loading, W, is equal to wS_B, where w is the weight per unit roof area (psi) and S_B is the beam spacing (in) as shown in the bottom sketch in Figure 3.7. According to FM Data Sheet 1-22 (1985), the sag, δ_c, is approximately equal to $0.07 L_b$ for open web joists, and equal to $0.09 L_b$ for heavy trusses and wide flange beams. A Similar equation is given in the NFPA firewall standard (NFPA 221-1994) with a safety factor of 1.25 and additional guidance for wood trusses. Alternatively, the sag can be estimated from the midspan deflection of a uniformly loaded simply supported beam. Before the elastic limit is exceeded, this deflection is

$$\delta_c = \frac{W L_b^4}{384 E_s I} \quad [3.4.3]$$

where I is the moment of inertia of the beam.

If there is continuous steel framework through the wall, the horizontal forces calculated from equations [3.4.1] and [3.4.2] should be resisted by the lateral strength of the steel on the other side. If it is a one-way fire wall, the wall itself will have to provide the resistance. FM Data Sheet 1-22 (1985) shows the recommended form of the ties for one-way fire walls and discontinuous roof trusses with purlins parallel or perpendicular to the fire wall.

Cantilevered fire walls

In the case of a cantilever fire wall, the spacing between the wall and adjacent steel framing should allow for steel expansion and/or wall deflection during the fire. The beam elongation, δ_b, is

$$\delta_b = \alpha_s L_b (T_b - T_0) \quad [3.4.4]$$

while the deflection at the top of a wall heated from one side is (Cooke, 1985)

$$\delta_w = \frac{\alpha_c H^2 (T_2 T_1)}{2d} \quad [3.4.5]$$

where δ_w is the horizontal deflection of top of wall (in), H is the wall height (in), d is the wall thickness (in), and $T_2 - T_1$ is the temperature difference across heated wall (°F).

Figure 3.8 illustrates these deflections when the wall bows away from the heated steel beam and when the fire occurs on the other side of the wall such that the wall bows toward the beam. When the fire is under the beam but not close enough to the wall to heat it, δ_w is negligible and the gap between the wall and the end of the beam should be greater than the beam displacement given by equation [3.4.4]. If $T_b - T_0 = 1000\,°F$ (538 °C), and α_s is given by equation [3.1.1], $\delta_b = 0.008 L_b$. The spacing recommended in FM Data Sheet 1-22 (1985) is approximately $0.0105 L_b$. This displacement corresponds to a temperature rise of 1240 °F (671 °C).

In the case of a fire on the other side of the wall, as illustrated in the bottom sketch in Figure 3.8, the wall-to-beam gap should exceed the deflection given by equation [3.4.5]. If $T_2 - T_1 = 500\,°F$ (260 °C) (T_1 should be less than 250 °F or 121 °C to pass the fire resistance test), $\delta_w = 0.002 H^2/d$. FM Data Sheet 1-22 (1985) specifies that H/d should not exceed 20 for hollow masonry walls and 30 for masonry walls that are at least 75% solid. At the upper limit for solid masonry walls, $\delta_w = 0.060 H$. This requirement can dictate larger gaps than that based on δ_b when $H > 0.13 L_b$.

Cantilever fire walls should also have a lateral strength sufficient to resist thermal stress induced by the fire, volumetric expansion and buoyancy pressures generated by fire gases (see Chapter 4), and the pull of flashing attached to the roof cover. The design basis lateral load specified in FM Data Sheet 1-22 (1985) for both cantilever walls and tied walls is 24 kg/m² (5 lb/ft²). This strength for cantilever walls usually entails use of reinforcing bars (with sizes and spacing determined

FIRE RESISTANT CONSTRUCTION 77

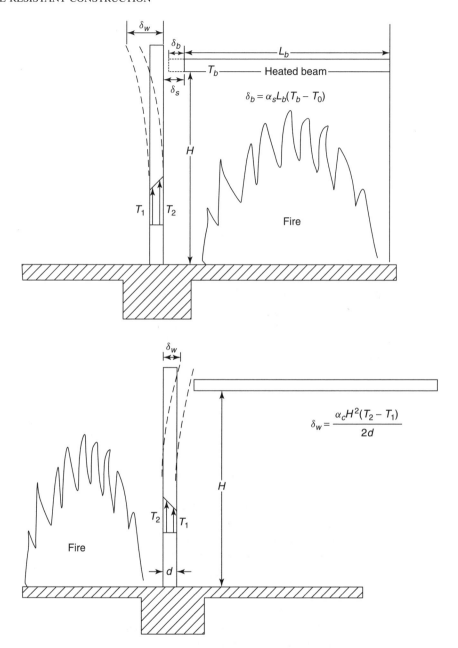

Figure 3.8. Deflections relevant to cantilevered fire wall

from concrete design manuals) extending from the concrete footing up to more than half the wall height as indicated in Figure 3.6. Pilasters with reinforcing bars or encased steel columns are also used to achieve lateral stability.

A double fire wall consists of two one-way fire walls situated back-to-back with a minimum spacing as specified above for cantilevered fire walls. The only connections between the walls should be the roof flashing which should consist of separate pieces with only frictional resistance holding them together. When constructed to these specifications, double fire walls are considered very reliable. They are often used when an addition to a plant requires a fire wall separation from the existing structure.

3.4.3 FIRE WALL LOSS EXPERIENCE

There have been numerous examples of fire walls providing effective isolation during severe fires. The General Motors Livonia fire described in Appendix B is one example where one fire wall isolated a relatively small part of the facility, and several other fire walls were needed. The Ford Cologne, Germany warehouse fire (also described in Appendix B) is a good example of a double brick fire wall confining a severe fire with the help of manual firefighting so that a large portion of the warehouse remained almost undamaged.

There are several examples of ineffective fire walls. One of the classic examples of fire wall failure is the K Mart fire also described in Appendix B. Failure of the K Mart fire walls has been attributed (Best, 1983) to the use of a tied wall with unprotected adjacent steel columns and framework, and the lack of sufficient reinforcement to resist the lateral forces developed by the collapsing steel (FM Data Sheet 1-22, 1985). Another important weakness was the use of water deluge curtains instead of fire doors at several openings in the fire wall.

The 1987 Sherwin-Williams warehouse fire also penetrated a fire wall, which in this case was a 16.5 cm (6.5 in) thick concrete cantilevered wall. Wall failure was attributed (Isner, 1988) to flammable liquid fire spread under the fire doors such that the wall was exposed to fire on both sides. This two-sided fire exposure caused massive spalling of the concrete wall panels and collapse of some panels. Both the Sherwin-Williams fire and the K Mart fire demonstrate the importance of providing effective fire doors to maintain fire wall integrity.

3.5 Fire doors

Personnel movement and material handling considerations dictate the need for openings in fire walls and other types of fire partitions. Fire doors, in principle, satisfy these needs without sacrificing the fire containment function of the wall/partition. They are designed in most cases to close automatically upon fire detection and to provide a minimum fire test endurance, which is usually three hours for fire walls and partitions. In practice, however, fire doors can compromise fire wall integrity via potential failure modes discussed in Section 3.5.3. Their overall effectiveness is probably best summarized by the Factory Mutual Loss Prevention Data Sheet 1-23 (1976), position that 'approved fire doors are the next best solution to no openings at all.'

3.5.1 TYPES OF FIRE DOORS

Industrial facility fire doors, as characterized by method of operation, include rolling steel doors, horizontal sliding doors, swinging fire doors, and to a lesser extent, vertical sliding doors and elevator/chute doors. Brief descriptions are provided here. Performance requirements, in terms of cycling operations, fire resistance, and durability, are prescribed in various approval and listing specifications such as FMRC Class 4100 (1988). Installation and maintenance guidelines are provided in various Codes and Standards such as NFPA 80 (1999).

A rolling steel door consists of an interlocking steel slat curtain, wall guides, a bottom bar, and an automatic release device that closes the slat curtain. Figure 3.9 illustrates a typical configuration with fusible link releases and a cylindrical hood for the rolled up slats and coil torsion spring mechanism located above the door opening. The spring tension must be reset after each door actuation. Operating instructions are provided by the manufacturers, and must be carried out

FIRE RESISTANT CONSTRUCTION 79

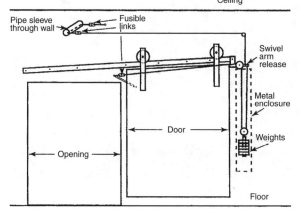

Figure 3.9. Rolling steel and horizontal sliding fire doors. Reprinted with permission from NFPA 80®, *Fire Doors and Windows*. Copyright © 1999 National Fire Protection Association, Quincy, MA 02269. This reprinted material is not the complete and official position of the National Fire Protection Association, on the referenced subject which is represented only by the standard in its entirety

Figure 3.10. Rolling steel fire door installed in fire wall

carefully since an improperly set rolling steel door is not readily apparent from casual observation. Figure 3.10 is a photograph of a raised rolling steel fire door with barriers near the wall opening to provide the needed clearance from the door guides.

As illustrated in Figure 3.9, horizontal sliding doors can be mounted on inclined tracks or on level tracks. The inclined track doors close by gravity usually when a fusible link detector (located on both sides of the fire wall) releases a counterweight. Level track doors usually close via a weighted pulley mechanism. Electric power operated doors are also used with the proviso that the door releases and closes automatically upon loss of power.

Sliding door construction materials include (NFPA 80, 1999) hollow metal doors, metal clad (Kalamein) doors, sheet metal doors, and composite doors. The metal clad doors usually have a wood core or interior insulation. Composite doors consist of wood, steel, or plastic sheets bonded to a solid core material. The insulating capability of a door can vary greatly depending on the type of construction and core material.

Swinging fire doors can be either self-closed or tripped with an automatic closing device. Opening and latching hardware for swinging fire doors have specifications (NFPA 80, 1999), which depend upon the door fire resistance rating and the type of construction. Material and construction designs are similar to those for horizontal sliding doors, including the use of binders.

FIRE RESISTANT CONSTRUCTION

Swinging doors approved/listed as fire exit doors are subjected to 'panic loading tests' (FMRC Class 4100, 1988) to verify that they can be easily opened and will not impede safe egress in an emergency.

Vertical sliding doors can be either single piece or two-section doors operated in vertical tracks via counterweights and/or helical springs. They are not as common as rolling steel doors because they usually require more vertical space for the raised door.

3.5.2 FUSIBLE LINKS AND DETECTORS

Fusible links are the predominant actuation trigger for industrial fire doors. Link actuation temperatures range from 51.6 °C to 260 °C (125 °F to 500 °F) and in varying load ratings to allow compatibility with door cable loads.

If a single door is installed on a fire wall opening, fusible links should be installed on both sides of the wall with a through wall connection so that operation of any one link will cause the door to close. One link on each side should be located near the top of the opening and another near the ceiling as shown in Figures 3.9 and 3.10. NFPA 80 (1990) specifies that the link not be situated in the stagnant air space defined as within 10 cm (4 in) of the wall-ceiling intersection.

Power operated doors can be closed from the actuation signal of a smoke detector or some type of optical flame detector. Smoke detectors can be advantageous in facilities that are vulnerable to smoke damage and/or smoldering fires, whereas optical flame detectors can be advantageous for rapid fire development such as a flammable liquid fire. Sometimes the detector also triggers fans, duct dampers, etc. as part of a smoke control system. In other cases, the detector triggers a fire alarm or a fire extinguishing agent such as carbon dioxide or Halon.

3.5.3 RELIABILITY ISSUES

Fire doors will not necessarily close properly when a fusible link or detector is actuated. Their propensity to hang up has been determined through an extensive series of automatic closure tests conducted annually or semiannually by the Factory Mutual Engineering Association (FMEA) on installed doors ('Fire Doors Closing the Safety Gap', 1988). Test results for the period 1984 to 1988 are shown in Table 3.5. The overall closure failure rate is 18.5%. Rolling steel doors have a significantly higher failure rate (21%) than the other types of fire doors. This is particularly noteworthy because rolling steel doors accounted for more than half the installed doors in the survey. Vertical sliding doors have the lowest reported closure failure rate (11%), but they only accounted for 2% of the doors tested.

Table 3.6 lists the most common causes of door closure failure. Improper spring tension is the leading cause of failure for rolling steel doors, accounting for one out of every three failures. Snagged chains or cables is the leading cause of failure (37%) for horizontal sliding doors, while damaged closers represent the most common cause of failure (38%) for swinging doors. In almost all cases, the doors were not defective when manufactured but were either installed incorrectly or

Table 3.5. Fire door closure test data (FMEA Surveys 1984–1988)

Type of door	# Failures	# Tests	% Failure
Rolling steel	1177	5587	21.1
Horizontal sliding	377	2463	15.3
Vertical sliding	17	156	10.9
Swinging	166	1183	14.0
Total	1737	9389	18.5

Table 3.6. Causes of fire door closure failure, % of failures in various types of doors (from FMEA Survey: 303 failures)

Cause of failure	Rolling steel	Horizontal sliding	Vertical sliding	Swinging	Overall
Spring tension	33%	5%	0	0	23%
Snagged chain	23%	37%	0	3%	23%
Opening blocked	10%	6%	25%	3%	9%
Damaged tracks	9%	17%	0	0	9%
Damaged closer	16%	3%	50%	38%	16%
Hood/curtain	5%	0	0	0	3%
Damaged binder	0	8%	0	21%	4%
Other	3%	24%	25%	35%	11%

were damaged after installation. Apparently, the closing mechanisms, including chains, cables, counterweights, etc. are mechanically complicated and damage prone when installed in heavily trafficked areas with lift trucks moving large materials. Until simpler, less damage prone door designs are forthcoming, special diligence is needed during installation, testing, and normal facility operations in the vicinity of the doors.

The basic laws of probability can be used to relate the single door failure rate data to the probability of at least one door remaining open during a fire near a multi-opening fire wall. The basic formula is

$$p_w = 1 - (1 - p_d)^n \qquad [3.5.1]$$

where p_w is the probability of at least one door remaining open, p_d is the single door probability of not closing properly, and n is the number of doors in the fire wall.

For example, if there are four openings in a fire wall and each opening is covered by a rolling steel door, then

$$p_w = 1 - (1 - 0.211)^4 = 0.612$$

However, if each of the four openings is covered by two doors, one on each side of the wall, then

$$p_w = 1 - (1 - p_d^2)^4$$

since both doors have to fail at each opening

$$p_w = 1 - (1 - (0.211)^2)^4 = 0.166$$

which is substantially lower than the failure probability of 0.61 calculated previously. If two horizontal sliding doors are used, than the failure probability becomes 0.09.

The preceding examples neglect two other potentially important fire door failure modes. First, there is a chance that the fusible link or detector will not actuate until after the fire has spread through the opening. This apparently occurred in the K Mart fire described in Appendix B because flaming aerosol cans rocketed through the wall openings while the doors were still open. Secondly, there is a chance that the fire will penetrate the closed door, as explained below.

Although fire door approval/listing entails passing a fire resistance test, Campbell (1986) has pointed out several differences between fire resistance ratings for fire doors and fire walls. One difference is the unlimited temperature rise of the unexposed side of most fire doors. Campbell notes that some steel doors become sufficiently hot to have unexposed surfaces glowing red during the furnace test. A second difference is allowable buckling and temporary flaming permitted in fire door resistance tests. A third difference is the presence of perimeter spaces that may not necessarily transmit flame or smoke in a furnace with a lower pressure than the surroundings, but could be a transmission path when a real fire generates a positive pressure on the fire side of the door. In view of these differences, there is a greater chance for fire to spread through a closed fire door than through the fire wall itself. Consequently, combustible materials should not be stored next to a fire door.

The 1987 Sherwin-Williams warehouse fire (Isner, 1988) provides some interesting applications of the fire door reliability issues described here. There were four vehicle openings in the center fire wall of the warehouse, with 3-hour rated horizontal sliding doors on both sides of each opening. There were also two personnel doors with 3-hour fire resistance ratings. The two personnel doors were fully closed during the fire, but two of the four horizontal sliding doors remained at least partially open. Thus, 20% of the 10 fire doors failed to close effectively. Since the two open doors were not located on the same opening, all the wall openings were closed during the fire. Isner (1988) attributes the fire spread across the wall to the flow of flammable liquid under the closed doors. Horizontal sliding doors can have a gap as large as 9.5 mm (3/8 in) between the bottom of the door and the sill, according to NFPA 80-1986.

Isner and others recommend the use of curbs and/or trenches alongside fire doors in flammable liquid warehouses to prevent this mode of fire penetration. The curbs and trenches are also desirable from the viewpoint of preventing sprinkler water runoff from flowing under the doors and potentially contaminating the surrounding environment. However, if the curbs/trenches are too high or steep they can become an obstacle for lift truck operation, such that they can actually lead to accidents.

Other examples of successful and unsuccessful fire door operation during fire incidents are described in 'Fire Doors: Closing the Gap', 1986.

3.6 Insulated metal deck roofing

Insulated steel decks are a common form of roofing at industrial facilities in the United States and elsewhere. Extensive fire spread due to the gassing of the roof insulation and the dripping of the asphalt adhesive was a major factor in the destruction of the General Motors Livonia, Michigan facility as described in Appendix B. Since the 1953 Livonia fire, it has become clear how to design, construct and test insulated metal decks so as to reduce their challenge to building fire resistance.

3.6.1 DESCRIPTION

The three primary components of an insulated steel deck roof are: (1) the steel deck itself; (2) the insulation and fasteners or adhesive; and (3) the weather resistant roof covering. In many cases a thin sheet vapor barrier is located between the deck and the insulation to prevent warm humid air from within the building from condensing on the insulation.

Figure 3.11 shows cutaway drawings of typical roof construction with adhesive (top drawing) and with mechanical fasteners (bottom drawing). The corrugated (also called ribbed) steel deck is made from 22 gage (0.76 mm) or thicker steel sheets. Deck sections are attached to the underlying purlins by welding or by special deck fasteners.

Vapor barriers described in Factory Mutual Loss Prevention Sheet 1-28 (1983) include asphalt saturated felt, vinyl plastic, Kraft paper, or laminated combinations thereof. Until recently, most vapor barriers were attached to the deck with a solvent adhesive or asphalt. The amount of adhesive used is a tradeoff between wind uplift resistance and fire resistance. FM Data Sheet 1-28 (1983) recommends the use of mechanical fasteners for attaching the vapor barrier to the deck.

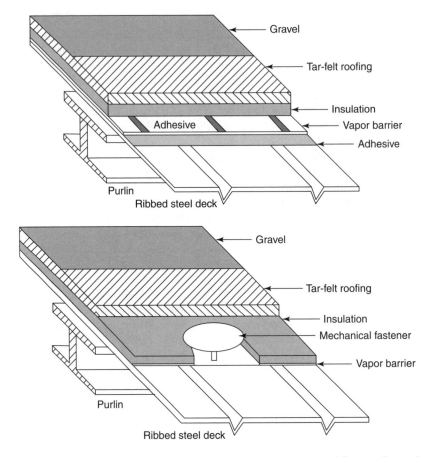

Figure 3.11. Insulated steel deck with adhesive (top) and mechanical fastener (bottom)

A wide variety of insulation materials have been used above the steel deck. Materials listed in FM Data Sheet 1-28 (1983) include glass fiber, wood fiber, perlite, paper-faced phenolic, and composite board containing polyurethane foam or isocyanurate foam. Sometimes multiple layers of insulation are used. As with the vapor barrier, adhesives or mechanical fasteners can be used. The mechanical fasteners have a large plate or washer to sit on the insulation, a steel shank, and a bottom tip to pierce the steel deck below.

The weather resistant outer layer of the insulated steel deck roof is usually multiple piles of roofing felts saturated with hot tar or asphalt and covered with gravel as shown in Figure 3.10. Alternatively, a single ply membrane can be adhered to the insulation.

3.6.2 WHITE HOUSE TESTS

The so-called White House tests were conducted after the 1953 General Motors Livonia fire to investigate the effect of roof construction on the propensity for flame spread under the roof. The tests ('Insulated Metal Roof Deck Fire Tests', 1955) were conducted in a 20 ft wide by 100 ft long by 10 ft high (6.1 m by 30.5 m by 3.05 m high) structure (Figure 3.12) enclosed on two sides and one end, with the other end open. The roof being tested was welded to the I beam purlins. The exposure fire at the closed end of the test structure was a gasoline spray fire with a flow

Figure 3.12. White house roofing fire tests. © 2002 Factory Mutual Insurance Company, with permission

rate increasing from 1.0 gpm (3.78 L/min) at the beginning of the test to 2.7 gpm (10 L/min) after 17 minutes. (Heat release rates increased from 2 MW or 2110 Btu/sec to 5 MW or 5275 Btu/sec). A blower provided a 5500 cfm (2.6 m^3/s) air supply for the gasoline burners.

Initially ('Insulated Metal Roof Deck Fire Tests', 1955), a series of five tests was conducted with a noncombustible roof, and with different combinations of combustible insulation, vapor barrier, and adhesives. Results demonstrated that the vapor barrier and adhesives under the insulation determined the extent of fire spread. Vapor barriers composed of asphalt impregnated felt caused flame to propagate the full length of the enclosure, as did asphalt adhesives containing at least 5.4 kg (12 lb) of asphalt per 9.3 m^2 (100 ft^2) of insulation. On the other hand, mechanically fastened insulation did not cause the fire to spread beyond the area of the exposure fire. Thus, any combustible material in or above the insulation did not contribute to the flame spread while combustible material between the insulation and the steel deck did exacerbate flame spread. Apparently, gases generated in the combustible adhesive and vapor barrier were forced down through the joints of the steel deck.

At least ten more White House tests were conducted with different insulation and asphalt loadings ('The Effect of Exposure Duration on Insulated Metal Roof Deck Fires', 1960). Furthermore, a similar test structure was constructed at another test facility to test proprietary roof assemblies. These additional tests showed that some combustible adhesives could be used with certain insulations without contributing to flame spread, so that separate tests were required for each new roof assembly.

3.6.3 SMALL-SCALE TESTS AND CLASSIFICATIONS

Two different small-scale tests have been used to assess under roof flame spread propensity without resorting to the expense of constructing a 6.1 m by 30.5 m (20 ft by 100 ft) roof structure for a White House test. The small-scale test used by Factory Mutual Research Corporation is the construction calorimeter. The small-scale test used at Underwriter's Laboratories is the Steiner Tunnel. Both tests have been purported to provide some correlation with the White House tests. The use and interpretation of both tests is described by Coursey (1988), and is summarized below.

The FMRC construction calorimeter uses a 1.2 m by 1.5 m (4 ft by 5 ft) section of roof. The test section is placed in the top of the furnace calorimeter and burned with a controlled flow of heptane as the exposure fire. Based on the convective flow in the furnace flue, the heat release rate contribution of the test section is calculated. If the heat release per unit test section area is no more than 76 kW/m² (410 Btu/min-ft²), the roof is given a Class I classification. This allows the roofing assembly to be installed in an otherwise noncombustible occupancy without requiring automatic sprinkler protection. If the heat release rate per unit roof area is greater than this critical value, the roof is designated as Class II and automatic sprinklers are needed regardless of the rest of the occupancy.

The UL Steiner Tunnel test uses a 0.3 m by 7.6 m (1 ft wide by 25 ft long) roofing section. The underside of the roofing assembly is exposed to an open flame generating 9 kW (510 Btu/min) with a 1.2 m/s (2.7 mph) air flow down the tunnel. If flame does not extend beyond 4.3 m (14 ft) down the tunnel during the 30 min test duration (and does not reach 10 ft or 3.05 m during the first 10 min), the assembly passes the test. UL listing of the roofing assembly entails passing a fire resistance test and the external flame spread test described below.

The external flame spread test used by both UL and FMRC, as well as several other fire testing laboratories, is the ASTM E-108 fire test fire test for roof coverings (ASTM E-108, 1983). There are five parts to the test, but the most relevant part for a severe external fire exposure subjects the top of the roof assembly to a gas flame in a 5.4 m/s (12 mph) air flow for 10 min. The roofing is given a Class A rating if the flame spread is not more than 1.8 m (6 ft). Lower ratings are given for more extensive flame spread.

Roofing assemblies that pass both the underside flame spread test and the external flame spread test are listed in the FMRC and UL directories for building construction materials.

3.7 Water spray protection of exposed structures

If a roof deck, wall, beam, or column cannot provide the desired level of fire resistance, can it be effectively protected with water spray nozzles or sprinklers? This question is addressed from both a theoretical viewpoint and from practical considerations in industrial facilities.

We begin with some simple heat transfer theory for water film coverage of a structural member exposed to a high heat flux from a nearby fire. If the water spray is to absorb via vaporization the impinging heat flux, the required water spray density is

$$m''_w = \frac{(q''_c + q''_r)}{\rho_w(L_v + c_p(100 T_0))} \qquad [3.7.1]$$

where m''_w is the required water spray density per unit exposed steel surface area (m³/m²-s), q''_c is the convective heat flux to steel surface (kW/m²), q''_r is the radiant heat flux to steel surface (kW/m²), ρ_w is the mass density of water (kg/m³), L_v is the heat of vaporization of water (kJ/kg), and T_0 is the initial temperature of water prior to heating (°C).

When $T_0 = 20$ °C (68 °F), the denominator in equation [3.7.1] is equal to 2590 kJ/liter (9280 Btu/gal). According to the specifications of the high intensity fire resistance tests in Section 3.3.3, the maximum heat flux for a small structure immersed in a pool fire is 173 kW/m² (914 Btu/min-ft²). The water application rate required for this exposure is, according to equation [3.7.1], $(173/2590)60 = 4.0$ liter/min-m² (0.1 gpm/ft²). Less intense exposures would, in principle, require proportionately lower water application rates.

The preceding theory is based on the critical assumption that the water spray can be applied in a manner that will allow it to absorb the entire heat flux at the surface of the exposed structure. This is extremely difficult, if not impossible, for the following reasons:

1. Gas or wind velocities may prevent the spray from reaching the surface of the structure.
2. Once the water droplets reach the surface of the structure, they may rebound or drip off it rather than cover it with a water film.
3. Flame radiation may be transmitted through the droplets or water film if the droplet diameter or film thickness is too small.

The first complication can be mitigated by locating the spray nozzle or sprinkler close to the exposed structure or by generating large drops moving at high velocity toward the exposed structure. The second complication depends strongly on the geometry and surface condition of the structure as well as the droplet impact parameters. The third complication can be analyzed knowing the spectral distribution of flame radiation and the spectral absorptivity of water. Lev and Strachan (1989) have calculated absorption fractions for black body emitters and have shown that a water film thickness of about 1 mm (0.04 in) is needed to absorb 90% of the radiation generated at flame temperatures. Thus it would be desirable to maintain a continuous film of water at least 1 mm thick on the surface of the exposed structure.

The thickness of water film generated by water spray application to a solid surface depends on the orientation of the surface. In the case of a vertical surface, the film thickness is given by (Lev and Strachan, 1989)

$$b = [(2.4\, m_w'' x \upsilon)/(\rho_w g)]^{1/3} \qquad [3.7.2]$$

where b is the water film thickness at a distance x below the top of the vertical surface (m), υ is the kinematic viscosity of water (m^2/s), and the other variables are as defined earlier.

It is clear from equation [3.7.2] that there will always be a region near the top of a vertical surface where the water film thickness will be too small to absorb the flame radiation. In fact, the entire vertical surface will have a film thickness under 1 mm when water application rates are of the order of the 4 liter/min-m^2 (0.1 gpm/ft^2) rate calculated above. Therefore vertical surfaces will probably not be able to sustain water films sufficiently thick to prevent significant heating at practical water spray application rates. By absorbing some of the incident heat flux, the water spray/film will delay the time needed for the fire to heat the steel to the point of structural failure. The calculation of that delay time, or effective fire resistance, of the structure is complicated by film or nucleate boiling considerations as well as the radiation transmissivity and two dimensional, time dependent heat conduction formulation.

The inability of water spray to provide complete protection of exposed steel structures has led to different practical approaches to water spray protection. In the case of ceiling steel, American sprinkler spray patterns don't attempt to provide much direct wetting of ceiling structures, whereas the British sprinkler spray patterns do. These differences reflect different priorities regarding whether the sprinkler spray can accomplish more by controlling the fire below or by providing at least partial wetting of the ceiling above.

References

Ahmed, G.N. and Hurst, J.P., Modeling the Thermal Behavior of Concrete Slabs Subjected to the ASTM E119 Standard Fire Condition, *J. of Fire Protection Engineering*, **7**, 125–132, 1995.

Alpert, R.L., Convective Heat Transfer in the Impingement Region of a Buoyant Plume, FMRC JI 0J0N1.BU, May 1985 (also in *Journal of Heat Transfer*, **109**, February 1987).

Alpert, R.L. and Ward, E.J., Evaluating Unsprinklered Fire Hazards, FMRC RC84-BT-9, *Fire Safety Journal*, **7**, 1984.

ASTM E-108, *Standard Methods of Fire Tests of Roof Coverings*, American Society for Testing and Materials, 1983.
ASTM E-119, *Standard Methods of Fire Tests of Building Construction and Materials*, American Society for Testing and Materials, Philadelphia, 1983.
ASTM P-191, *Proposed Test Methods for Determining Effects of Large Hydrocarbon Pool Fires on Structural Members and Assemblies*, American Society of Testing and Materials, 1986.
Barnett, J.R., Use and Limitations of Computer Models in Structural Fire Protection Engineering Applications, *Fire Safety Journal*, **9**, 137–146, 1989.
Best, R., Fire Walls that Failed: The K Mart Corporation Distribution Center Fire, *Fire Journal*, **77(3)**, 74, May 1983.
BS 476 Part 8, *Test Methods and Criteria for the Fire Resistance of Elements of Building Construction*, British Standards Institution, 1972.
Campbell, J.A., Confinement of Fire in Buildings, *Fire Protection Handbook*, Section 7/Chapter 9, National Fire Protection Association, 1986.
Cooke, G.M.E., Fire Engineering of Tall Fire Separating Walls, Fire Research Station Paper 46/85, IFSEC Conference on Flexible Approaches to Fire Resistance and Passive Protection, April 1985.
Coursey, R., Do Roof Fire Tests Assure Performance? *Plant Engineering*, 54–57, April 14 1988.
Design Guide Structural Fire Safety Workshop, CIB W14, *Fire Safety Journal*, **10**, 1986.
Drysdale, D., *An Introduction to Fire Dynamics*, John Wiley & Sons, 1985.
Fitzgerald, R., Structural Integrity During Fire, Section 7, Chapter 8, *Fire Protection Handbook*, National Fire Protection Association, 1986.
Factory Mutual Loss Prevention Data Sheet 1-1, 'Building Construction and Materials,' 1983.
Factory Mutual Loss Prevention Data Sheet 1-21, '*Fire Resistance of Building Assemblies*,' Factory Mutual Engineering Corp., 1977.
Factory Mutual Loss Prevention Data Sheet 1-22, '*Criteria for Maximum Foreseeable Loss Fire Walls and Space Separation*,' Factory Mutual Research Corporation, September 1985.
Factory Mutual Loss Prevention Data Sheet 1-23, '*Protection of Openings in MFL Fire Walls*,' Factory Mutual Research Corporation, 1976 (also Technical Advisory Bulletin on Fire Doors, September 1987).
Factory Mutual Loss Prevention Data Sheet 1-28, '*Insulated Steel Deck*', Factory Mutual Research Corporation, May 1983.
Factory Mutual Research Corporation Approval Standard for Fire Door and Frame Assemblies, Class Number 4100, Factory Mutual Research Corporation, October 1988.
'*Fire Doors: Closing the Safety Gap*,' Record, Factory Mutual Training Resource Center, Sept/Oct 1986 and March/April 1988.
'Fire Resistance Tests – Elements of Building Construction,' International Standards Organization Ref. No. ISO 834-1975/A1-1979/A2-1980 (E).
Fleischmann, C., 'Analytical Methods for Determining Fire Resistance of Concrete Members,' Section 4/Chapter 10, *SFPE Handbook of Fire Protection Engineering*, NFPA, 1995.
Gandhi, P.D., Correlations of Steel Column Fire Test Data, *Fire Technology*, 20–32, February 1988.
'*Insulated Metal Roof Deck Fire Tests*,' Factory Mutual Engineering Division, 1955.
Isner, M.S., $49 Million Loss in Sherwin-Williams Warehouse Fire, *Fire Journal*, March/April 1988.
Latham, D.J., Kirby, B.R. and Thomson, G., The Temperatures Attained by Unprotected Structural Steelwork in Experimental Natural Fires, *Fire Safety Journal*, **12**, 139–152, 1987.
Lev, Y. and Strachan, D.C., A Study of Cooling Water Requirements for the Protection of Metal Surfaces Against Thermal Radiation, *Fire Technology*, 213–229, August 1989.
Lie, T.T., Fire and Buildings, Applied Science, London, 1972.
Lie, T.T., Fire Temperature-Time Relations, *The SFPE Handbook of Fire Protection Engineering*, Section 4/Chapter 8, the National Fire Protection Association, 1988.
Milke, J., Analytical Methods for Determining Fire Resistance of Steel Members, Section 4/Chapter 9, *SFPE Handbook of Fire Protection Engineering*, NFPA, 1995.
NFPA 80, '*Fire Doors and Windows*,' National Fire Protection Association, 1999.
NFPA 221, '*Fire Walls and Fire Barrier Walls*,' National Fire Protection Association, 1994.
Odeen, K., Fire Resistance of Wood Structures, *Fire Technology*, **21(1)**, 35, February 1985.
Pettersson, S.E., Magnusson, S.E. and Thor, J., *Fire Engineering Design of Steel Structures*, Swedish Institute of Steel Construction, 1976.
'*Standard Calculation Methods for Structural Design for Fire Conditions*,' American Society of Civil Engineers and Society of Fire Protection Engineers, 1995.
Taloma, D., Kruppa, J., Franssen, J. and Recho, N., Factors Influencing the Behavior of Steel Columns Exposed to Fire, *J. of Fire Protection Engineering*, **8**, 31–43, 1996.
'The Effect of Exposure Duration on Insulated Metal Roof Deck Fires,' Factory Mutual Engineering Division, Serial No. 12652-S5/RES, 1960.

Warren, J.H. and Coronna, A.A., This Method Tests Fire Protective Coatings, Hydrocarbon Processing, 1978.
White, R.H., Analytical Methods for Determining the Fire Resistance of Timber Members, Chapter 4-11, *SFPE Handbook of Fire Protection Engineering*, National Fire Protection Association, 1988.
You, H.Z. and Faeth, G.M., Ceiling Heat Transfer During Fire Plume and Fire Impingement, *Fire and Materials*, **3**, 140–147, 1979.

4 SMOKE ISOLATION AND VENTING

The concept of fire containment via compartmentation is extended here to considerations of smoke control via isolation and venting. The usual objective in an industrial setting is to control smoke damage to susceptible equipment. Another objective, which is common to all fire protection applications, is to prevent smoke from threatening safe egress and personnel survivability.

Three different types of smoke isolation are envisaged. The greatest degree of isolation would be to confine the smoke to a particular piece of equipment or work station so as to avoid contaminating nearby equipment in the same compartment as illustrated in the top sketch in Figure 4.1. This could entail providing a dedicated ventilation system and suppression system for the equipment or work station as in a clean room in an electronics plant, a supercomputer in a computer room, or a special hazard in a chemical or biotechnology laboratory. The second type of isolation would be the confinement of smoke to the compartment in which the fire originated such that other areas in the building remain relatively smoke-free as illustrated in the middle sketch in Figure 4.1. An example might be a fire in an office building or in a cutoff room adjacent to a computer room or a telephone exchange or a manufacturing or storage area. The third type of isolation would be the confinement of smoke to the area of fire origin via the use of curtain boards and roof vents in an otherwise undivided large work or storage area as illustrated in the bottom sketch in Figure 4.1. Each situation is treated separately in this chapter.

4.1 Isolation and halon suppression within ventilated equipment

If smoke damage is to be confined to a burning piece of equipment, detection and suppression must take place locally, i.e. before the smoke spreads throughout a large area of the room. For example, a plastic wet bench fire in a clean room in a microelectronics assembly plant should be detected and suppressed before the fire overtaxes the clean room ventilation system and causes widespread contamination. This particular scenario has been simulated in a recent series of fire tests that illustrate the challenge and potential solution associated with this type of smoke isolation.

Figure 4.2 shows the simulated clean room work station and ventilation used in the fire tests conducted by Fisher *et al.* (1986). The air supply to the polypropylene wet bench (overall dimensions of $5.8 \times 2.5 \times 1.7$ ft or $1.77 \times 0.76 \times 0.52$ m) was directed downward through a High Efficiency Particle Airfilter (HEPA) and by a laminar flow vinyl curtain. The laminar air flow is exhausted through both the bench exhaust duct and a finger wall return plenum behind the bench. Various types of fire/smoke detectors were located in the bench exhaust duct, the return plenum, and the 9 ft high ceiling of the simulated clean room. Automatic sprinkler and Halon suppression systems were also incorporated into the tests.

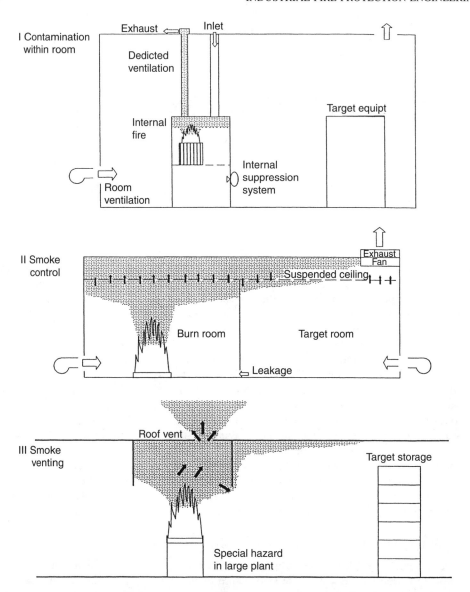

Figure 4.1. Smoke isolation and venting scenarios

The most common wet bench fire scenario is an exposed immersion heater igniting the polypropylene wall of a bench tub/tank. Upon melting, the polypropylene begins dripping and spreading flame into the interior of the bench. Furthermore, the downward air flow often entrains smoke into the bench interior and exhaust. Most of the tests reported in Fisher *et al.* (1986) involved fire initiated within the bench itself, under the tub. The timetable shown in Table 4.1 is indicative of the fire development and detector/sprinkler response reported by Fisher *et al.* for a fire ignited within a wet bench containing a pressurized water line. Presumably, a similar sequence of events occurring at somewhat later times would result from a fire initiated in the tub rather than within the bench.

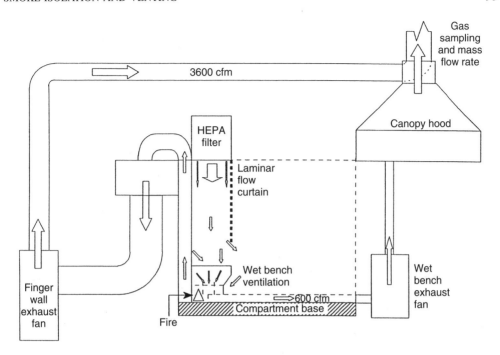

Figure 4.2. Side elevation of the simulated clean room section indicating the ventilation conditions, and location of fire in wet bench

Table 4.1. Observations in wet bench fire test 5, (data from Fisher et al., 1986)

Time (min)	Observation
0	Ignition of insert in poly propylene (PP) bench
4:25	Molten PP drips onto floor of bench
5:40	IR Detector in bench alarms
11:28	Photoelectric detector in duct alarms
14:55	Water line in bench ruptures; fire intensifies
16:40	Flame drawn into exhaust duct
17:05	Fire breaches exhaust duct
17:08–18:50	Plenum and ceiling detectors alarm
22:30–24:44	Ceiling sprinkler heads open
23:20	Curtain falls; smoke escapes
26:35	Sprinklers extinguish fire

The critical event with regard to widespread smoke contamination is the breaching of the bench exhaust duct at approximately 17 minutes after ignition. Detectors within the bench and exhaust duct alarmed well before that time, but ceiling detectors and sprinklers (fast response and conventional response) did not respond in time. In two tests, the bench IR detector tripped a 3.6 kg (8 lb) Halon 1301 suppression bottle within the bench.

The Halon 1301 (CF_3Br) rapidly extinguished the bench fire, but the excessive discharge rate from the oversize extinguisher generated an overpressure that blew the insert tub out of the bench.

Hotta and Horiuchi (1986) and Grant (1995) offer relevant detection and suppression guidelines, respectively, for fires within enclosed ventilated equipment such as the wet bench. Hotta and Horiuchi (1986) tested the response of ionization and photoelectric smoke detectors to smoldering fires in a smoke tunnel with air flow velocities in the range 0.5–8 m/sec (1.6–26.2 ft/sec). Velocities in this range are needed for cooling modern electronic equipment such as computers using LSI and VLSI microprocessors. Their data show ionization detectors respond earlier than photoelectric detectors when air velocities exceed about 1 m/sec (3.3 ft/sec). The relative ineffectiveness of photoelectric detectors at these velocities is believed to be due to the short particle residence times not allowing time for agglomeration and other particle growth processes. Based on these results, a miniature ionization detector (4×4 cm or 1.6×1.6 in) has been commercialized for ventilated equipment fire applications. Response time is reportedly under 15 seconds.

If the ionization detector is to shut off the ventilation system and trigger a discharge of either Halon or some other gaseous suppressant in the isolated equipment, the quantity of suppressant needed to achieve a concentration c is (Grant, 1995)

$$W = Vc/[s(1-c)] \qquad [4.1.1]$$

where W is the mass of suppressant (Halon) required, kg (lbm), V is the volume of equipment enclosure, m³ (ft³), c is the required suppressant (Halon 1301) volume fraction, s is $1/\rho$ which is the Halon vapor specific volume, m³/kg (ft³/lbm), s is $0.1478 + 5.67 \times 10^{-4}T$ m³/kg for T in °C, and s is $2.206 + 5.046 \times 10^{-3}T$ ft³/lb for T in °F.

More generally, s is $RT/(148.9P)$, where R is the universal gas constant (8.314×10^3 J/kmol-°K), T is temperature (°K), and P is the absolute pressure in the enclosure (Pa).

The value of c specified for Halon 1301 extinguishment of most flaming fires is 0.05. Deep-seated smoldering fires and extinguishment using other gaseous agents require higher concentrations. In the case of the wet bench used in the Fisher et al. (1986) fire tests, V was about 0.3 m³ (10 ft³), T was about 21 °C (70 °F) prior to the fire, and according to equation (4.1), W should be about 0.10 kg (0.22 lb) for Halon 1301.

If the ventilation air flow through the equipment cannot be shutdown, an extended discharge of suppressant (Halon) is needed to achieve and maintain the required concentration. In particular, the discharge rate needed to develop a concentration, c, at any given time after initiating discharge into the ventilated enclosure is (Grant, 1995)

$$R = \frac{cE}{s(1-c)(1-\exp(Et/V))} \qquad [4.1.2]$$

where R is the Halon 1301 discharge rate, lbm/sec (kg/s), E is the enclosure ventilation rate, ft³ per s(m³/s), and t is the time after start of discharge (s).

After the discharge is completed, the concentration decay is as follows:

$$c = c_0 \exp(-E(t-t_d)/V) \qquad [4.1.3]$$

where c_0 is the concentration at end of discharge, vol%, and $t - t_d$ is the time from the end of discharge.

According to the NFPA 12A standard for Halon 1301 (1989), the desired soak time (during which Halon concentrations remain at or above 5 vol%) depends upon how deep-seated the smoldering fire is at the time of Halon application. A 10 minute soak time is usually specified for room total flooding discharges. In the case of ventilated equipment with a sensitive ionization detector, the soak time can presumably be significantly shorter than 10 minutes. However, it would also be prudent to provide automatic sprinklers or a gaseous agent total flooding system as a second level of protection for the surrounding room.

According to the NFPA standard for carbon dioxide (NFPA 12, 1989), a design concentration of 50% is recommended for electrical installations. There is no explicit carbon dioxide hold time requirement for surface fires, but a 20 minute minimum hold time is specified for deep-seated fires. Required flame extinguishing concentrations for Halon replacement agents, as specified in the Appendix of NFPA 2001 (1996), are between those cited here for Halon 1301 and for carbon dioxide. NFPA 2001 states that hold time requirements for these replacement gases should be sufficient to allow effective emergency action by trained personnel.

The pressure developed in the enclosure during the discharge of a gaseous suppressant stored as a pressurized liquid is governed by several thermodynamic and fluid dynamic phenomena including:

1. Flash vaporization of a fraction of the liquid discharged into the equipment/room and cooling of the remaining liquid such that liquid and vapor fractions are both at the atmospheric pressure boiling point (for Halon: $-72\,°F$, $-58\,°C$).

2. Heating of the vaporized agent and vaporization of the remaining fraction of liquid upon mixing with the air in the enclosure.

3. Condensation of water vapor in the enclosure and warming of the vapor-air mixture by the heat released during condensation.

4. Pressurization of the enclosure from the additional volume of vapor (compared to the pre-discharge volume of air) that exists in the Halon-air mixture at the temperature produced at the end of 3.

5. Heating of the cool vapor-air mixture by the walls and equipment in the enclosure.

6. Leakage of halon-air mixture from the enclosure due to either forced ventilation, enclosure pressurization, or density differences within and outside the enclosure (see Section 4.2.1).

Saum *et al.* (1988) have compiled quantitative representations of these six phenomena assuming that they occur sequentially following halon discharge. For example, they report that a design basis Halon discharge into an enclosure initially at $21\,°C$ ($70\,°F$), will produce a temperature of $-2.5\,°C$ ($27.5\,°F$) prior to water vapor condensation, and a mixture temperature of $3\,°C$ ($38\,°F$) after condensation of 50% relative humidity air.

The pressure in the enclosure during and following discharge can be calculated from the ideal gas equation of state in the form

$$p = \frac{(W_0 + W)RT}{VM_{H-a}} \qquad [4.1.4]$$

where p is the enclosure pressure, psfa (Pa), V is the enclosure volume, ft^3 (m^3), W_0 is the initial mass (weight) of air in enclosure, kg (lb), W is the mass (weight) of Halon in enclosure from equation [4.1.1], R is the universal gas constant, 1545 ft-lbf/lbmole °R (8314 J/kgmole °K), M_{H-a} is the molecular weight of halon-air mixture (34.8 lb/lbmole (kg/kgmole) at 5 vol% Halon), and T is the temperature of halon-air mixture, °R (°K).

If equation [4.1.4] is applied immediately after Halon discharge into an isolated (unventilated) enclosure with minimal leakage, the pressure depends strongly on the value of T. In the case of the Saum *et al.* (1988), calculation of $3\,°C$ ($38\,°F$) after thermodynamic processes 1 through 4 above, equation [4.1.4] yields a pressure that is about 4 psf (190 Pa) less than the initial atmospheric pressure. However, if the halon-air mixture warmed up to its $21\,°C$ ($70\,°F$) initial temperature without any leakage, the pressure would be about 130 psf (6200 Pa) higher than its initial atmospheric pressure. Leakage (and the relatively slow wall heating) in most practical situations prevents the pressure from increasing that much. Measurements reported in Saum *et al.* (1988) and other studies indicate that the enclosure pressure initially decreases by 1.6–10 psf (77–480 Pa), and then increases to a maximum of 1–15 psf (50–720 Pa).

In the simpler case of a continuous halon discharge into a leaky enclosure, the pressure increase in the enclosure can be approximated by the following equation from the 1977 edition of NFPA 12A Halon Standard:

$$\Delta p = bR/A_L \qquad [4.1.5]$$

where Δp is the pressure increase in enclosure, psfg (kPa), R is the Halon discharge rate from equation (4.2), A_L is the enclosure leakage area, in^2 (cm^2), and b is the 13.2 for English units and 130 for metric units above.

Equations [4.1.2] and [4.1.5] can be used to determine the appropriate discharge rate to maintain concentrations above 5 vol% without jeopardizing the enclosure structural integrity by generating pressures beyond the allowable strength. If the equipment ventilation is shutdown at agent discharge, but there is an opening through which Halon can leak out of the enclosure, additional agent beyond the quantity given by equation [4.1.1] will be needed. The dilution of agent-air mixture in the enclosure can be estimated and accommodated with a higher initial concentration. NFPA 12A provides curves for estimating the initial concentration that will produce a 5% concentration at the end of the designated soak time.

Figure 4.3 shows a new wet bench equipped with a commercially available internal Halon suppression system. Similar systems for electronic equipment are being developed for US Air Force usage.

In the case of a carbon dioxide system intended for use in enclosed equipment, similar phenomena occur during discharge. The calculation of temperatures in the enclosure are particularly important for CO_2 systems because at atmospheric pressure pure CO_2 exists as a solid at $-79\,°C$ ($-110\,°F$), so there is concern about possible cold shock damage to sensitive electronic equipment. A series of mathematical models to predict CO_2-air mixture temperatures, pressures, and concentrations during and following discharge in unventilated leaky enclosures was recently developed by Cheng (1991). Similar models and test data will be needed for the various candidate Halon replacement gases intended for total flooding applications in enclosed equipment.

4.2 Isolation within rooms–building smoke control

Control of smoke movement across room boundaries is accomplished via differential pressurization. This usually involves using either the building Heating Ventilating and Air Conditioning (HVAC) system or special fans, blowers, and ducting capable of functioning in the high temperatures and pressures generated during a fire. In high-rise building applications, pressure differences associated with different air temperatures within and outside the building are also substantial and significantly influence intercompartment air flows. General design guidelines for smoke control are presented in Klote and Fothergill (1983), Klote and Milke (1992), and NFPA 92A (1988).

A prerequisite for the effective design of smoke control systems is knowledge of room pressures generated by the fire itself. The fire affects room pressures by generating buoyancy induced pressure differences relative to adjacent ambient temperature enclosures, and by the volumetric expansion associated with the rate of energy addition to room gases. Both sources of room overpressure are discussed here along with the corresponding smoke propagation velocities and pressure differences across doors and dampers.

4.2.1 BUOYANCY PRESSURE DIFFERENCES

The buoyancy pressure difference between the room of fire origin and an adjacent room is due to the different gas densities in the two rooms. If velocities are small, and gas densities are uniform

SMOKE ISOLATION AND VENTING

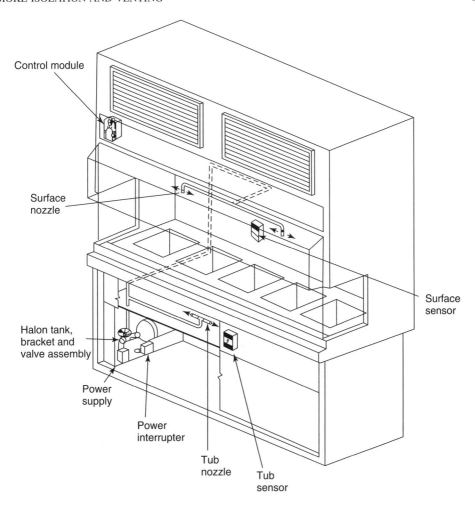

Figure 4.3. Halon suppression system for wet bench (from Ansul Co H-1000 Spec Sheet)

in each room (top sketch in Figure 4.4), pressures are approximately equal to the following static distributions:

$$p_0 = p_{np} - \rho_0 g(z - z_{np}) \qquad [4.2.1a]$$

$$p_f = p_{np} - \rho_f g(z - z_{np}) \qquad [4.2.1b]$$

where p_0 is the pressure at elevation z in the adjacent room, p_f is the pressure at elevation z in the fire room, p_{np} is the pressure at a neutral plane where $p_0 = p_f$, z_{np} is the neutral plane elevation, ρ_0 is the air density in the adjacent room, ρ_f is the gas density in the fire room, and g is the acceleration due to gravity (9.81 m/s^2).

The pressure distributions represented by equations [4.2.1a] and [4.2.1b] are depicted in Figure 4.4. The pressure difference between the rooms is

$$p_f - p_0 = g(\rho_0 - \rho_f)(z - z_{np}) \qquad [4.2.2]$$

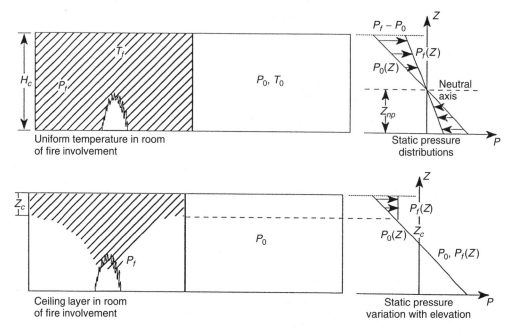

Figure 4.4. Buoyancy pressure differences

Both densities can be represented by the ideal gas equation of state taking the absolute pressure to be approximately constant and equal to p_{np}. Equation [4.2.2] becomes

$$p_f - p_0 = (p_{np}g/R)(T_0^{-1} - T_f^{-1})(z - z_{np}) \quad [4.2.3]$$

where R is the universal gas constant for air, R/28.8.

According to NFPA 92A (1988), $p_{np}g/R = 7.64$ in water/°R-ft, and the recommended NFPA 92A smoke control pressure to overcome the buoyancy pressure is

$$\Delta P = 7.64(T_0^{-1} - T_f^{-1})(2H_c/3) + 0.03 \text{ in water} \quad [4.2.4]$$

where H_c is the ceiling height in the room of fire involvement, and the 0.03 in water is a safety factor. The value of T_f specified in paragraph 2-2.1 of NFPA 92A for nonsprinklered rooms is 927°C (1700°F). The corresponding pressure difference given by equation [4.2.5] for $T_0 = 21$°C (70°F) and $H_c = 2.7$ m (9 ft) is 0.095 in water (24 Pa).

A similar result can be obtained for the case of a well defined ceiling layer in the fire room as shown in the bottom sketch in Figure 4.4. If the gas density below the ceiling layer is equal to the density in the adjacent room, the maximum pressure difference is

$$\Delta P = \rho_0 g z_c (T_f - T_0)/T_f \quad [4.2.5]$$

where z_c is the ceiling layer depth.

Use of equation [4.2.5] instead of equation [4.2.4] will produce smaller calculated pressure differences when z_c is less than $2H_c/3$. When z_c is larger than $2H_c/3$, the assumption that the gas density (and temperature) below the ceiling layer is equal to the adjacent room density (and temperature) is no longer valid because of heat transfer from the ceiling layer to the lower layer.

4.2.2 VOLUMETRIC EXPANSION PRESSURES

Most of the literature data on volumetric expansion pressures generated by an enclosure fire were obtained in enclosures with large natural ventilation areas in the form of open doors and windows. Pressures generated in these naturally ventilated fires are usually limited to about 4–15 Pa (Drysdale, 1985, p 367) for mass burning rates of about 5 kg/min (11 lb/min) (heat release rate on the order of 1 MW). The range of volumetric expansion pressure (in Pascals) reported by Drysdale based on Fung's data is

$$0.25(m_b/T_0 A)^2 T_f < \Delta P < 1.0(m_b/T_0 A)^2 T_f \qquad [4.2.6]$$

where m_b is the mass burning rate (kg/min), and A is open area of the enclosure (m^2).

If the doors and windows are closed, such that buoyant flow into and out of the enclosure is restricted, volumetric expansion pressures will be the principle driving force for mass flow out of the enclosure through small gaps around the nominally closed doors and windows. The mass flow rate corresponding to Bernoulli's equation for the velocity through these small enclosure openings is

$$m = C_D (2\rho_f \Delta P)^{1/2} A \qquad [4.2.7]$$

where m is the mass flow rate (kg/s) through enclosure leakage area A (m^2), P is the pressure drop across enclosure openings (Pa), and C_D is the orifice discharge coefficient (dimensionless). The convective energy flow out of the enclosure is $mc_p(T_f - T_0)$, and at steady state conditions,

$$YQ = mC_p(T_f - T_0) \qquad [4.2.8]$$

where Q is the fire heat release rate (kW), Y is the fraction of Q transferred to outflowing gases, and C_p is the specific heat for enclosure gases (kJ/kg °K).

Substituting equation [4.2.7] into [4.2.8], and solving for ΔP,

$$\Delta P = 0.50 \rho_f^{-1} [YQ/(C_D A c_p (T_f - T_0))]^2 \qquad [4.2.9a]$$

or, equivalently,

$$\Delta P = 0.50 (T_f/T_0) \rho_0^{-1} [\gamma Q/(C_D A c_p (T_f - T_0))]^2 \qquad [4.2.9b]$$

where ΔP is in Pascals when ρ_0 is kg/m^3 and A is in m^2.

Calculated volumetric expansion pressure differences associated with equations [4.2.9a,b] are sensitive to the assumed value of Y. In the case of enclosures with large open doorways, Y can be as large as 0.60 or 0.70. However, in the case of small gaps around closed doors and windows, Y is much smaller. A rough estimate for Y for small enclosure openings can be ascertained from the results of computer model calculations such as those obtained with the HAZARD I code developed at the National Institute of Standards and Technology. Two of the examples presented for HAZARD I (Bukowski and Peacock, 1989) include enclosure fires with peak heat release rates of about 250 kW in rooms with only a 2 cm (0.8 in) high gap at the bottom of a closed door. The HAZARD I calculated overpressures of about 30 Pa for these examples of small gap enclosures would also be obtained from equation [4.2.9] if a value of approximately 0.20 was used for Y.

Hinkley (1995) has suggested the following equation for expansion pressure increases in fires with specified rates of temperature rise rather than specified heat release rates:

$$\Delta P = 500(H_c A_c dT_f/dt)^2/[A^2 T_f^3] \qquad [4.2.10]$$

where A_c is the enclosure surface area (m^2), and dT_f/dt is the rate of temperature rise in compartment (°C/sec).

As an example, equation [4.2.10] would predict for $H = 3$ m (9.8 ft), $A_c = 50$ m^2 (538 ft^2), $T_f = 1143\,°$K (1600 F), and $dT_f/dt = 4\,°$C/sec (7 °F/sec), and the other parameters as specified above, $\Delta P = 3.7$ Pa.

There is recent evidence that substantially higher pressures can develop in enclosure fires with balanced supply/exhaust ventilation systems, and with venting into a confined area such as a corridor, particularly at high heat release rates. For example, pressure differences as high as 0.8 in water (200 Pa) have been measured for propylene burner fires with heat release rates increasing as t^2 in an enclosure vented into a corridor (Heskestad and Hill, 1987). Similarly, fire induced overpressures in the range 0.07 to 0.7 in water (18 to 180 Pa) have been measured at FMRC when forced air supply ventilation was provided at a rate of 5–15 volume changes per hour. If indeed pressures can reach approximately 100 Pa during severe forced ventilation enclosure fires, they may exceed HVAC system pressure capabilities (Milewski, 1985) and produce uncontrolled smoke transport through a large portion of the building. Additional research is needed to determine the fire intensities and room/ventilation conditions required for these excessive fire induced room pressures.

Anomalously high volumetric expansion pressures can also be produced in certain other special situations where fuel rich vapor-air mixtures are generated. One such situation involves an elevated fire source immersed in the ceiling layer of oxygen vitiated gases. This situation is prone to pulsating combustion as the ceiling layer is vented, fresh air rushes in and suddenly allows rapid burning of both the primary fuel and fuel rich vapors remaining in the ceiling layer, until the oxygen concentration is again reduced in the vicinity of the fire source. Another special situation is the prolonged burning in a ventilation limited enclosure in which pyrolysis products have been accumulating. The sudden admission of air from the opening of an enclosure door can cause combustion to accelerate rapidly enough to be called a 'smoke explosion.' This has been a cause of many casualties among both amateur and professional firefighters.

4.2.3 ISOLATION VIA VENTILATION EXHAUST

One smoke control configuration that is particularly relevant to many industrial applications is the use of exhaust fans to maintain negative pressures in the burn room (relative to adjacent rooms) and in plenums above suspended ceilings. The exhaust ventilated burn room situation is depicted in Figure 4.5, and the ventilated plenum above a suspended ceiling is shown in the middle sketch in Figure 4.1. Note that the supply air damper to the burn room is closed in Figure 4.5. Klote (1982) has compared smoke obscuration levels in a burn room with exhaust only ventilation as shown in Figure 4.5 to those obtained with combined supply-exhaust ventilation corresponding to an open supply air damper. Smoke accumulation in the burn room was considerably reduced with exhaust only ventilation, and the burn room pressure was held about 1 Pascal below the adjacent unventilated space for most of the test (Test 10 in Klote, 1982).

Proper design of the smoke isolation system sketched in Figure 4.5 should include guidance on the ventilation rates required to prevent smoke propagation into Room 2 for various intensity fires in Room 1 and leakage areas between rooms. Although data for this configuration are not yet available, data correlations are presented in Section 4.2.4 for preventing upstream smoke propagation. When the duct dampers are set as shown in Figure 4.5, Room 2 is effectively upstream of Room 1 in the absence of fire. Thus, the data correlations in Section 4.2.4 should be applicable providing heat release rates are sufficiently small to neglect volumetric expansion pressures.

A limited amount of data is available for the case of a ventilated plenum above the burn and target rooms. Klote (1982) has conducted both low heat release rate and high heat release rate fire tests with this configuration as illustrated in Figure 4.6. The low heat release rate fire was conducted with a fire retardant cotton innerspring mattress, while the high heat release rate fire involved a polyurethane foam mattress. Klote's results indicate that exhausting the space above

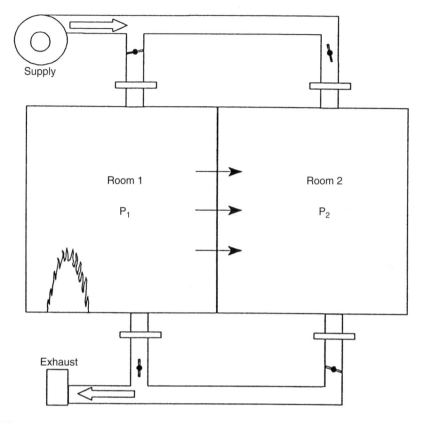

Figure 4.5. Smoke isolation via ventilation exhaust (Smoke control $\Rightarrow P_2 - P_1 > \Delta P_{\text{Buoyancy}} + \Delta P_{\text{expansion}}$)

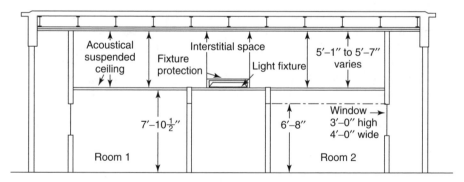

Figure 4.6. Section of test facility used by Klote (1982) to investigate smoke control via exhaust ventilation of burn room and area above suspended ceiling

the suspended ceiling at a rate of two air changes per hour (110 cfm or 3.1 m³/min) is sufficient to preclude downward smoke flow through the ceiling to the target room when there is a low heat release rate fire in the burn room. However, he did observe some downward smoke flow into the target room (Room 2 in Figure 4.6) when there was a flaming polyurethane foam mattress fire in the burn room. The peak pressure difference between the burn room and the ventilated plenum

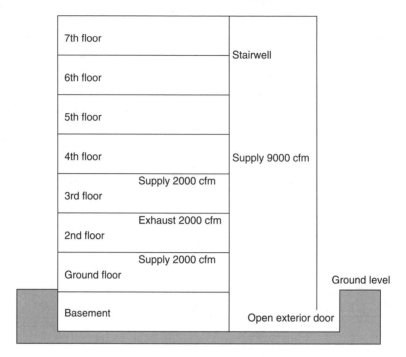

Figure 4.7. Schematic of the smoke control system in Plaza Hotel tests. (Notes: (1) The second floor is the smoke zone, and it is exhausted at about six air changes per hour; (2) The first and third floors are pressurized at about six air changes per hour; (3) The stairwell is pressurized by 9000 cfm, and the exterior stairwell door remains open throughout pressurization)

was about 9 Pascals for both fires. Unfortunately, heat release rates were not measured so it is difficult to generalize the results from Klote (1982).

The concept of smoke confinement to the room of fire origin can be extended to zoned smoke control where the smoke zone includes more than a single room. One example of this expanded concept of zoned smoke control is the confinement of smoke to one floor in a multistorey building. Fire tests designed to demonstrate this application of zoned smoke control were conducted in 1989 in the seven story Plaza Hotel building in Washington, DC (Klote, 1990). The smoke control system used in the tests is shown conceptually in Figure 4.7. It consisted of a 2000 cfm (0.94 m^3/s, six volume changes per hr) exhaust fan for the fire floor, two other 2000 cfm supply fans to pressurize the floors immediately above and below the fire floor, and a 9000 cfm (4.2 m^3/s) fan to pressurize the stairwell at the fire floor. The stairwell doors at the fire floor and at the seventh floor were closed in some tests and open ($\frac{1}{2}$ in or 1.3 cm gap at fire floor; completely open on 7th floor) in other tests.

Wood crib fires with estimated heat release rates of 1.5 MW in the Plaza Hotel second floor corridor produced significant smoke obscuration on upper floors (particularly the top floor) in the absence of any smoke control. When the tests were repeated with the installed smoke control system activated, there was virtually no smoke on the upper floors. Carbon monoxide levels on the 7th floor were also reduced substantially by smoke control. When the smoke control was actuated at ignition, CO levels were negligible or the 7th floor; when smoke control was delayed four minutes, 7th floor CO levels were higher, but were still limited to about 30% of the baseline levels without smoke control (Figure 4.8). Tests with a ceiling sprinkler near the fire also produced negligible smoke and CO levels on the seventh floor.

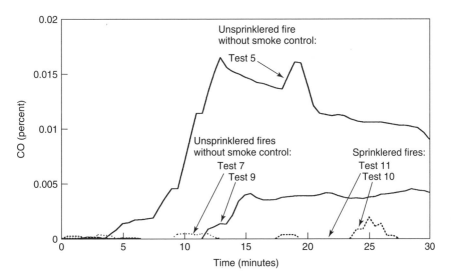

Figure 4.8. CO concentrations on the seventh floor in the Plaza Hotel tests

Zoned smoke control systems operated in situations where the mechanical ventilation system (and in the case of multi-storey buildings, stack effects and wind loading), rather than fire induced pressures, is responsible for smoke transport can be evaluated using available computer codes. For example, the smoke control system for the Plaza Hotel tests was designed with the aid of the ASCOS computer code described in Klote and Fothergill (1983) and currently being distributed by the Society for Fire Protection Engineers.

One first order approximation to combine the various sources of pressure difference across compartment boundaries would be to linearly superimpose them. Thus,

$$\Delta p = \Delta p_{HVAC} + \Delta p_b + \Delta p_{exp} \qquad [4.2.11]$$

where Δp_{HVAC} is the HVAC system (fan) induced pressure difference, Δp_b is the buoyancy induced pressure difference (equations [4.2.3], [4.2.4], or [4.2.5]), and Δp_{exp} is the volumetric expansion induced pressure difference (equations [4.2.9] or [4.2.10]).

There is no fundamental reason why equation [4.2.11] should be valid because the three effects may be interactive and therefore nonlinearly superimposed on each other. However, an approach equivalent to equation [4.2.11] (except that volumetric expansion pressures were computed from mass flow rate differences) was used by Klote to analyze the pressure differences measured in the Plaza Hotel smoke control tests. His comparisons with data indicated good agreement, i.e. to within 0.02 in H_2O (5 Pa).

Situations involving significant interactions between the fire and mechanical ventilation systems are much more difficult to evaluate. One potential pitfall in the latter situations is the possibility of unstable fan operation when the static pressure increase across the fan is outside of its design range. This can lead to flow reversal across the fan and smoke propagation into the intended smoke free zone. Klote and Cooper (1989) have developed a computer model, called FANRES, to account for the possibility of flow reversal due to excessively high pressure increases across a fan. The FANRES model is intended for use with enclosure fire models in which the fire compartment pressure is calculated as part of the solution. The simpler situation in which there is a potential for upstream smoke propagation via buoyancy and diffusion, rather than flow reversal, is described in the following Section.

4.2.4 UPSTREAM SMOKE PROPAGATION

Buoyant gravity flow causes the leading edge of a hot smoke layer to propagate away from the fire source in a duct or corridor (Heskestad and Hill, 1987). Thomas (1968) performed wind tunnel tests to measure the critical air velocity required to prevent upstream smoke propagation. His data correlation is

$$U_c = [gQ/\rho_0 C_p w T_0]^{1/3} \quad [4.2.12]$$

where U_c is the minimum duct flow velocity to prevent upstream smoke propagation (m/s), Q is the heat release rate into the duct (kW), w is the duct/corridor width (m), g is gravitational acceleration (m/s^2), ρ_0 is the upstream air density (kg/m^3), C_p is the specific heat of air (kJ/kgK), and T_0 is the upstream air temperature (°K).

Klote (1995) has provided a graphical representation of equation [4.2.12] for $\rho_0 = 1.3$ kg/m^3, $C_p = 1.005$ kJ/kg-°C, $T_0 = 27$ °C, which is reproduced here as Figure 4.9.

Thomas' results can be expressed in the form of a critical Froude number defined as follows:

$$Fr_c = U_c/[gH_D(T_{HL} - T_0)/T_0]^{1/2} \quad [4.2.13]$$

where Fr_c is the critical Froude number to prevent upstream smoke propagation, H_D is the duct (or corridor) height (m), T_{HL} is the hot layer (smoke) temperature (°C), and T_0 is the air temperature in duct or corridor (°C).

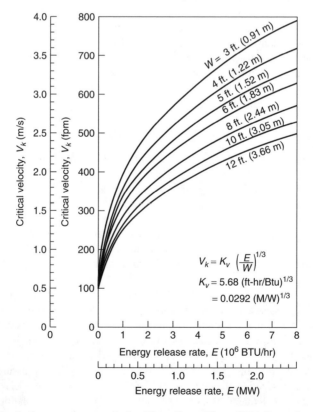

Figure 4.9. *Critical velocity to prevent smoke backflow* (from Klote, 1988. Reproduced by permission of Society of Fire Protection Engineers)

Thomas' wind tunnel data correlation suggested that Fr_c is approximately equal to 1 (the actual value depends on the ratio of the convective heat transport rate to the chemical heat release rate). Heskestad and Spaulding (1991) conducted experiments to determine the minimum air inflow velocity to prevent smoke from escaping through wall and ceiling openings in an enclosure in which a steady-state fire is burning. They correlated their data in the form of a critical Froude number of the form

$$Fr'_c = \frac{U_c}{[2gL(T_{HL}T_0)/T_{HL}]^{1/2}} \quad [4.2.14]$$

where U_c is the minimum inflow velocity to prevent upstream smoke propagation out of the enclosure, L is the height of wall aperture or width of ceiling aperture, and T_{HL} is the temperature of hot layer at top of wall aperture or at ceiling aperture.

Heskestad and Spaulding found that the critical Froude numbers were in the following ranges depending on whether the enclosure opening was in the wall or the ceiling:

$$0.32 \leq Fr'_c \leq 0.43 \text{ for wall openings} \quad [4.2.15a]$$

$$0.23 \leq Fr'_c \leq 0.38 \text{ for ceiling openings} \quad [4.2.15b]$$

In the case of wall openings, the lower end of the Fr'_c range corresponds to highly stratified hot layers as would be produced at low heat release rates, while the upper end (0.43) corresponds to nearly uniform hot layers as would be produced at relatively high heat release rates. In the case of ceiling openings, the low end of the Fr'_c range corresponds to large enclosures (more precisely, large Grashof numbers as defined in Heskestad and Spaulding, 1991), while the upper end (0.38) corresponds to small enclosures.

Consider the implications of equations [4.2.14] and [4.2.15] for the double enclosure situation depicted in Figure 4.5, neglecting the presence of the filters. First, imagine the case when the dampers above and below Room 2 are both closed, the dampers above and below Room 1 are both open, and there is negligible leakage between the rooms. In other words, the air flows in through the ceiling and out through the floor of Room 1 and Room 2 is isolated if there is no damper leakage. We further specify a ceiling aperture width of 0.5 m (1.6 ft) and a hot layer ceiling temperature of 1000 °K (1340 °F). Since this is a relatively large opening, we choose a value of Fr'_c toward the lower end of the range given by equation (4.20b); perhaps 0.25. The minimum air velocity to prevent smoke propagation out through the ceiling opening in this case is

$$U_c = 0.25[2(9.81)(0.5)(707)/1000]^{1/2} = 0.66 \text{ m/sec} \ (2.2 \text{ ft/sec})$$

If the ceiling opening is square, the required air flow rate is $0.66(0.5)^2 = 0.16 \text{ m}^3/\text{sec} = 350$ cfm.

Now suppose the dampers are set as shown in Figure 4.5 and there is a 0.5 m (1.6 ft) high opening in the wall separating the two rooms. In this case, the critical Froude number for air flow through the wall into Room 1 is given by equation [4.2.15a], with a value toward the lower end (perhaps 0.34) being more appropriate for high heat release rates and high values of T_{HL}. Thus,

$$U_c = 0.34[2(9.81)(0.5)(707)/1000]^{1/2} = 0.90 \text{ m/sec} \ (3 \text{ ft/sec})$$

If the wall opening is square, the required air flow rate now is $0.22 \text{ m}^3/\text{sec} = 480$ cfm.

The pressure drop required to produce the critical velocity, U_c, through the enclosure opening is

$$\Delta P_c = \rho_0 U_c^2 / 2C_D^2 \quad [4.2.16]$$

where ΔP_c is the critical pressure drop across enclosure opening to prevent upstream smoke propagation, and C_D is the discharge coefficient for flow through enclosure opening.

Substitution of equation [4.2.14] into equation [4.2.16] yields

$$\Delta p_c = (Fr'_c/C_D)^2 \rho_0 g L (T_{HL} - T_0)/T_{HL} \qquad [4.2.17]$$

Discharge coefficients to be used in equations [4.2.16] and [4.2.17] are given in Heskestad and Spaulding (1991). In the case of wall openings, C_D is equal to 0.64, which is approximately the value for cold flow through sharp edged orifices. In the case of ceiling openings, C_D ranges from a low of about 0.20 at the critical Froude number to about 0.60 for Froude numbers seven times as large as Fr'_c.

Application of equation [4.2.16] and a discharge coefficient of 0.20 to the preceding case of smoke flow through a ceiling opening 0.5 m (1.6 ft) wide with $T_{HL} = 1000\,°K$ (1340 °F), results in $P_c = 6.5$ Pa. If the enclosure opening is in the side wall, such that $U_c = 0.90$ m/sec (3 ft/sec) and $C_D = 0.64$, $P_c = 1.2$ Pa. Thus, higher velocities but smaller pressure drops are required for wall openings than for ceiling openings.

The prevention of upstream smoke propagation can be an important aspect of manifolded ventilation ducting serving a multiple clean room facility such as that shown in Figure 4.10. The exhaust duct air velocity required to prevent smoke infiltration from the burn room (Room 2 in Figure 4.10) to an upstream room (Room 1) can be estimated from equation [4.2.13] or Figure 4.9. However, the exhaust fan capacity should account for the expansion of combustion products in the burn room. Thus, the ratio of exhaust flow rate to air supply rate in Room 2 should be approximately equal to the temperature ratio T_f/T_0. This ratio should be at least a factor of 2 and more likely a factor of 3 (corresponding to $T_f = 900\,°K$ or 1160 °F).

One other smoke control issue applicable to clean room ventilation is the appropriate location for smoke detectors. Since clean room air flow is usually supplied from the ceiling and removed from the floor as shown in Figure 4.10, the ventilation air flow is directly opposite to the buoyant rise velocity of the smoke. Sugawa et al. (1987) have conducted experiments and calculations to determine whether a small downward ventilation velocity can prevent the buoyant smoke from reaching the clean room ceiling. They found that a downward velocity of about 22 cm/sec (8.7 in/sec) was sufficient to prevent smoke from a 6 kW fire from reaching most of the ceiling. They recommend placing the smoke detector in the clean room exhaust duct. Their results are (at least) qualitatively consistent with the large-scale simulated clean room fire tests described in Fisher et al. (1986).

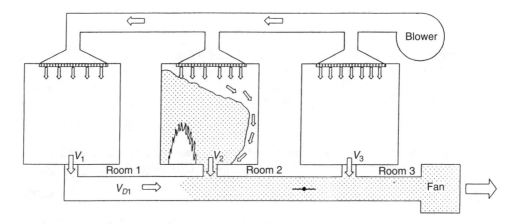

Figure 4.10. Smoke transport and control for clean rooms with manifolded ventilation ducts

4.2.5 DOOR AND DAMPER SMOKE LEAKAGE

If a smoke control installation is to be effective, there should be reliable data on the smoke leakage characteristics of smoke dampers in HVAC systems and intercompartment doors. Cooper (1985) has reviewed the status of available test methods for measuring smoke leakage rates through doors. These methods can be categorized (ISO DP 5925, 1977) according to the temperature of the leaking gases as: (1) ambient temperature air leakage tests (both laboratory tests and in-situ tests on installed doors); (2) medium temperature (100–250 °C or 212–482 °F) air leakage tests; and (3) high temperature gas leakage tests. These three categories would be applicable to doors (1) far away from the fire compartment, (2) in a compartment adjacent to the fire source, and (3) in a compartment in which there is a fully developed fire, respectively.

The International Organization for Standardization (ISO) has established test methods for all three categories of smoke leakage (ISO DP 5925, 1977). Cooper's review (1985) includes a synopsis of the ISO tests and proposed modifications. The high temperature leakage test has been especially criticized, and there are several proposed revisions. All versions involve installing the door on the furnace used for fire resistance testing as described in Section 3.3.1, and operating the furnace with controllable pressure differences (up to about 100 Pa) across the door. The different versions involve various methods for measuring and reporting gas leakage rates. Once these different versions are reconciled, it should be possible to have standard smoke leakage ratings for doors.

If a door is subjected to high pressure differences needed for smoke control, there is a question as to its ability to be opened to allow emergency egress. According to NFPA 92A and NFPA 101, Life Safety Code, the maximum force required to open a door in a fire situation should be 30 lbs (133 N). The corresponding maximum pressure difference across the door is 0.21–0.45 in water (53–112 Pa), depending on door closing force and door width. These pressures are higher than the buoyant pressure differences discussed in Section 4.2.1 and the expansion pressures discussed in Section 4.2.2 for natural ventilation. However, they could be less than the expansion pressure differences for balanced supply-exhaust ventilation in the burn room. Therefore, an effective smoke control system may require shutting down the air supply to the burn room so as to eliminate the need to generate excessively high pressure differences across the burn room door.

4.3 Heat and smoke roof venting

Besides serving as a smoke isolation method, heat and smoke roof vents are intended to facilitate manual fire fighting and/or evacuation by preventing smoke accumulation and smoke layer descent in the fire area. Two key design parameters for roof vents are the required vent area and the allowable depth of the smoke layer. The following synopsis of the theoretical design basis in NFPA 204M (1982) explains the relationship between these parameters. A similar theory is used in the equivalent European standards cited by Hinkley (1995). A fundamental assumption in both theories is that the fire is not ventilation limited, i.e. there is ample ventilation at low elevations such that there is no air inflow through the roof vents, just outflow. The situation that results when there is not adequate low level air inflow is discussed at the end of this section.

The basic premise in NFPA 204M is that the total vent area should be sufficient to allow the smoke/gas flow rate through the vent to equal the fire plume flow rate entering the smoke layer. The plume flow rate is dependent on the fire heat release rate and flame height, as illustrated in Figure 4.11. A convective heat release rate that produces a flame height equal to the height of the bottom of the smoke layer is called the critical heat release rate, Q_c. The Heskestad flame height correlation utilized in NFPA 204M corresponds to the following equation for Q_c:

$$Q_c = C(H - d)^{5/2} \qquad [4.3.1]$$

where H is the roof height, d is the depth of the smoke layer, C is 11.3 for H, d in ft and Q_c in Btu/sec, and C is 233 for H, d in m and Q_c in kW.

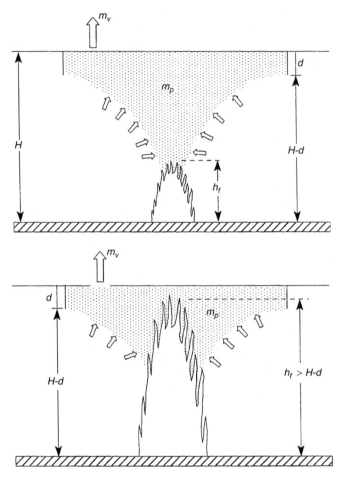

Figure 4.11. Fire plume and smoke layer for flames below draft curtain (top) and flames entering ceiling layer (bottom).

When $Q < Q_c$ ($h_f < H - d$), the plume mass flow rate, m_p, entering the smoke layer is

$$m_p = 0.022 Q^{1/3}(H-d)^{5/3}[1 + 0.19 Q^{2/3}(H-d)^{-5/3}] \qquad [4.3.2]$$

where m_p is in lb/sec, Q is in Btu/sec, and H and d are in ft.

When $Q > Q_c$ ($h_f > H - d$),

$$m_p = 0.097(H-d)^{5/2}(Q/Q_c)^{3/5} \qquad [4.3.3]$$

where m_p is in lb/sec, and H and d are in ft.

The mass flow rate, m_v, through the vent area, A_v, due to the buoyant rise velocity of the hot smoke layer is

$$m_v = (2g\rho_0^2 T_0 \Delta T/T^2)^{1/2} A_v d^{1/2} \qquad [4.3.4]$$

A good approximation for smoke layer temperatures in the range 300–1000 °F (422–811 °K) is

$$(T_0 \Delta T/T^2)^{1/2} \approx 0.50 \qquad [4.3.5]$$

SMOKE ISOLATION AND VENTING

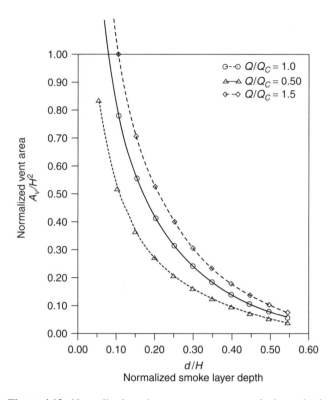

Figure 4.12. Normalized smoke vent area versus smoke layer depth

Substitution of equation [4.3.5] into [4.3.4], and setting

$$m_v = m_p \qquad [4.3.6]$$

with m_p from equation [4.3.2], leads to the following solution for the vent area, A_v:

$$A_v = 0.075 Q^{3/5}(H-d)/d^{1/2} \qquad [4.3.7]$$

where A_v is in ft², Q is in Btu/sec, and, H and d are in ft.

Equation [4.3.7], which is approximately valid for subcritical heat release rates as well as supercritical values of Q, dictates that small values of d will require very large values of A_v. A nondimensional version of equation [4.3.7], can be obtained by using equation [4.3.1] to normalize Q. The result is

$$\frac{A_v}{H^2} = \frac{0.322(Q/Q_c)^{3/5}(1-d/H)^{5/2}}{(d/H)^{1/2}} \qquad [4.3.8]$$

Figure 4.12 is a graph of A_v/H^2 versus d/H based on equation [4.3.8]. It shows that A_v/H^2 is less than 0.10 when $d/H \geq 0.50$, i.e. when the smoke layer extends at least halfway from the ceiling to the floor. If L represents the horizontal distance between smoke curtains, such that L^2 is the ceiling area within the curtained and vented roof, the percent roof area vented is $(A_v/H^2)(H/L)^2$. When $H/L = 0.5$, A_v/H^2 being equal to 0.10 implies that $(0.10)(0.5)^2 = 2.5\%$ of the roof area needs to be vented (for a smoke layer equal to half the building height). If it is desired to keep $d/H \leq 0.20$, Figure 4.12 indicates that $Av/H^2 \geq 0.40$ for $Q/Q_c \geq 1.0$. This would correspond to a vent/roof area ratio of at least $(0.40)(0.5)^2 = 10\%$ when $H/L = 0.50$.

NFPA 204M suggests that $A_v < 2d^2$ in order to allow the smoke layer to fill the volume between draft curtains. Furthermore, the selection of an appropriate draft curtain depth should be made with considerations of the fire spread potential due to radiation from the hot smoke layer.

NFPA 204M provides design guidance for both constant heat release rate fires and fires with unlimited growth rates. In the case of fuel area limited fires, Q can be estimated from tables of heat release rate per unit surface area (e.g. in Appendix A and in Chapter 7). In the case of unlimited fuel surface area, NFPA 204M suggests use of the following form of a second power (parabolic) growth rate:

$$Q = 1000(t/t_g)^2 \qquad [4.3.9]$$

where t is time (sec) from effective ignition (after accounting for any incipient fire development), and t_g is the characteristic fire growth time (sec) for the combustible fuel array, such that $Q = 1000\,\text{Btu/sec}$ when $t = t_g$. Representative values of t_g are shown in Table 4.2 for various warehouse commodities. Values of the parameter $a = 1000/t_g^2$ for other warehouse commodities are given in Chapter 5.

The suggested vent area design basis in NFPA 204M for a fire with a heat release rate satisfying equation [4.3.9] is obtained by designing for Q when

$$t = t_d + t_r \qquad [4.3.10]$$

where t_d is the time required to detect the fire, and t_r is the time to adequately respond to the fire either with manual fire fighting or by personnel evacuation.

Substitution of equations [4.3.9] and [4.3.10] into [4.3.7] yields

$$A_v = 4.8[(t_d + t_r)/t_g]^{6/5}(H - d)/d^{1/2} \qquad [4.3.11]$$

The value of A_v calculated from equations [4.3.11] or [4.3.7] is the aerodynamic vent area equal to the actual vent area times the vent discharge coefficient, C_D. Typical values of C_D are in the range 0.60 to 0.80.

Design bases similar to NFPA 204M have been developed by researchers at the Fire Research Station and adopted in several European heat/smoke vent guidelines. These correlations and analyses are reported in Thomas et al. (1963), Thomas and Hinkley (1964) and summarized in Hinkley's chapter in the SFPE Handbook (1995).

If it is not practical to provide the amount of roof area given by the preceding equations, powered ventilators can be used. The design basis for these powered smoke extractors is that they provide a volumetric flow rate equivalent to the plume mass flow rates given in equations [4.3.2] or [4.3.3], and that they generate a pressure sufficient to provide this flow rate against a back pressure corresponding to the wind induced adverse pressure on the roof. Hinkley (1995) provides elaboration and examples of powered extractor design calculations. One additional consideration discussed by Hinkley (1995) is the prevention of 'plugholing,' which is the entrainment of cold air beneath the smoke layer into the vent flow such that the bulk of the smoke layer is not vented. This detrimental phenomenon occurs when the vent outflow velocity is greater than a critical value corresponding to a critical vent outflow Froude number.

If there is not sufficient air inlet beneath the smoke layer, to provide a mass inflow equal to the plume/ceiling layer outflows, the smoke layer will descend toward the floor and the preceding equations for required roof vent area are not valid. The resulting roof vent flow pattern in this case consists of a concentric outflow through the central portion of the vent and an inflow through the outer area of the vent. Clearly the venting efficiency is greatly reduced. Epstein (1992) has presented a theory to calculate the reduced outflow for a steady-state fire in a building with

Table 4.2. Maximum heat release rates (data from NFPA 72)

Warehouse materials	Growth time (tg) (sec)	Heat release density (q)	Classification (s = slow, m = medium, f = fast)
1. Wood pallets, stack, 1 1/2 ft high (6–12% moisture)	150–310	110	f-m
2. Wood pallets, stack, 5 ft high (6–12% moisture)	90–190	330	f:m
3. Wood pallets, stack, 10 ft high (6–12% moisture)	80–110	600	f
4. Wood pallets, stack, 16 ft high (6–12% moisture)	75–105	900	f
5. Mail bags, filled, stored 5 ft high	190	35	m
6. Cartons, compartmented, stacked 15 ft high	60	200	f
7. Paper, vertical rolls, stacked 20 ft high	15–28	–	–[a]
8. Cotton (also PE, PE/Cot, Acrylic/Nylon/PE), garments in 12-ft high rack	20–42	–	–[a]
9. Cartons on pallets, rack storage, 15–30 ft high	40–280	–	f-m
10. Paper products, densely packed in cartons, rack storage, 20 ft high	470	–	s
11. PE letter trays, filled, stacked 5 ft high on cart	190	750	m
12. PE trash barrels in cartons, stacked 15 ft high	55	250	f
13. FRP shower stalls in cartons, stacked 15 ft high	85	110	f
14. PE Bottles, packed in Item 6	85	550	f
15. PE bottles in cartons, stacked 15 ft high	75	170	f
16. PE pallets, stacked 3 ft high	130	–	f
17. PE pallets, stacked 6–8 ft high	30–55	–	f
18. PU mattress, single, horizontal	110	–	f
19. PE insulation board, rigid foam, stacked 15 ft high	8	170	–[a]
20. PS jars, packed in Item 6	55	1200	f
21. PS tubs nested in cartons, stacked 14 ft high	105	450	f
22. PS toy parts in cartons, stacked 15 ft high	110	180	f
23. PS insulation board, rigid, stacked 14 & high	7	290	–[a]
24. PVC bottles, packed in Item 6	9	300	–[a]
25. PP tubs, packed in Item 6	10	390	–[a]
26. PP and PE film in rolls, stacked 14 ft high	40	350	–[a]
27. Distilled spirits in barrels, stacked 20 ft high	23–40	–	–[a]
28. Methyl alcohol	–	65	–
29. Gasoline	–	200	–
30. Kerosene	–	200	–
31. Diesel oil	–	180	–

For SI Units: 1 ft = 0.305 m
NOTE: The heat release rates per unit floor area are for fully involved combustibles, assuming 100% combustion efficiency. The growth times shown are those required to exceed 1000 Btu/sec heat release rate for developing fires assuming 100% combustion efficiency.
(PE = polyethylene; PS = polystyrene; PVC = polyvinyl chloride; PP = polypropylene; PU = polyurethane; FRP = fiberglass-reinforced polyester.)
[a]Fire growth rate exceeds design data

no low-level air inlet. He also presents results for the maximum possible mass burning rate in a fire that is controlled by the rate of air inflow through the roof vent. Since the fire heat release rate in this case increases with increasing roof vent area, there is a question of the overall desirability of using roof vents in conjunction with a fire suppression system such as a sprinkler system.

4.4 Heat and smoke venting in sprinklered buildings

The use of heat and smoke vents in sprinklered industrial buildings is a controversial subject for which there is widely differing opinions and a noncommittal position in NFPA 204 (1998). The following is a brief synopsis of the current understanding as described in various papers and reports published between 1974 and 1994 and included in the reference list.

Potential benefits of vents in sprinklered buildings are qualitatively the same as those in unsprinklered buildings; namely, they can delay loss of visibility and maintain tenable conditions for manual firefighting and cleanup. It has also been argued that venting can serve as a backup in the sense that it can reduce ceiling temperatures in case the sprinkler system does not function or is ineffective against a high challenge fire.

Potential drawbacks of vents in sprinklered buildings are: (1) they may increase the burning rate by providing an unlimited supply of oxygen, and thereby (depending on the draft curtain layout in relation to the fire) cause additional sprinkler heads to open away from the fire and possibly overtax the available water supply; and (2) the cooling of the hot ceiling layer by sprinkler spray can decrease its buoyancy and reduce the effectiveness of the vents to the point where visibility cannot be maintained even in their presence.

4.4.1 TESTING

Relevant test programs conducted prior to 1980 are summarized in Heskestad (1974), Ward (1982), and FM Data Sheet 1-10 (1978). Some early testing performed at Factory Mutual in 1956 utilized a gasoline spray fire in a 120 × 60 ft (36.6 × 18.3 m) test building. Sprinklers were used with as much as $3\,m^2$ ($32\,ft^2$) of vent area in the $212\,m^2$ ($2280\,ft^2$) curtained (5 ft or 1.5 m deep) area. The draft curtain was beneficial in reducing the sprinkler water demand compared to the unvented configuration without a curtain. However, the presence of vents and the vent area did not appreciably affect the water demand. Ceiling layer temperatures were slightly reduced due to the vents.

A comprehensive series of model tests was conducted by Heskestad (1974) to examine vent area and location and draft curtain effects as well as fire intensity. Results indicated that water demand was decreased with vents directly over the fire, but was increased in several other configurations.

FMRC also occasionally explored venting effects in its rack storage tests at the FM Test Center. One noteworthy test involving rubber tires was initially conducted without any venting, but doors and windows were opened about one hour into the test when the sprinklers appeared to have controlled the fire. Soon after the building was vented, the fire flared up and opened more than twice as many sprinklers (95). Miller (1984) maintains that the timing of this flare up is only coincidentally related to vent opening. However, there are numerous fire incidents (e.g. the Chatham mattress fire described by Heseldon, 1984) in which sudden venting of intense, oxygen deficient fires have led to rapid flare up and explosions due to accumulated pyrolysis products.

The 1982 IITRI test program (Waterman *et al.*, 1982) is perhaps the most controversial. Waterman *et al.* claim the IITRI tests in a 7.6 × 23 × 5.2 m (25 × 75 × 17 ft) high building demonstrate that automatic roof vents do not impair sprinkler control of growing fires. They attribute the occasional increased water demand in some vented tests to the normal data scatter. Miller (1984) supports the IITRI conclusion and further asserts that some of the data suggest "that venting may aid sprinklers during the critical period when marginal fire control begins to deteriorate." Heskestad (1983), on the other hand, maintains that the results are inconclusive for several reasons including the relatively large effective vent area in the nominally unvented tests.

Another fire test project involving sprinklers and vents was conducted in Ghent, Belgium (1990) in a building with overall dimensions of 53 × 22 × 11.3 m (174 × 72 × 37 ft) to the apex of the roof. Steady-state and growing hexane pool fires were used with varying amounts of roof

SMOKE ISOLATION AND VENTING   113

Table 4.3. Measured average first-sprinkler opening times and fire heat release rates at first-sprinkler actuation for growing fires

	Time of first sprinkler opening (s)	Heat release rate at first sprinkler opening (MW)	Number of sprinklers opened[a]
Without vents	148	9.7	41
With vents	158	11.3	10

[a]Based on growing fire tests in which the fire heat release rate was reduced by 20% at the time of first sprinkler opening

vent area. There were 55 ceiling sprinklers. The specific objectives of the tests include validation of mathematical model predictions (Section 4.4.3) of the effects of venting on sprinkler actuation times, and determination of any significant reduction in venting effectiveness due to sprinkler spray cooling of the hot ceiling layer. Table 4.3 shows the measured average first-sprinkler opening times and fire heat heat release rates at first-sprinkler actuation for the growing fires.

It is clear from the results in Table 4.3 that the vents caused a minor delay in the first sprinkler opening, and that they provided a significant reduction in the number of sprinklers opened. The latter is due to the reduction in ceiling gas temperature away from the axis of the fire when the roof vents were employed. The roof vents also allowed a clear area beneath the smoke layer, overcoming the tendency of the sprinkler spray to drive the smoke layer downward. These results are encouraging, but they are not necessarily applicable to solid fuels because the solid fuels may be more readily controlled by the sprinkler spray in the absence of the vents.

4.4.2 LOSS EXPERIENCE

Fire brigade experiences described by Heseldon (1984) are entirely pro-venting including several incidents in which venting was required to extinguish fires only partially controlled by sprinklers. However, several contrary examples are also noteworthy. The K Mart fire (Appendix B.3) was a situation in which the sprinkler system was overcome and yet the roof vents were also ineffective in relieving a sufficient quantity of heat to prevent roof collapse. This also occurred in the Sherwin-Williams flammable liquid warehouse fire (Isner, 1988 from Chapter 3). An example of a much smaller industrial fire occurred in a room outfitted with two 5 ft (1.5m) diameter normally open vents. The vents were considered to be detrimental in that they delayed the opening of a sprinkler located over a plastic injection molding machine fire. Furthermore, the vents did not prevent the room from becoming smoke logged and inaccessible to firefighters, because closed fire doors prevented the air inflow required for effective venting. Thus, there is an inevitable conflict between the need for air inflow for venting and the need to close all openings in fire walls.

4.4.3 MATHEMATICAL MODELING

Several researchers are beginning to develop mathematical models of a sprinklered, vented fire in order to better understand the interactions between sprinklers and heat/smoke vents. Hinkley (1986) has described one such model that includes the cooling effect of the sprinkler spray on the hot ceiling gases, but doesn't include either the effect of the sprinklers or the local oxygen concentration on the prescribed fire heat release rate. Calculations with Hinkley's model indicate that venting is unlikely to delay the actuation of sprinklers in a large enclosure. In small enclosures or in buildings with deep draft curtains, venting of the hot layer can produce a minor delay in sprinkler actuation, but Hinkley points out that first sprinkler actuation time is far more sensitive to heat release rate, ceiling height, sprinkler link temperature, etc. than to the presence of venting.

He concludes that there is no reason to delay the actuation of vents until sprinklers start actuating, as recommended in some guidelines.

The National Institute of Standards and Technology (NIST) has developed a mathematical model (Cooper, 1988; Davis and Cooper, 1989) that can be used to calculate actuation times for automatic heat/smoke vents installed on sprinklered roofs. The NIST model includes the effect of sprinkler and vent link depth under the ceiling. Calculations for a situation in which the sprinklers are located nearest the fire, but further below the ceiling than the vent link (1.0 ft versus 0.3 ft, or 30 cm versus 9 cm), showed the vents actuating prior to the sprinklers even though all links fuse at the same link temperature. The explanation for the earlier vent opening is the higher gas temperatures nearer the ceiling and the lower thermal inertia (as characterized by Response Time Index) of the vent links in the example in Davis and Cooper (1989).

Additional calculations with the Hinkley model show the effects of vents on additional sprinkler actuations, and the effect of sprinkler spray cooling on smoke layer depth (Hinkley, 1989). As for additional sprinkler (after the first) actuations, Hinkley's results show vents either having no significant effect or resulting in fewer sprinkler actuations in large buildings without draft curtains. In small or curtained buildings, the effect of vents could not easily be generalized because it depended on the fire growth rate, sprinkler link thermal inertia, and the assumed effect of the sprinkler spray on the fire development. The latter assumption remains to be more fully evaluated before the effect of venting on sprinkler water demand can be answered.

As for the effect of vents on smoke layer depth, calculations with the Hinkley model (1989) for a 15 m (49 ft) high building with 3% of the roof area vented show the vents either substantially delay or prevent the descent of the smoke layer to elevations at which visibility and breathing would be impaired. The calculated benefit of vents in this regard is particularly noteworthy for slowly developing fires that do not actuate ceiling sprinklers until the layer descends to the 2–3 m (6.6–9.8 ft) elevation in the unvented building.

4.4.4 CLOSING REMARKS

My assessment of the current situation is that automatically actuated heat and smoke vents might be used judiciously in special situations where the need to maintain visibility and delay smoke logging outweighs possible detrimental effects on sprinkler performance. These situations might include the need for extended evacuation times and the need to avoid smoke damage to delicate electronic equipment. Another situation might be an isolated hazard (such as a dip tank in a metalworking facility) for which draft curtains can be installed to aid venting. Unlike the unsprinklered building, there is no valid design basis for selecting vent area in a sprinklered building. A conceptual approach would be to use the test data cited previously to indicate how sprinkler spray might reduce the hot layer temperature rise in the vent flow rate equation (4.26), and how the free burn values of Q in the preceding equations might be reduced by the spray. The sprinkler spray effect on the hot layer can be assessed using the methodology developed by Cooper (1991). A preliminary estimate of the sprinkler spray reduction on heat release rate can be developed from the commodity classification fire products collector tests described in Chapter 5, but further research is needed to provide a formal methodology. In carrying out these extensions to smoke venting theory, it will probably be important to account for the differences between the spray patterns produced by European sprinklers (designed to emphasize ceiling cooling) and American sprinklers (designed to emphasize fire control or suppression).

References

Bouchard, J.K., Venting Practices, *NFPA Handbook*, Section 5, Chapter 10, 1982.

Bukowski, R.W. and Peacock, R.D., 'Example Cases for the HAZARD I Fire Hazard Assessment Method,' *NIST Handbook 146*, Vol III, June 1989.

Cheng, W. H., 'Modelling of Carbon Dioxide Total Flooding Discharge Tests,' MS Thesis, Worcester Polytechnic Institute, 1991.
Cooper, L.Y., 'The Need and Availability of Test Methods for Measuring the /smoke Leakage Characteristics of Door Assemblies,' ASTM STP 882, pp 310–329, 1985.
Cooper, L.Y., Estimating the Environment and the Response of Sprinkler Links in Compartment Fires with Draft Curtains and Fusible-Link-Actuated Ceiling Vents – Theory, *Fire Safety Journal*, **16**, 137–163, 1990. (Also as NBSIR 3734, April 1988.)
Cooper, L.Y., 'Interaction of an Isolated Sprinkler Spray and a Two-Layer Compartment Fire Environment', National Institute of Standards and Technology, NISTIR 4587, 1991.
Davis, W.D. and Cooper, L.Y., 'Estimating the Environment and the Response of Sprinkler Links in Compartment Fires with Draft Curtains and Fusible-Link-Actuated Ceiling Vents – Part II: User Guide for the Computer Code LAVENT,' NISTIR 89-4122, August 1989.
Drysdale, D., *An Introduction to Fire Dynamics*, John Wiley & Sons, 1985.
Epstein, M., Maximum Air Flow Rate Into a Roof-Vented Enclosure, *J. Heat Transfer*, **114**, 535–538, 1992.
Factory Mutual Data Sheet 1-10, 'Smoke and Heat Venting in Sprinklered Buildings,' December 1978.
Fisher, F.L., Williamson, R.B., Toms, G.L. and Crinnion, D., Fire Protection of Flammable Work Stations in the Clean Room Environment of a Microelectronic Fabrication Facility, *Fire Technology*, **22(2)**, 148, May 1986.
Fung, F., 'Evaluation of a Pressurised Stairwell Smoke Control System for a Twelve Storey Apartment Building', National Bureau of Standards, NBSIR 73-277, 1973.
Grant, C., Halon Design Calculations, Chapter 4–6, *The SFPE Handbook of Fire Protection Engineering*, SFPE, NFPA, 1995.
Heselden, A.J.M., 'The Interaction of Sprinklers and Roof Venting in Industrial Buildings: the Current Knowledge,' Building Research Establishment Report, 1984.
Heskestad, G., 'Model Study of Automatic Smoke and Heat Vent Performance in Sprinklered Fires,' FMRC JI # 21933, 1974.
Heskestad, G., 'Review of IITRI Report on Fire Venting of Sprinklered Buildings,' FMRC Memo submitted to NFPA Committee on Heat and Smoke Venting, 1983.
Heskestad, G. and Hill, J.P., 'Experimental Fires in Multiroom/Corridor Enclosures,' FMRC JI 0J2N8.RU, National Bureau of Standards, October 1985 (also 'Propagation of Fire Smoke in a Corridor,' *Proceedings of the 1987 ASME-JSME Conference*, vol 1, ASME, 1987).
Heskestad, G. and Spaulding, R.D., 'Inflow of Air Required at Wall and Ceiling Apertures to Prevent Escape of Fire Smoke,' *Proceedings of the Third International Symp. on Fire Safety Science*, pp 919–928, 1991.
Hinkley, P.L., The Effects of Vents on the Opening of the First Sprinklers, *Fire Safety Journal*, **11**, 211–225, 1986.
Hinkley, P.L., The Effect of Smoke Venting on the Operation of Sprinklers Subsequent to the First, *Fire Safety Journal*, **14**, 221–240, 1989.
Hinkley, P.L., Smoke and Heat Venting, *The SFPE Handbook of Fire Protection Engineering*, Chapter 3–9, SFPE, NFPA, 1995.
Hotta, H. and Horiuchi, S., Detection of Smoldering Fire in Electrical Equipment with High Internal Air Flow, *Fire Safety Science – Proceedings of the First Intl Symp.*, Hemisphere Publishing, p 699, 1986.
ISO DP 5925, Fire Tests – Smoke Control Door and Shutter Assemblies, Part 0 – Commentary, ISO/TC 92/WG 3N204, International Organizations for Standards, 1977.
Klote, J.H., 'Smoke Movement Through a Suspended Ceiling System,' NBSIR 81-2444, February 1982.
Klote, J.H. and Fothergill, J.W., Design of Smoke Control Systems, *NBS Handbook 141*, National Bureau of Standards, 1983.
Klote, J.H. and Cooper, L.Y., 'Model of a Simple Fan-Resistance Ventilation system and its Application to Fire Modeling,' NISTIR 89-4141, September 1989.
Klote, J.H., 'Fire Experiments of Zoned Smoke Control at the Plaza Hotel in Washington DC,' NISTIR 90-4253, 1990.
Klote, J.H. and Milke, J.A., *Design of Smoke Management Systems*, American Society of Heating, Refrigeration, and Air-Conditioning Engineers and the Society of Fire Protection Engineers, 1992.
Klote, J.H., Smoke Control, *The SFPE Handbook of Fire Protection Engineering*, Chapter 4–12, SFPE, NFPA, 1995.
Milewski, L., Control Safety Considerations for HVAC Smoke Management Techniques, *Fire Safety Science and Engineering*, ASTM STP 882, pp 301–309, 1985.
Miller, E.E., Venting Sprinklered Buildings, *The Sentinel*, Third Quarter 1984.
NFPA 12, 'Carbon Dioxide Extinguishing Systems,' National Fire Protection Association, 1989.
NFPA 12A, 'Halogenated Fire Extinguishing Agent Systems,' National Fire Protection Association, 1989, also Factory Mutual Data Sheet 4-8N, 1989.
NFPA 92A, 'Recommended Practice for smoke Control Systems,' National Fire Protection Association, 1988.
NFPA 204M-1982, 'Guide for Smoke and Heat Venting,' National Fire Protection Association, 1982, updated in 1998 Edition of NFPA 204.

NFPA 2001, 'Clean Agent Fire Extinguishing Systems,' Proposed 1993 Edition, National Fire Protection Association, 1993.

Saum, D., Saum, A., Hupman, J. and White, M., Pressurization Air Leakage Testing for Halon 1301 Enclosures, *Conference of Substitutes and Alternatives to CFCs and Halons*, Washington, DC, January 1988.

Sugawa, O., Oka, Y. and Hotta, H., Fire Induced Flow in a Clean Room with Downward Vertical Laminar Flow, *9th UJNR Panel Meeting*, May 1987.

Thomas, P. et al., 'Investigation into the Flow of Hot Gases in Roof Venting,' Fire Research Station Technical Paper No. 7, HMS0, London, 1963.

The Ghent Fire Tests, Colt International Ltd, 1990.

Thomas, P. and Hinkley, P.L., 'Design of Roof-Venting Systems for Single Storey Buildings,' Fire Research Station Technical Paper No. 10, HMSO, London, 1964.

Thomas, P.H., 'The Movement of Smoke in Horizontal Passages Against an Air Flow,' Fire Research Station Note No. 723, September 1968.

Ward, E., 'Design Requirements and the Need for Standards,' *IFSSEC*, London, 1982.

Waterman, T.E. et al., 'Fire Venting of Sprinklered Buildings,' IITRI Project J08385, July 1982.

5 WAREHOUSE STORAGE

If there is one mainstream topic in industrial fire protection, it is storage in general and warehouse storage in particular. Traditional concerns about large quantities and concentrations of combustible material have been compounded by new challenges and new solutions to classical problems. The new challenges are (1) recent large loss fires in seemingly well protected warehouses, and (2) new advances in material handling capability stimulating storage height increases beyond the protection limits of conventional ceiling sprinklers. New solutions available to the fire protection engineer include: (1) new test methods to provide a better measure of commodity protection requirements; (2) new, less flammable, packaging materials; and (3) new, more effective, sprinkler systems for warehouse storage.

This chapter sets out to provide a working understanding of these issues by:

1. Providing some statistical and illustrative perspective on large loss warehouse fires.
2. Describing how commodity characteristics and storage height and configuration influence the fire hazard and protection requirements.
3. Documenting heat release rate characteristics and associated sprinkler protection requirements for representative ordinary combustible and plastic commodities and typical, storage configurations.
4. Reviewing the design basis and applicability of Early Suppression Fast Response Sprinklers to new warehouse construction.

5.1 Warehouse fire losses

A NFPA task force concerned with a series of multi-million dollar warehouse fires in the 1980s compiled loss statistics to provide a broader perspective on storage facility fires (NFPA Ad Hoc Task Force, 1988). Three leading Highly Protected Risk insurers (Factory Mutual, Industrial Risk Insurers, and Kemper) provided the loss data shown in Table 5.1 for fires in storage facilities and other insured properties.

Thus storage fires incur an average property damage three times as high as the average for all insured facilities ($262,100 versus $88,800). Furthermore, $419 million of the $515 million (81%) in storage property damage was due to 167 fires (8.5% of all storage fires) with a loss of at least $500,000 per fire. The large loss potential of storage fires is due in part to the large quantities of combustible material and vulnerable products present in storage facilities. Inadequate sprinkler protection for stored commodities is often listed as the primary factor responsible for the extent of the loss as indicated in Table 5.2 for storage fires.

Table 5.1. Insurance loss statistics for the period 1982–1986 (data from NFPA Ad Hoc Task Force, 1988)

	Storage facilities	All facilities
Number of fire losses	1961	13,845
Property damage	$514.8 million	$1229.5 million
Business interruption	$54.9 million	$256.2 million
Total losses	$569.6 million	$1485.7 million
Property damage per loss	$262,100	$88,800
Total $ per loss	$290,500	$107,300

Table 5.2. Sprinkler system status in storage fires during 1983–1987

	Adequate protection	Sprinklers needed
Number of fires	164	168
Total loss	$25.2 million	$130.2 million
Average loss per fire	$153,000	$775,000

The category 'Adequate protection' covers fires in which at least one sprinkler operated, no additional sprinklers were needed, and there were no sprinkler system or water supply defects or shut valves. The category 'Sprinklers needed' includes fires with inadequate sprinkler coverage as well as those without any sprinklers installed. The average loss for a storage fire in which sprinklers were needed is five times as large as the average storage fire loss with adequate sprinkler protection. An even larger average loss (albeit a smaller number of fires) was incurred in the category sprinkler system/water supply defects. This category includes fires with obstructed sprinkler piping, a water supply impairment, fire pump not started promptly, inappropriate sprinkler link temperature (too high or too low), excessive distance between storage and sprinklers, and/or excessive dry pipe trip time.

Table 5.3 provides a listing of twelve of the largest warehouse fires during the period 1977–1987. Two of these losses, the K Mart fire and the Ford Cologne, West Germany fire, are described in Appendix B. Most of the warehouses listed in Table 5.3 were sprinklered with relatively good water supplies. Nevertheless, the sprinkler protection was not adequate for the 4.3–9.1 m (14–30 ft) high rack storage in most cases. In each incident the warehouse was virtually destroyed and losses exceeded $20 million. Effective sprinkler protection design for warehouse storage requires an understanding of how storage configuration, storage height, and storage commodity influence fire development, as discussed in the following sections.

5.2 Storage configurations

Most warehouse storage is either solid pile on floor, on pallets, or on racks. A small but growing fraction of warehouse storage is in open top bins or wire baskets. Automated Storage and Retrieval is another small but important storage configuration. Fire protection considerations inherent in each configuration are described here.

Solid piled storage of cartoned commodities is storage without any spaces between cartons in each stack. The only exposed surfaces in solid piled storage are the outer walls and top surface of the storage pile. Large, structurally stable commodities can be solid piled quite high, but many smaller cartoned commodities cannot be solid piled higher than about 3.7 m (12 ft). The relatively

WAREHOUSE STORAGE

Table 5.3. Representative very large loss warehouse fires (at least $20 million per loss)

Year	Company	Location	Stored items	Storage area (1000 sq ft)	Building height (ft)	Storage height (ft)	Storage configuration	Property damage	Contributing factors
1977	Ford Motor	Cologne, W.G.	Auto Parts	1000	30	17–20	Baskets In Racks	>$100M	Aisle Storage, Plastics, Motor Oil
1978	Montgomery Ward	Bensonville, IL	Gen. Merchandise	200	?	?	?	$30M	Aisle Storage
1979	Supermarket General	Edison, NJ	Gen. Merchandise	290	?	20	Rack	$30M	Aerosols
1981	K Mart	Falls Township, PA	Gen. Merchandise	1200	30	15	Rack, Palletized	>$100M	Aerosols, Ineffective Fire Wall
1984	Hanworth	London	Computer Equipment	78	?	?	Rack	$65M	No Sprinklers
1984	Cricklewood Trading	London	Gen. Merchandise	750[a]	25–50[a]	20–30	?	$100M	Partially Sprinklered, Cocoa Butter
1983	Multi-Occupancy	Bradford, England	Gen. Merchandise	330	Four Stories	?	Rack	$25M	No Sprinklers, LPG Cylinders
1983	British Army	Donnington, England	Army Supplies	440	33	30	Rack	$330M	No Sprinklers, Roof Vents
1985	MTM (Mitsui)	Elizabeth, NJ	Aerosols, Gen. Merchandise	500	24	17	Rack	$150M	Aerosols, Ineffective Fire Walls
1986	Sandoz	Basel, Switzerland	Chemicals	50	26–40	15–20	Palletized Drums	$20M	No Sprinklers, Flammable Liquids
1987	Service Merchandise	Garland, TX	Gen. Merchandise	200	25	24	Rack	$52M	Delayed Detection, High-Piled Storage
1987	Sherwin-Williams	Dayton, OH	Paint, Solvent	180	30	14–16	Palletized, Rack	$49M	Flammable Liquids, Aerosols

[a]The Cricklewood warehouse complex fire involved nine buildings

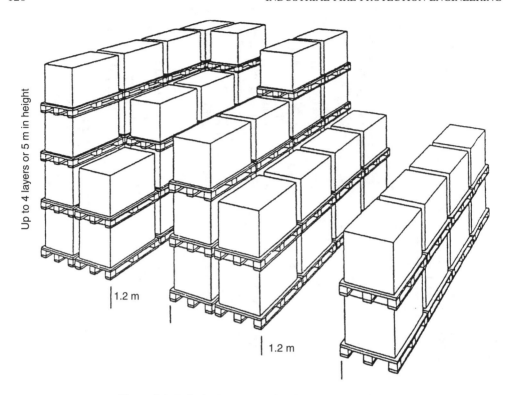

Figure 5.1. Palletized storage with 1.2 m (4 ft) wide aisles

small exposed surface area and restricted air access in solid piled storage restrict burning rates compared to other storage modes.

Palletized storage entails stacking and handling of pallet loads on top of each other as illustrated in Figures 5.1 and 5.2. The pallets are typically $1.1–1.5\,\text{m}^2$ ($12–16\,\text{ft}^2$) and about 15 cm (6 in) deep with slatted upper and lower surfaces. Pallet surface areas as well as commodity wall surfaces are exposed to flame, and the pallets allow air access to burning horizontal surfaces. Fire growth rates in palletized storage are comparable to those in solid piled storage. This implies that the early fire development is primarily governed by the exposed vertical surfaces available for flame spread rather than the exposed horizontal surfaces. Indeed, tests have shown that doubling the number of vertical exposed surfaces by providing transverse flue spaces as well as longitudinal flue spaces between storage stacks, does cause the rate of fire growth to increase substantially.

Palletized storage is usually more stable than solid piled storage and has somewhat greater air access during burning because of the pallet spacing. These comparisons, which imply an enhanced flammability of rack storage, are summarized in Table 5.4.

Rack storage entails placing each unit load on a structural steel rack with open, slotted, or solid shelves. The rack structures provide greater load carrying capacity and stability than either solid piled or palletized storage. Consequently, greater storage heights can be achieved with rack storage. Figure 5.3 illustrates how narrow aisle, side loading high lift trucks allow access to upper tier storage at heights up to 12.2 m (40 ft). Since rack storage tiers are often 1.5 m (5 ft) high, this corresponds to unit load storage up to eight tiers high. Rack supported building structures can go significantly higher than the eight tiers.

Figure 5.2. Warehouse palletized storage

Table 5.4. Comparison of storage configuration burning characteristics

	Solid piled	Palletized	Rack storage
Exposed surfaces per unit load	4	4 + pallet	5 + pallet
Air access	Restricted	Less restricted	Least restricted
Load stability	Least stable	More stable	Most stable
Water spray access	Depends on longitudinal and transverse flues		

Storage racks are termed single-row, double-row, or multiple-row, depending on the row spacing. Single-row racks have aisle spacing of at least 1.1 m (3.5 ft) on both sides of each row. Double-row racks are separated by a relatively narrow longitudinal flue on one side and by a wider aisle on the other side as illustrated in Figure 5.3. Multiple-row racks are racks greater than about 3.7 m (12 ft) wide or single-row racks or double-row racks with aisle spacing less than 1.1 m (3.5 ft). The relatively narrow spacing between rows in multiple-row racks usually promotes more rapid fire spread across rows.

Unlike solid piled and palletized storage, the unit load in each tier of rack storage has an exposed top surface. This exposed top surface may not necessarily increase the early fire development rate, but it does contribute to the burning rate eventually and allows for greater air access in ventilation limited fires. (If the racks contain solid shelves, sprinkler spray access to the lower burning tiers will be obstructed unless in-rack sprinklers are employed.) The overall effect is that rack storage is a greater fire protection challenge than either solid piled or palletized storage. This increased challenge has resulted in the average property damage for rack storage fires being more than twice as large as the average property loss in solid piled and palletized storage.

Most warehouse storage racks have open or slatted shelves to support the stored commodities. This is a critical feature for protection by ceiling sprinklers because it allows water to reach

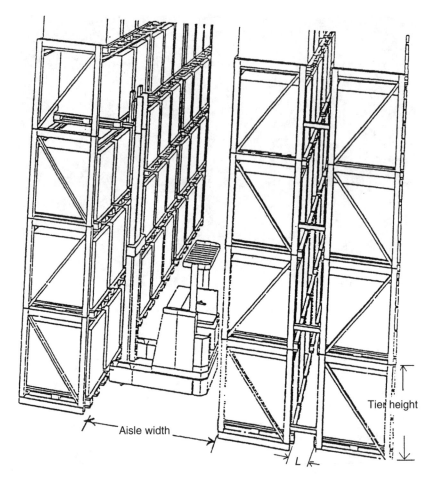

Figure 5.3. Palletized rack storage with indicated aisle width, tier height, and longitudinal flue space, L

the lower storage tiers. Racks with solid or slatted shelves provide a much greater challenge, and often need in-rack sprinklers to supplement ceiling sprinkler protection. One example of the effect of solid shelving is a pair of comparison fire tests conducted with 20 ft high rack storage of plastic commodity protected by $K = 11.4$ gpm/psi$^{1/2}$ ceiling sprinklers at a discharge density of 0.60 gpm/ft^2 (Troup, 1994). In the test with slatted wood shelves, only four sprinklers actuated and controlled the fire with only three pallet loads of commodity consumed. The test with solid wood shelves resulted in the fire spreading to the ends of the storage array, with 15 pallet loads consumed and 15 ceiling sprinklers opening.

Storage in open top bins is far less prevalent than solid piled, palletized, and rack storage, but it is gaining in popularity, particularly for small parts storage in electronic assembly facilities. Fires in the open top bins are usually slow in developing but can eventually become quite intense. This occurred in the fire tests conducted at the British Fire Research Station (Field and Murrell, 1988) using a bin storage configuration with 1.2 m wide by 2.3 m high (3.9 ft wide by 7.5 ft high) aisles covered with a chipboard. There were a variety of bins fabricated from fiberboard, steel boxes, metal shelf trays, cardboard boxes, and timber and cardboard. Storage within the bins included small metal parts, paper, leather and plastic. Test results indicated that automatic

WAREHOUSE STORAGE 123

Figure 5.4. Large open automated storage and retrieval system

Figure 5.5. Enclosed ASRS photo with CO_2 suppression system. Reproduced by permission of White Systems, Inc.

sprinklers installed centrally beneath the walkway at a maximum spacing of 2 m (6.6 ft) would provide the best available fire protection.

Bins are often used in small Automated Storage and Retrieval Systems (ASRS). In-rack sprinklers are recommended by Field and Murrell (1988) for ASRS with combustible containers stored over 3.6 m (12 ft) high. The larger ASRS consist of a large unenclosed rack storage system with a motorized picker installed on rails in the aisle as shown in Figure 5.4. Fire protection choices for these systems are equivalent to those for high-rise rack storage. The smaller ASRS are sometimes comprised of enclosed modular units such as the one shown in Figure 5.5. The particular ASRS shown in Figure 5.5 is protected by a carbon dioxide system, where the CO_2 is stored outside the unit and piped to discharge nozzles within the module.

5.3 Effect of storage height, flue space, and aisle width

How does the fire severity and corresponding fire protection requirements vary with storage height? Data obtained by You (1989) for rack storage of the 'standard plastic commodity' (described in Section 5.4) indicate that the heat release rate early in the fire is directly proportional to the number of storage tiers. This data is shown here in Figure 5.6 and can be represented by the third power curve fit

$$Q_{con} = 0.0448 N(t - t_0)^3 \quad t - t_0 < 26 \text{ s}, 1 < N < 6 \quad [5.3.1]$$

where Q_{con} is the convective heat release rate in kW at time t after ignition, t_0 is incubation time between ignition and self-sustained burning, and N is the number of tiers of storage.

Equation [5.3.1] is only valid early in the fire when $t - t_0$ is less than about 26 seconds. Convective heat release rates beyond this time increase less rapidly. Measurements of mass burning rates obtained by Kung et al. (1984) and shown in Figures 5.7 and 5.8 indicate that the mass burning rates for both the standard Class II commodity and the standard plastic commodity (the commodities are defined in Section 5.4) are directly proportional to the number of rack storage tiers for $t - t_0$ less than or equal to at least 180 sec in the case of the Class II commodity

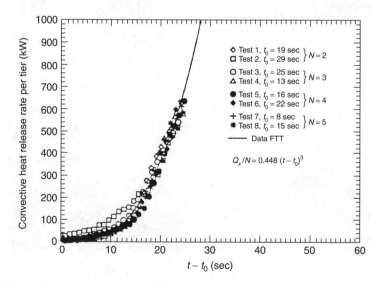

Figure 5.6. Heat release rate data for 2, 3, 4, 5-tier storage of Group A plastic commodity. © 2002 Factory Mutual Insurance Company, with permission

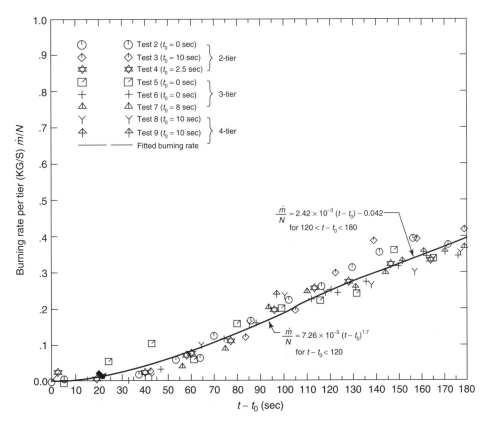

Figure 5.7. Mass burning rates for 2-, 3-, and 4-tier storage of Class II commodity. © 2002 Factory Mutual Insurance Company, with permission

and 150 sec in the case of the plastic commodity. Thus it is tempting, albeit still premature, to generalize the conclusion that the mass burning rate and possibly the convective heat release rate is proportional to the storage array height for many commodities. (The convective heat release rate can be obtained by multiplying the burning rate by the effective convective heat for a specific commodity; values of effective convective heat for various commodities are discussed in Section 5.4.)

Mass burning rates for wood pallet piles are also proportional to pile height according to data obtained by Krasner and reviewed by Babrauskas (1995). The peak heat release rate for a fully involved stack of pallets 1.22 × 1.22 m (4 × 4 ft) was represented by the correlation

$$Q_{\max} = 1450(1 + 2.14 h_s)(1 - 0.027 M_w) \qquad [5.3.2]$$

where $Q_{\max}$ is the theoretical peak heat release rate in kW based on mass burning rate and a heat of combustion of 12 kJ/g, h_s is the height of pallet stack (m), $h_s > 0.5$ m, and M_w is the pallet moisture content (wt %).

Babrauskas and others (for example, data in Table A.4) suggest that equation [5.3.2] can be generalized to other pallet sizes by normalizing on the basis of pallet floor area. The time required to reach the peak heat release rate given by equation [5.3.2] can be estimated from the characteristic 1 MW growth times listed for pallet stacks in Table 4.2. These characteristic growth times

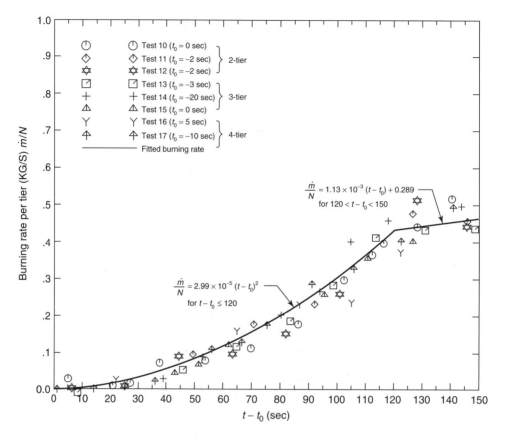

Figure 5.8. Mass burning rates for 2-, 3-, and 4-tier storage of plastic commodity. © 2002 Factory Mutual Insurance Company, with permission

generally decrease with stack height, implying that a given heat release rate is reached sooner in tall stacks than in short stacks.

In the case of a single wall, as represented for example by solid pile or palletized storage with minimal flue spaces burning on the aisle face, the peak heat release rate per unit wall width seems to vary as the wall height to the 1.25 power according to the data listed in Table A.3. Thus, peak heat release for storage stacks seem to vary as stack height to a power between 1.0 and 1.25 depending on the stack configuration and flue space.

A pair of comparison tests reported by Dean (1980) demonstrates how the placement of an additional tier of storage on the top of a four-tier high rack storage array can overcome the design basis sprinkler protection. The baseline test with four-tier high storage of Class II Metal Lined Double Tri-wall Cartons opened 31 ceiling sprinklers and had a maximum ceiling steel temperature of 521 °C (970 °F). When an extra tier of storage was inserted on the top tier of the rack, 36 sprinklers opened and the maximum steel temperature increased to 655 °C (1211 °F). The data are summarized in Table 5.5.

The amount of sprinkler water needed to suppress rack storage fires of different stack height increases with stack height in a manner investigated by Lee (1984). Lee measured the Required Delivered Density (RDD), i.e. the water spray flow rate per unit area at the top of the storage array; for the standard plastic commodity stacked three, four, and five tiers high. The RDD for

Table 5.5. Effect of an extra tier and of mixed commodity storage on sprinkler protection effectiveness for rack storage (data from Dean, 1980)

	Test 69R2	Test C1	Test F1	Test F2
Commodity	Class II	Class II	II + Plastic	II + Plastic
Storage height	19 ft	23 ft	19 ft	19 ft
First sprinkler open (min:sec)	2:56	2:43	2:25	2:40
Total sprinklers open	31	36	51	88
Total water flow rate (gpm)	895	1085	1515	2700
Maximum Ceiling Steel Temp (C)	521	655	894	712

Notes: Ceiling Height was 30 ft and Sprinkler Discharge Density was 0.30 gpm/ft^2 from 286 °F $\frac{1}{2}$-in orifice heads in all tests. Plastic commodity was placed in the top tier for Test F1 and the bottom tier for Test F2

Table 5.6. RDD varies with stock height

Number of tiers	Stack height (ft)	RDD (gpm/ft^2)	Redevelopment fraction
3	14.5	0.30	1/6
4	19.5	0.40	1/3
5	24.5	0.50	2/4

each stack height was determined by applying water when the fire in a two pallet wide by two pallet deep by (3, 4, or 5) tier high array reached a predetermined convective heat release rate in the range 350–1900 kW (331–1800 Btu/sec). The RDD corresponding to the boundary between fire suppression and fire redevelopment was independent of water application heat release rate in this range and varied with stack height, as shown in Table 5.6.

Thus, the RDD is roughly, but not exactly, proportional to stack height. The nominal RDD values specified for suppression are greater than those listed above by 0.05 gpm/ft^2 (2 L/min-m^2) and 0.10 gpm/ft^2 (4 L/min-m^2) for five tier high because the fire redeveloped in a certain fraction of the tests with the listed RDD, as indicated by the last column above.

Insurance loss data of the type described in Section 5.1 generally reflect the increased challenge of higher storage heights. For example, the average loss for 5.5 m (18 ft) high rack storage is about $1.7 million, which is about three times the value for 3.05 m (10 ft) high rack storage. However, the data is too sparse to develop a quantitative correlation with storage height.

The Swedish National Testing and Research Institute (SP) has conducted tests to determine the effect of rack storage flue space on the rate of fire development and the maximum heat release rate. As the longitudinal and transverse flue spaces get larger, there is less re-radiation from carton to carton, and the convective heat transfer to adjacent cartons decrease. However, when the flue space becomes too small, air access to the burning surfaces becomes limited. Data reported by Ingason (2001) for the Standard Class II commodity tested at four different flue spaces in the range 7.5–30 cm, show that the optimum spacing from the standpoint of rapid fire growth rate, i.e. time for the free burn heat release rate to reach 2 MW; is 15 cm (6 in). The tests with a 30 cm flue space required two minutes longer to reach 2 MW; however, the maximum heat release reached with the 30 cm flue space (greater than 10 MW) was higher than the maximum heat release rates with the narrower flue spaces. Furthermore, the time for the fire to grow from 2 MW to 7 MW was shorter with the 30 cm flue than with the narrower flues. Thus, the worst-case flue space depends on the time period and range of heat releases of interest. In most applications, the range of free burn heat release rates is limited to less than 2 MW because the ceiling sprinklers should actuate before that value is reached.

Aisle widths are important from the viewpoint of fire spread to adjacent storage rows. Since flame spread occurs via radiant heating of the exposed surface of the target row, the effect of aisle width show depend on how the radiant view factor decreases with distance from the flame to the exposed target. Sample calculations have been conducted to explore this effect using the methodology described in Section 5.7. In the case of 6.1 m (20 ft) high rack storage of the prototypical Group A plastic commodity, the calculated time to jump increased from 81 seconds to 94 seconds when the aisle width was increased from 1.83 m to 3.66 m (6 ft to 12 ft). These calculated times do not account for the effects of any sprinkler discharge, which is the primary defense against aisle jump. Warehouse sprinkler protection guidelines discussed in Section 5.6 give credit for larger aisle widths.

One other advantage of wide aisles is the improved access for locating and manually responding to a warehouse storage fire at an early stage of development. Similarly, fire cleanup operations can be conducted more expeditiously with the improved access.

5.4 Commodity effects

A warehouse storage commodity consists of the basic product, its packaging, and its container. Flammability properties of product, packaging, and container materials need to be evaluated in establishing appropriate fire protection. These evaluations can either be in the form of generic material classifications, laboratory flammability testing, or larger scale fire testing involving one or more warehouse unit loads of commodity. Each approach is discussed in this section.

5.4.1 GENERIC COMMODITY CLASSIFICATION

The National Fire Protection Association Standard 231 (1998) for General Storage is based on a generic classification system involving seven categories of commodities. The seven categories in NFPA 230 (1999) and in the Factory Mutual Commodity Classification Data Sheet 8-1 (1998) are Classes I, II, III, and IV, and Group A, B, and C Plastics. There is significant overlap between Class III and Group C plastics, and between Class IV and Group B plastics. Class I commodities are the least flammable, while Group A plastics are considered to be the most flammable class of general storage. There are also numerous categories for special commodities discussed in Chapters 6, 7, and 8.

Table 5.7 summarizes the classification definitions for Class I, II, III, and IV commodities. A Class I commodity is a noncombustible product in a container of ordinary combustibility, such as an ordinary cardboard carton. Examples of Class I products include glass bottles either empty

Table 5.7. Generic commodity classification in NFPA 230

Class	Product	Packaging	Plastic content
I	Noncombustible	None; on pallets. Single wall carton, or paper wrap.	Negligible
II	Noncombustible	Multi-wall carton, or wood crate or wood box.	Negligible
III	Wood, Paper, Leather, Natural fiber textile, or Group C plastic.	None or ordinary combustible.	Negligible ($\leq 5\%$) Group A or Group B
IV	Class I, II, or III with 5-15 weight % or 5-25 vol % of Group A plastic; or Group B plastic.	Anything except Group A plastic.	Either Group B or an appreciable amount of Group A as defined for product.

or filled with nonflammable liquids, metal pots and pans, bags of cement, and ceramic products without any packing material.

A Class II commodity is a Class I product in a slatted wooden crate, a solid wood box, or a multiple thickness corrugated carton. Examples of Class II commodities include beer or wine in (up to 20% alcohol) in wood crates or barrels, lightbulbs in multiple corrugated containers, and large appliances such as washing machines in triwall cardboard cartons.

A NFPA 230 Class III commodity consists of a combustible product with a negligible amount of Group A or Group B plastics (Table 5.8) in either the product or the packaging. Examples include shoes, books, cotton or wool clothing, wood cabinets or furniture (with only a negligible amount of plastic padding), combustible food products, and Group C plastic products such as polyvinyl chloride insulated cable on metal or wood reels.

A NFPA 230 Class IV commodity is defined either as a Class I, II, or III product containing an appreciable amount of Group A plastics in an ordinary cardboard carton, or a Class I, II, or III product in cartons with Group A plastic packaging materials, or a Group B plastic product in a cardboard carton. The NFPA 230 and Factory Mutual definition of an appreciable amount of Group A plastic is that a pallet load of Class IV commodity (including the pallet) should not contain more than 25% by volume of expanded (i.e. foamed) plastic, or more than 15% by weight of unexpanded plastic.

Examples of Class IV commodities cited in NFPA 230 include fiber glass insulation rolls, empty PET beverage bottles in cartons, pharmaceuticals (pills) in cartoned plastic bottles, wood furniture with plastic coverings, wax-coated paper cups and plates, and electrical wire on plastic spools. NFPA 230 also defines a free flowing Group A plastic, comprised of pellets, powder, or flakes, as a Class IV commodity because the free flowing bulk plastic tends to flow out of the container and into the flue space where it can have a smothering effect on the fire.

The Group A, B, and C classification system for plastics is based on the recognition that the flammability properties of plastics and other polymers varies from almost negligible (as in polytetrafluoroethylene) to substantially more severe than cellulosic materials. Examples of Group A, B,

Table 5.8. Plastic classification examples in NFPA 230 (data from Tewarson, 1995)

Polymer	Heat of combustion (kJ/g)	Heat of gasification (kJ/g)
Group A Plastics		
Acrylonitrile-Butadiene-Styrene (ABS)	30.8	3.23
Fiberglass Reinforced Plastic	12.9–26.0	1.4–6.4
Polycarbonate	29.7	2.1
Polyethylene	43.6	1.8(LD), 2.3 (HD)
Polymethyl Methacrylate (PMMA)	25.2	1.6
Polypropylene	43.4	2.0
Polystyrene	39.9	1.3–1.9
Polyurethane Foam	23–28	1.2–2.7
Group B Plastics		
Cellulose Acetate	17.7	–
Nylon	30.8	2.4
Silicone Rubber	21.7	–
Group C Plastics		
Phenolic	10–36.4	1.6–3
Polytetraflouroethylene (PTFE)	5.3	–
Polyvinyl Chloride (PVC, rigid)	16.4	2.5
Urea Formaldehyde	14.6	–

and C plastics as specified in NFPA 230 are listed in Table 5.8. Heats of complete combustion and effective heats of gasification are also listed in Table 5.8. The corresponding values for cellulose are a heat of combustion of 16.1 kJ/g (7000 Btu/lb) and a heat of gasification for corrugated paper of 2.2 kJ/g or 1000 Btu/lb (from Table 3–4.4 of the *SFPE Handbook*, 1995). According to the data listed in Table 5.8, Group A and B plastics generally have heats of combustion significantly higher than cellulose/paper, while Group C plastics have heats of combustion comparable or significantly less than cellulosic materials. The data also shows that there can be wide variations for certain generic polymers.

The distinction between Group A and Group B plastics is different in NFPA 230 than in Factory Mutual Data Sheet 8-1. Several of the NFPA 230 Group A plastics are considered to be examples of Group B plastics in Factory Mutual Data Sheet 8-0S. These include polycarbonate, polyethylene, polypropylene, and thermosetting polyesters. The Group A, B, and C plastics listings in NFPA 230 are apparently based on the original series of small-scale and large-scale plastic commodity fire tests described in Sections 5.4.3 and 5.4.4. The groupings in Factory Mutual Data Sheet 8-1 are based on material heats of combustion and horizontal flame spread rates as well as various sprinklered fire tests. Factory Mutual Research Corporation maintains a computer database of heats of combustion and flame spread rates of commodity samples submitted for classification. The listing of examples of commodity classifications in NFPA 230 has increased significantly in recent editions, but there are still ambiguities, particularly with regard to the use of various types of flame retardants and reformulated polymers.

The Class I, II, III, IV and Group A, B, C classification implicitly assume that the commodity is stored on a wood pallet. If a plastic pallet is used, NFPA 230 and NFPA 13 require that the classification be increased to the next higher (more challenging) classification unless the plastic pallet has been shown to be equivalent to a wood pallet through testing and corresponding certification. The type of testing conducted for this certification is equivalent to the Commodity Classification tests described in Section 5.4.3.

According to laboratory flammability data (Section 5.4.2), the steady-state burning rate per unit horizontal area is proportional to the ratio of the heat of combustion to the heat of gasification. Values of this ratio calculated from the data for the various polymers in Table 5.8 are listed in Table 5.9.

The ratios listed in Table 5.9 suggest that a possible demarcation between Group A, B, and C plastics might be set at ratios over 20, between 10 and 20, and under 10. However, the fact that the range of ratios shown for polyurethane foam is from under 10 to over 20 indicates that some

Table 5.9. Ratios of heat of combustion to heat of gasification

Polymer	Ratio $\Delta H_c / \Delta H_g$
Polystyrenex	21–31
Polypropylene	21
Polyethylene	19(h.d.), 24 (l.d.)
Polyurethane Foam	8.5–23
PMMA	16
Polycarbonate	14
Nylon 6/6	13
ABS	9.5
Wood (Douglas Fir)	8.9
Corrugated Paper	7.3
Rigid PVC	6.6

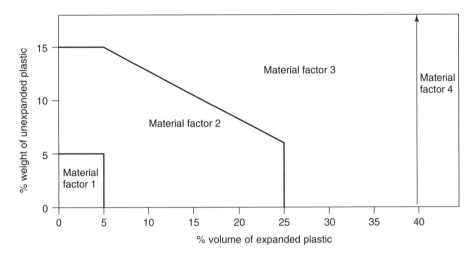

Figure 5.9. CEN prEN 12845 commodity material factor categories (from CEN, 2001)

plastics may fall into any of the three groups depending on density, the presence of plasticizing or flame retardant additives, and physical form. This illustrates the futility and inappropriateness of any rigid generic commodity classification scheme. Both Factory Mutual and NFPA realize that their generic classification schemes are more valuable for providing a preliminary indication of relative flammability than a firm irrefutable determination.

The European Committee for Standardization has a generic commodity classification scheme in its draft standard for automatic sprinkler systems (CEN prEN 12845, 2001). CEN has four commodity categories, which depend on the commodity material and storage configuration. The material factor depends on the expanded plastic and unexpanded plastic content of the commodity according to the regions delineated in Figure 5.9. If there is less than 5% plastic (both by weight and by volume) in the product and packaging, the commodity is designated as a Material Factor 1. As the percent plastic is increased, the Material Factor increases to 2, 3, or 4. Cartoned commodities are designated as Category I, II, III, or IV according to their Material Factor. Exposed plastic commodities are categorized according to their contents as explained in Annex B of the draft CEN standard. The physical nature of the material is also a factor if it is a solid block, a powder, or an open (low volume fraction) material. Annex C of the CEN standard is a listing of the categories of numerous specific commodities. Although the same factors are utilized in the NFPA commodity classification scheme, there is no direct relationship between a CEN Category IV commodity and a NFPA/FM Class IV commodity, and likewise for the other three categories.

5.4.2 LABORATORY FLAMMABILITY TESTING

Warehouse storage fires involve a combination of flame spread over the storage array and burning into the storage array. Since flame spread is the propagation of an ignition boundary, laboratory flammability tests measuring flame spread and heat release rates per unit area usually also measure sample ignition characteristics. Several test methods, described, by Tewarson (1995) and Drysdale (1998), have been developed for this purpose. The ignition tests involve measuring ignition delay times, t_{ig}, during exposure of material samples to a series of radiant heat fluxes.

Hamins and McGrattan (1999) have measured ignition times of 4 mm thick corrugated paper samples exposed to radiant heat fluxes in the LIFT apparatus (designated as ASTM E 1321 and ISO 5658) and the Cone Calorimeter (designated as ASTM E 1354 and ISO 5660). Their data

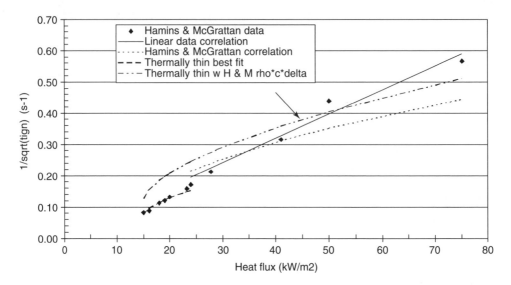

Figure 5.10. Inverse time to ignition versus heat flux for corrugated paper

are plotted in Figure 5.10 in the form of $t_{ig}^{-1/2}$ versus radiant heat flux. A linear relationship on this plot implies that the material is behaving as a thermally thick material, for which the ignition time should satisfy the following theoretical equation (Drysdale, 1998, p. 218)

$$t_{ig} = \frac{\pi}{4} k\rho c \frac{(T_{ig} - T_0)^2}{q_e''^2} \qquad [5.4.1]$$

where k, ρ, and c are the thermal conductivity, density, and specific heat of the sample material, T_{ig} is its ignition temperature, T_0 is the ambient temperature, and q_e'' is the radiant heat flux impinging on the sample (and assumed to be absorbed by the sample). A best linear fit has been obtained over the heat flux range 28–75 kW/m² for the data in Figure 5.10.

If the material responds to the imposed heat flux as if it is thermally thin (uniform temperature), and if heat losses at the exposed surface are neglected, the ignition time should vary as:

$$t_{ig} = \rho c \delta \frac{(T_{ig} - T_0)}{q_e''} \qquad [5.4.2]$$

where δ is the sample thickness. If convective heat losses at the exposed and unexposed surface are included, the equivalent thermally thin equation is (Drysdale, 1998, p. 214)

$$t_{ig} = \frac{\rho c \delta}{2h} \ln\left(\frac{q_e''}{q_e'' - 2h(T_{ig} - T_0)}\right) \qquad [5.4.3]$$

where h is the convective heat transfer coefficient. Introducing the critical heat flux, q_{cr}'' (minimum heat flux for ignition at infinite exposure time) is introduced, equation [5.4.3] can be rewritten as

$$t_{ig} = \frac{\rho c \delta}{2h} \ln\left(\frac{q_e''}{q_e'' - q_{cr}''}\right) \qquad [5.4.4]$$

Hamins and McGrattan (1999) fit the high heat flux portion of their corrugated paper ignition data to equations [5.4.3] and [5.4.4], and found $q_{cr}'' = 14.5$ kW/m2, $T_{ig} = 370\,°C$, $h = 0.042$ kW/m², and $\rho c\delta = 0.98 \text{ kJ/m}^{2\circ}\text{K}$. In a later paper (McGrattan *et al.* 2000), they report $\rho c\delta = 1.5 \pm 0.4 \text{ kJ/m}^{2\circ}\text{K}$. Both of their curve fits are shown in Figure 5.10. It is clear that from Figure 5.10 that the thermally thick approximation fits the high heat flux range of the data much better than equation [5.4.4]. This is because the ignition times at these high heat fluxes are small, and the corresponding heat conduction thickness, $(kt_{ig}/\rho c)^{1/2}$, is smaller than the corrugated paper thickness.

Equation [5.4.4] has also been fit to the portion of the data in Figure 5.10 corresponding to heat fluxes less than 25 kW/m^2. Since the curve does seem to fit the data well in this range, the thermally thin approximation appears to be valid for these small heat fluxes and corresponding long ignition times. Thus, the corrugated paper can be treated as either thermally thin or thermally thick, depending on the heat flux range of interest. This is consistent with the observation of Silcock and Shields (1995) that many sample materials are intermediate between thermally thin and thermally thick in terms of best-fit correlations to inverse ignition time data.

One popular flame spread test is the ASTM E 162 (1983) test in which the flame spread is downward on an inclined sample opposite a radiant heat source. The result of the ASTM E 162 test is a flame spread index which is a product of a flame spread factor (proportional to the flame spread rate) and a heat generation factor. Underwriters Laboratories recently measured flame spread indices for a variety of warehouse container materials used by the US Air Force ('Flammability Test Method/Requirements for Packaging Materials', 1988). Downward flame spread rates were greatest for fiberboard materials (10–15 in/min or 25.4–38.1 cm/min) and smallest for medium density polyethylene (about 5 in/min, or 12.7 cm/min). Since results in this test configuration are influenced by the melting and dripping of thermoplastics such as polyethylene, other test configurations may be more relevant to warehouse commodities.

Time to ignition data for treated and untreated cardboard carton materials have been obtained by Khan (1987) and Tewarson (1995) using the FMRC Flammability Apparatus. Tewarson (1995) correlated all his data with the following version of equation [5.4.1].

$$t_{ig}^{-1/2} = \frac{\sqrt{4/\pi}}{TRP}(q_e'' - q_{cr}'') \qquad [5.4.5]$$

in which $TRP = (k\rho c)^{1/2}(T_{ig} - T_0)$ is called the *thermal response parameter*. Tewarson and Khan reported values of q_{cr}'' for ordinary corrugated paper sheet and for flame retardant corrugated sheet of 10 kW/m^2 (0.88 Btu/sec-ft^2) and 15 kW/m^2 (1.32 Btu/sec-ft^2), respectively. The untreated paper critical flux of 10 kW/m^2, which is 4.5 kW/m^2 smaller than the value measured by Hamins and McGrattan. The difference may be due to the black paint applied to the surface of the samples tested by Tewarson and Khan in order to increase the surface absorptivity. The value of TRP reported by Tewarson for lightweight corrugated paper is $152 \text{ kW-s}^{1/2}/\text{m}^2$.

Khan correlated the time to ignition data for ordinary corrugated sheets by

$$t_{ig}^{-1} = 10^{-4} q_e''\{2 + 0.45 q_e''\} \quad \text{for } q_e'' > 20 \text{ kW/m}^2 \qquad [5.4.6]$$

and the data for fire retardant corrugated sheets by

$$t_{ig}^{-1} = 0.001 q_e'' \quad \text{for } q_e'' > 30 \text{ kW/m}^2 \qquad [5.4.7]$$

Equation [5.4.7] is an attempt to account for both the thermally thin and thermally thick behavior of the corrugated sheets at small and large heat fluxes, respectively. Equation [5.4.7] is more appropriate for a thermally thin solid, which the fire retardant sheets apparently approximate because of their relatively long ignition times.

Figure 5.11 is a plot of $t_{ig}^{-1/2}$ versus q_e'' for the various types of corrugated paper tested by Khan and Tewarson. The corrugated sheets treated with fire retardant coatings have substantially longer

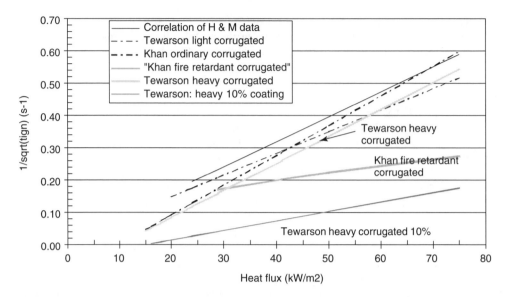

Figure 5.11. Correlations for inverse time-to-ignition for different grades of corrugated paper

ignition times (smaller $t_{ig}^{-1/2}$ values) than the ordinary corrugated sheets, based on Tewarson's reported values of TRP. This implies that flame spread rates over the fire retardant corrugated cartons will be significantly slower than those over ordinary cartons.

Laboratory measurements of fire heat release rates are being made by a variety of methods incorporating externally applied radiant heat fluxes. The most commonly used apparatus for this purpose is the cone calorimeter (ASTM E1354). The standard test sample for cone calorimeter testing is a 10×10 cm horizontal surface exposed to the heat flux. The FMRC flammability tests (Tewarson, 1982) using a similar apparatus have generated heat release rate data for 80 cm² (12.4 in²) by 2–5 cm (0.8–2 in) thick horizontal samples of numerous polymers and packaging materials exposed to external heat fluxes and sometimes to enhanced oxygen atmospheres. Data for the asymptotic (with increasing heat flux) heat release rate for various polymers span the range 120–1200 kW/m² (10.6–106 Btu/sec-ft²).

Hamins and McGrattan (1999) have conducted cone calorimeter tests with a $10 \times 10 \times 10$ cm miniature corrugated paper cell containing a polystyrene cup used in fabricating the prototype Group A plastic commodity. The heat release rate per unit area at an imposed radiant heat flux of 50 kW/m² was 400–500 kW/m². They later report (McGrattan *et al.* 2000) that their computer model provides a better match to large-scale test data when they use a specific heat release rate of 600 kW/m² for the prototype Group A plastic. One reason for the higher heat release per unit area in large-scale tests is that the heat flux impinging on the vertical surface of the cardboard boxes from the flames below is typically 90–100 kW/m².

Another laboratory flammability test is the ASTM E 906 apparatus (ASTM E-906, 1983) in which a 10×15 cm (3.9×6 in) horizontally oriented sample or a 15×15 cm (6×6 in) vertically oriented sample is introduced into an environmental chamber with a radiant heat source. Underwriters Laboratories ('Flammability Test Method/Requirements for Packaging Materials', 1988) used the ASTM E 906 apparatus to measure heat release rates for various Air Force warehouse container and cushioning materials exposed to a radiant heat flux of 35 kW/m² (3.1 Btu/sec-ft²) on the vertical exposed surface of the sample. Fiberboard and wood container materials had the lowest heat release rates (less than 220 kW/m² or 19.4 Btu/sec-ft²), while polyethylene had the highest heat release rates (it melted and burned in a pool). Among the cushioning materials tested, flexible polyethylene foam had the highest heat release rates while fire retarded rubberized hair had the lowest heat release rates.

UL also measured heat release rates for 2 by 2 by 2 ft (61 by 61 by 61 cm) Air Force warehouse containers by testing under the UL product calorimeter ('Flammability Test Method/Requirements for Packaging Materials', 1988). The 30 sec average peak heat release rates were correlated to the corresponding ASTM E 906 bench scale heat release rates. A linear correlation seemed to represent most of the data except for the bins with polyethylene foam cushioning material which dripped and generated relatively large heat release rates in the ASTM E 906 tests, but was confined to the bins in the product-scale tests. This inconsistency notwithstanding, Underwriters Laboratories recommended using the ASTM E 906 heat release rates and smoke release rates as a basis for classifying warehouse commodities for use in un-sprinklered Air Force warehouses. However, there have not been any large scale tests to confirm the UL recommended classification scheme. Moreover, there is much more interest in classifying commodities for sprinklered warehouses than un-sprinklered warehouses.

Factory Mutual Research Corporation has embarked on a long-range research program to develop suitable bench-scale flammability tests for classifying commodities for sprinklered warehouses. Toward this end, Tewarson (1995) and his co-workers have recently extended the conventional horizontal sample measurements to vertically oriented samples and vertically stacked miniature packaging boxes with upward flame propagation. They have also started conducting tests with water sprays and films applied to the commodities. Data from both the horizontal samples and the vertically oriented configurations are currently used as a screening tool to suggest appropriate materials for testing in the small unit load storage array tests described in Section 5.4.3. Laboratory cone calorimeter tests with water application are also being conducted now by several other investigators (for example, Hietananiemi *et al.* 1999). Results to date indicate that the relative effect of the water application depends both on the material and the method of application (nozzle versus perforated pipe), as well as the applied water density.

5.4.3 SMALL ARRAY TESTS

Fire tests described here involve 8–12 pallet arrays of various industrial warehouse commodities. The tests were conducted at Factory Mutual Research over a period of 20 years beginning with the original plastics storage test program (Dean, 1975; Delichatsios, 1983) and evolving into the recent Fire Products Collector Commodity Classification tests (FMRC Update, 1990) and Required Delivered Density Tests (Yu, 1989). The tests are intended to provide a basis for classifying warehouse commodities on the basis of heat release rate data with, and in some cases without, water application.

The original small array in the stored plastic test program consisted of a two pallet load wide by two pallet load deep by three pallet load high palletized storage configuration with the stacks butted together in one direction and separated by a 6 in (15 cm) flue in the other direction. The array occupied an 8×8.5 ft (2.4×2.6 m) floor area and extended to a height of 13–16 ft (4–4.9 m) depending on the commodity. The array was situated on a weighing platform to obtain weight loss data during the burns. A pair of tests was conducted for each commodity. A free burn test was conducted to measure weight loss rates and temperatures at a 18.3 m (60 ft) ceiling above the burning commodity. A sprinklered test was conducted under a 9.1 m (30 ft) ceiling equipped with 1/2 in orifice, 138 °C (280 °F) sprinklers on 7.4 m^2 (80 ft^2) spacing and discharging 0.45 gpm/ft^2 (18.3 L/min-m^2) upon actuation. The number of sprinklers operated, the maximum ceiling air temperatures, and the time of first sprinkler activation were the primary data considered from the sprinklered tests.

A total of 22 pairs of freeburn and sprinklered tests were conducted with different commodities in the original small array plastics storage test program. The complete set of data and commodity descriptions has been reported by Dean (1975). Delichatsios (1983) has provided a convenient summary shown here as Table 5.10 for the key results for 15 commodities. Besides a cursory commodity description, Table 5.10 shows the fuel volume fraction (the rest of the storage volume being air), the percent weight of plastic and cardboard (ignoring the percent weight of the pallet), a power law curve fit to the cumulative mass loss history, the maximum freeburn theoretical heat release rate (the peak mass burning rate times the theoretical heat of combustion based on the weighted fraction of plastic and cardboard), the time of first sprinkler operation, the corresponding theoretical heat release rate at sprinkler activation, the number of sprinkler rings opened (each ring being an imagined circle centered at the ignition site), and the total number of sprinklers opened.

The commodities listed in Table 5.10 are grouped into three categories: commodities in compartmented cartons, loosely packed commodities, and foamed plastic insulation boards in cartons. Within the grouping of commodities in compartmented cartons, the polystyrene jars produced the highest freeburn heat release rate and as many or more opened sprinklers as the other plastics. The peak heat release rates for the five commodities in this group decreased in the order

$$\text{polystyrene} > \text{polypropylene} > \text{polyethylene} > \text{PVC} > \text{emptycarton}.$$

Each of these five commodities in compartmented cartons had t^2 cumulative weight loss histories implying a linearly increasing mass loss rate and heat release rate. Surprisingly, the empty compartmented cartons produced the most rapid rate of weight loss early in the fire. Delichatsios attributes this to the relatively small volume fraction of fuel in the empty cardboard cartons, such that empty cartons and other fuels with small volume fractions are more rapidly heated to their ignition temperature.

Quantitatively, Delichatsios offers the following equation based on his model of flame spread through a porous fuel assembly (Appendix A of Delichatsios, 1983):

$$dm/dt = m''[u_f A_f t \phi (Sfe/Vfe)] \qquad [5.4.8]$$

where m'' is the specific burning rate per unit surface area burning at time t, u_f is the flame spread rate over exposed fuel surface, A_f is the flame surface area within fuel array, ϕ is the fuel volume fraction in storage array, and Sfe/Vfe is the exposed surface area of fuel per unit volume of storage.

Delichatsios' model assumes that the flame spread rate, u_f, varies inversely as ϕ, and he derives an equation for the specific burning rate based on natural convection of air through the open air spaces in the fuel array. His model shows that

$$dm/dt \propto m''(1-\phi)^{3/4} \qquad [5.4.9]$$

Although equation [5.4.9] is roughly consistent with the small array data, the data cover too limited a range to verify the exponent in equation [5.4.9].

The foamed insulation boards listed at the bottom of Table 5.10 produced the fastest rate of fire development as represented by the exponents (4 and 12) in the weight loss history curve fits. The loosely packed commodities produced several fires with t^3 mass loss histories; of these, the foamed plastic meat trays had the most rapid mass loss rates and opened the greatest number of sprinklers. The rapid development of expanded/foamed plastic commodities (in some cases after an incipient period to get the plastic involved) warrants there being treated as a more challenging class of commodities in NFPA 230.

Table 5.10. Summary of small array plastic storage tests palletized storage: $2 \times 2 \times 3$ high (data from Delichatsios, 1983)

Commodity description	Volume fraction of fuel (ϕ) (%)	Weight percentage of plastic and cardboard	Approximate ($\pm 20\%$) weight loss history[b] t: in min	Maximum HRR[a] (Btu/min)	Time of 1st sprinkler operation (min)	HRR[c] at 1st sprinkler operation (Btu/min)	Sprinkler rings[d] opened	No. of sprinklers
A. Commodities in compartmented cartons								
Test 4, jars, polystyrene 16 oz in compartment cartons	15	75% plastic 25% cardboard	m = 15 t²	7.06×10^6	1.82	8.75×10^5	1 + 3	13
Test 6 bottles PVC 32 oz in compartmented cartons	16.4	50.3% plastic 49.7% cardboard	m = 14 t²	2.38×10^6	1.68	6.07×10^5	1	4
Test 15, bottles 16 oz polyethylene In compartmented cartons	20	57% plastic	m = 8 t²	2.58×10^6	1.68	4.03×10^5	1 + 3	13
Test 19 tubes polypropylene 16 oz in compartmented cartons	15	53% plastic 47% cardboard	m = 8.5 t²	3.26×10^6	2.25	5.51×10^5	1 + 3	12
Test 41; Cardboard boxes without any plastic commodity	13	100% cardboard	m = 17.5 t²	1.41×10^6	1.6	4.48×10^5	1	4
B. Commodities loosely packed (including meat trays)								
Test 30A toy parts polystyrene	14	91% plastic	m = 7 t²	2.19×10^6	3.43	7.94×10^5	1	4
Test 13 meat trays wrapped in plastic sheet	2.7	100% plastic	m = 19.5 t³	4.36×10^6	0.93	1.10×10^6	1 + 3	13
Test 23 meat trays wrapped in paper	31	100% plastic	m = 24 t²	4.5×10^6	0.95	9.13×10^5	1 + 3	12

(*continued overleaf*)

Table 5.10. (continued)

Commodity description	Volume fraction of fuel (ϕ) (%)	Weight percentage of plastic and cardboard	Approximate (±20%) weight loss history[b] t: in min	Maximum HRR[a] (Btu/min)	Time of 1st sprinkler operation (min)	HRR[c] at 1st sprinkler operation (Btu/min)	Sprinkler rings[d] opened	No. of sprinklers
Test 44 tubes 16,24 32 oz polystyrene	–	92% plastic 8% cardboard	$m = 9.2 t^2$	2.78×10^6	–	–	–	–
Test 5 bottles 16 oz Polyethylene	9.2	81% plastic 19% cardboard	$m = 1.1 t^{2.5}$	–	5.90	–	1	4
Test 7 bottles assorted 1/2 12 oz polyethylene	–	–	$m \approx 5 t^2$	1.44×10^6	5.28	4.81×10^5	1	3
Test 22. trash barrels polyethylene	5.8	87% plastic 13% cardboard	$m = 4.35 t^3$	2.72×10^6	1.9	7.72×10^5	1	3
Test 43 bottles 64 oz polyethylene	6	–	$m = 3.97 t^3$	2.46×10^6	–	–	1	4
C. Insulation boards								
Test 34 polyurethane foam with paper facing	–	71% plastic	$m = 64 t^4$ (t < 1.5 min) $m = 65 t$ (t > 1.5 min)	1.09×10^6	0.56	5.04×10^5	1	4
Test 35 polystyrene board no wrapping	–	100% plastic	$m \approx 4.41 t^{12}$ 5 min < t < 7 min	9.6×10^6	2.38	–	–	10

[a]HRR: Heat Release Rate (estimated)
[b]See equations 5.4.8 and 5.4.9
[c]See page 136 for sprinkler details
[d]1 means first ring of sprinklers, 3 means third ring of sprinkler

WAREHOUSE STORAGE 139

Table 5.11. Fire severity – comparison of arrangement of bottles in cartons

Commodity	Total sprinklers opened	First sprinkler operation (min:sec)	Maximum ceiling temperature over ignition (F)	Duration above 1000°F on ceiling over ignition (min)	Flames reach ceiling (min)	Bottles begin to fall (min)	Initial collapse (min)
Compartmented PE 16 oz Bottles (Test 17)	13	1:41	1545	9.8	0.7	–	18.4
Ordered PE 16 oz Bottles (Test 8)	4	5:54	540	0.0	Did Not Reach	2.0	7.2

Polystyrene foamed insulation boards generated the highest peak heat release rate of any of the commodities (about nine times as high as the polyurethane foam paper faced insulation boards). Within the group of loosely packed commodities in Table 5.10, the foamed polystyrene meat trays wrapped in plastic sheet and wrapped in paper produced the highest peak heat release rates and opened the most sprinklers. Thus, within all three groupings of commodities, polystyrene produced the most severe fires of the various plastics tested. This is consistent with the listing in Table 5.9 indicating that polystyrene has the largest ratio of heat of combustion to effective heat of gasification.

One additional interesting result from the small-array stored plastic tests, was the effect of loosely packed plastic parts within the commodity. For example, Table 5.11 compares the results of a sprinklered test with polyethylene cups loosely packed in an orderly fashion within the cartons to an equivalent test with polyethylene cups in compartmented cartons. There were dramatically higher ceiling temperatures and greater number of sprinklers operated in the test with the cups in compartmented cartons. This is due to the fact that the cups remained in the compartmented cartons and fell out of the cartons without any dividers when the carton sides burned out. After falling from cartons, the cups tended to smother the fire such that flames never reached the ceiling. This tendency for some commodities to spill out of their container during a fire and reduce the resulting fire intensity has been observed in numerous fire tests and is now recognized as a factor to be considered in either classifying commodities, or in specifying sprinkler protection within a given class of commodities (NFPA 230 and FM Data Sheet 8-9).

The conclusion that polystyrene cups in compartmented cartons represent the most severe fire hazard of the high density (unexpanded) plastics tested in the small array tests has led to this being selected as the prototype Group A plastic commodity. It is shown in Figure 5.12, along with the commodity selected as the prototype for Class II. Note that the cups are tested upside down so that they will not collect sprinkler water when the tops of the cartons burn out. The prototype Class II commodity is a metal lined double triwall carton that produced a much lower peak heat release rate (0.95 million Btu/min or 16.7 MW) than the empty compartmented cardboard cartons (1.4 million Btu/min or 24.6 MW) in the small palletized-array tests.

Although the small palletized-array tests produced a wealth of data that form the current basis for generic commodity evaluations, the test configuration itself had several drawbacks. First, the palletized array is more prone to the vagaries of pile collapse and commodity spillage than a rack storage array. Second, the sprinkler configuration involving only 80 ft^2 per sprinkler contributed to the skipping of the second ring such that the number of operating sprinklers was not a reliable indication of the relative water demand. Third, the sprinkler model used in the original tests is no longer commercially available. In view of these drawbacks, another series of small array tests was conducted in 1979 with several of the same commodities.

The 1979 test series was conducted with a two-tier high rack storage array. The sprinklered tests were conducted using different sprinklers with 9.3 m^2 (100 ft^2) per sprinkler. The sprinklered

Figure 5.12. (a) Prototype Group A plastic commodity; (b) Prototype Class II commodity. © 2002 Factory Mutual Insurance Company, with permission

and freeburn test results for four commodities are described by Delichatsios (1983), who found that all four freeburns could be represented by second power heat release rates of the form

$$Q = a(t - t_0)^2 \quad \text{for } Q < 0.6 Q_{\max} \qquad [5.4.10]$$

where a, t_0, and $Q_{\max}$ are empirical parameters determined by curve fits. The values for these parameters obtained by Delichatsios (1983) were based on theoretical heats of combustion as well as weight loss histories. More accurate values are now available based on combustion calorimeter tests.

One important difference between the rack storage and small palletized array tests with the same commodities is the form of equation [5.4.10]. The palletized array tests resulted in only a first order linearly increasing heat release rate with time, while the rack storage array produced a second power variation. This is because flame spread was limited to a single narrow flue in the palletized array, and was allowed to encompass all exposed sides (including carton tops and bottoms) in the rack storage array. The peak freeburn heat release rates in the three tier high palletized-array were higher than the corresponding peaks in the two tier high rack storage tests.

Advances in heat release rate calorimetry and in modeling of sprinkler operating times now make it possible to obtain more accurate and repeatable data from both freeburn and water application tests with small storage arrays. These advances have been utilized in tests using the FMRC 10 MW (9500 Btu/sec) capacity Fire Products Collector (FPC) and special water applicator at the West Gloucester, Rhode Island Test Center. Thus the two by two by two tier high rack storage array tests described above are now conducted using the test configuration shown in Figure 5.13.

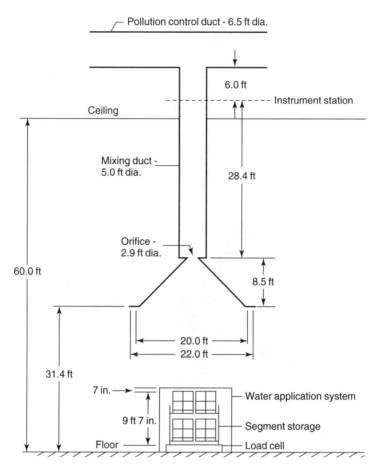

Figure 5.13. Commodity classification fire product collector test setup. © 2002 Factory Mutual Insurance Company, with permission

Fire Products Collector data have been obtained for several commodities in addition to the prototype Class II and Group A plastic commodities described earlier. The additional commodities include the prototype Class I, II and IV commodities shown in Figure 5.14. Each commodity consists of eight compartmented, single-wall, corrugated paper cartons, with each carton containing 125 cups or glasses made of the appropriate materials for that commodity class. The appropriate cup/glass materials are as follows: glass for the prototype Class I commodity, paper for the prototype Class III material, and a mixture of polystyrene and paper for the prototype Class IV commodity. The content of the commodities (including the wood pallet) is listed in Table 5.12 along with the total mass and theoretical combustion energy per pallet. The heat release rates are determined from oxygen consumption rates and combustion energy per unit mass of oxygen consumed (13.1 kJ/g or 5600 Btu/lb).

Values of the coefficient a in the second power heat release rate histories (equation (5.12)) are listed in Table 5.13, along with the peak freeburn total, convective, and radiative heat release rates. If the values of a in Table 5.13 are normalized by the value for Metal Lined Double Triwall (prototype Class II), the normalized values are about 4.4 for the polystyrene cups (prototype Group A plastic), and about 2.1 for the polyethylene terepthalate (PET) cups (representative

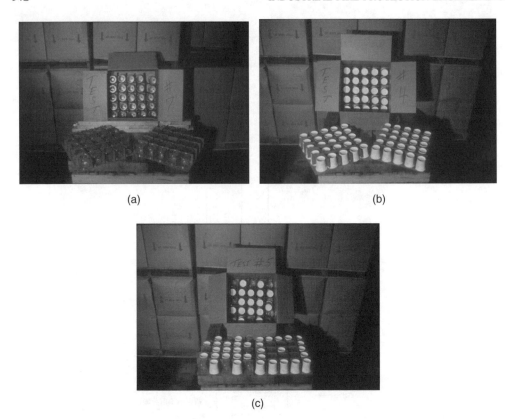

Figure 5.14. (a) Prototype Class I commodity: glass jars in compartmented cartons; (b) Prototype Class III commodity: paper cups in compartmented cartons; (c) Class IV commodity: paper and polystyrene cups in compartmented cartons. © 2002 Factory Mutual Insurance Company, with permission

Table 5.12. Composition of commodities tested under fire products collector

Commodity	NFPA class	Cardboard/paper (kg)[a]	Wood (kg)[a]	Plastic (kg)[a]	Total (kg)[a]	Combustion energy (MJ)[a]
Polystyrene cups	Group A plastic	19	26	31	76	2070
Polyethylene cups	Group A plastic	19	26	27	72	1950
PET cups	Group A plastic	19	26	41	86	1680
Polyurethane foam	Group A plastic	12	26	5.5	43	810
Modified PU foam	Tests required	12	26	6.8	45	850
Polystyrene (25 wt%)+ paper cups	Class IV	30	26	9.0	65	1300
Paper cups	Class III	34	26	0	60	1060
ML double triwall	Class II	36	26	0	85[b]	1100
Glass jars in carton	Class I	19	26	0	320[c]	800

[a] Tabulated quantities per pallet load of commodity
[b] Includes 23 kg metal liner inside double triwall commodity
[c] Includes 275 kg of glass jars

Table 5.13. Key results from fire products collector free burn tests (from Lee, 1987, and Spaulding, 1988)

Commodity	a (kW/s^{-2})	Q_{max} (MW)	Q_{cmax} (MW)	Q_{rmax} (MW)
Polystyrene cups	0.90	22.5	16.5	6.0
Polyethylene cups	0.57	17.8	11.8	6.0
PET cups	0.45	13.3	8.8	4.5
Polyurethane foam	0.87	16.0	11.8	4.2
Modified PU foam	0.98	10.6	7.5	3.1
ML double triwall	0.21	8.1	6.2	1.9

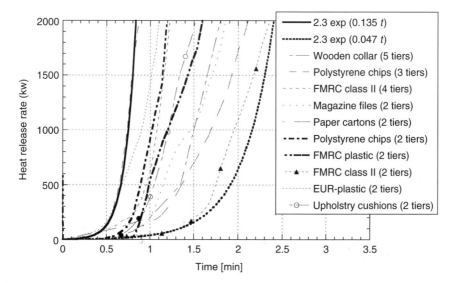

Figure 5.15. Commodity free burn heat release rates measured at SP. Reproduced by permission of H. Ingason

Class IV commodity). The values of a for the two foamed polyurethanes are comparable to that for the unexpanded polystyrene cups. However, the values of Q_{max} for the polystyrene cups polyethylene cups (6 MW) are substantially larger than those for the polyurethane foams (3–4 MW), and about three times as large as the prototype Class II commodity Q_{max}.

The Swedish National Testing and Research Institute (SP) has also constructed a large calorimeter for warehouse commodity testing. SP has used the calorimeter to conduct comparative tests of representative European commodities and the FM prototype Class II and Group A plastic commodities. Free burn heat release rate plots for various commodities are shown in Figure 5.15 for heat release rates up to 2 MW. Most of the commodities were tested in the 2-tier high rack storage small arrays used in the FM test configuration shown in Figure 5.13. The commodity with the most rapidly growing heat release rate curve in Figure 5.15 is polystyrene packing chips in corrugated paper cartons. The FM prototype Class II commodity had the slowest growing heat release rate. Ingason (2001) has fit the following two exponential curves to the heat release rate data for these maximum and minimum fire growth rates:

$$Q = 2.3e^{0.135t} \quad \text{for cartoned polystyrene chips 3 tiers high} \quad [5.4.11a]$$

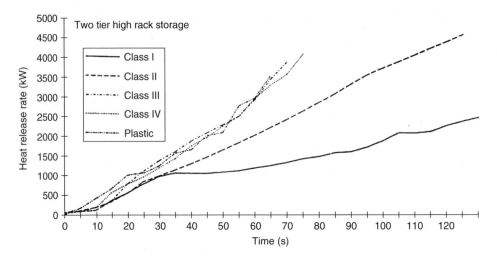

Figure 5.16. Commodity free burn heat release rates for 2 × 2 × 2 pallet array (FM)

$$Q = 2.3e^{0.047t} \quad \text{for prototype FM Class II 2 tiers high.} \qquad [5.4.11b]$$

where Q is <2000 kW and t is in seconds.

The t^2 and exponentially growing heat release rates expressed by equations [5.4.10] and [5.4.11] are only applicable for a limited period of time early in the fire. Beyond this period, the heat release rate develops less rapidly, perhaps with a linear increase in time. Heat release rate histories measured in the two tier high rack storage commodity classification tests are shown in Figure 5.16 for times from effective ignition (heat release rates of 30 ± 10 kW, or 28.4 ± 9.5 Btu/sec) of at least 60 seconds. Note that at times in the range 30–60 seconds, the Class III, IV, and Plastic commodity free burn heat release rates are all approximately the same. Class II and Class I prototype commodities do have distinctly lower heat release rates. One interpretation of the data in Figure 5.16 is that free burn tests are not appropriate for reliable commodity classification determination.

Fire Products Collector water application tests are conducted by calculating the time of simulated sprinkler operation for the specified sprinkler location and link actuation parameters. The tests reported by Lee (1987) Spaulding (1988) and the FMRC Update (1990) for commodity classification purposes applied water based on an assumed 6.1 m (20 ft) high ceiling 3.05 m (10 ft) clearance above the array) with sprinklers having a link temperature rating of 141 °C (286 °F) and a Response Time Index of 276 (m-s)$^{1/2}$ (500 (ft-s)$^{1/2}$). The fire plume is assumed to be centered between four sprinklers with 10 × 10 ft spacing and with links situated 20 cm (8 in) below ceiling level. A nonsteady plume and ceiling layer model similar to that described in Section 5.6 was used along with a sprinkler actuation model equivalent to the one in Appendix D. Calculated convective heat release rates at sprinkler actuation are shown here in Table 5.14.

The actual time of water application varied slightly from test to test for the same commodity because of differences in fire incubation time to achieve self-sustained burning. Water application rates corresponded to delivered densities in the range 0.10–0.36 gpm/ft^2. Figure 5.17 shows the convective heat release rate histories for five different cartoned commodities with a water application density of 0.21 gpm/ft^2 (8.5 L/min-m^2). The data indicate the delivered density of 0.21 gpm/ft^2 (8.5 L/min-m^2) is sufficient to suppress the prototype Class I and II commodity fires but the fire either redevelops or continues burning at the same intensity for Class III, IV, and Group A plastic commodities. These higher challenge commodities are controllable/suppressible

Table 5.14. Calculated values of Q_c at sprinkler activation for two tier racks

Commodity	Q_c (MW)
Polystyrene cups in cartons	4.5
Polyethylene cups in cartons	3.8
PET cups in cartons	3.7
Polyurethane foam in cartons	4.1
Modified PU foam in cartons	4.1
Polystyrene (25 wt%) and paper cups in cartons (IV)	3.6
Paper cups in cartons (Class III)	3.3
Metal lined double triwall (Class II)	2.9
Glass jars in cartons (Class I)	1.9

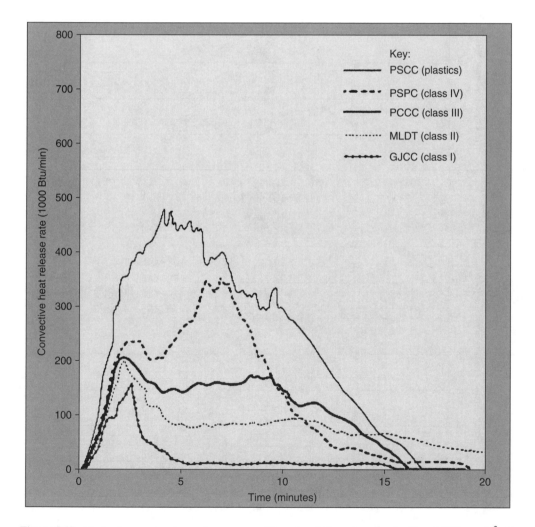

Figure 5.17. Heat release rates for various commodities at a delivered water density of 0.21 gpm/ft^2. © 2002 Factory Mutual Insurance Company, with permission

at the higher water application rates. Similar data for other commodities and water application rates are presented in Lee (1987) and FMRC Update (1990).

The Fire Products Collector measured peak (one-minute-average) convective heat release rates and peak chemical heat release rates for each water application test are plotted against water application density in Figure 5.18. Data for four of the five prototypical commodities show decreasing peak convective heat release rate with increasing water application density. The Class I commodity has a virtually constant value of Q_{cmax} which corresponds to the value of Q_c at water application.

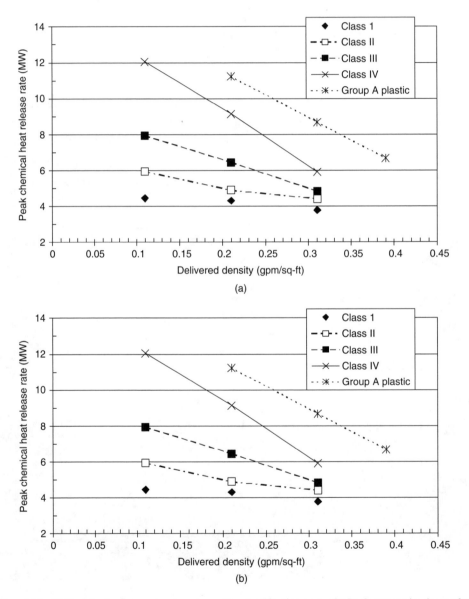

Figure 5.18. (a) Peak heat release rates in commodity classification tests; (b) Peak convective heat release rates in commodity classification tests

A formalized methodology for establishing commodity classification for unknown commodities on the basis of Fire Products Collector water application tests is described in the FMRC Update (1990). The tests entail using three different water application rates for each commodity, and comparing results to those for the prototypical commodities described above. The classification criteria used include the peak heat release rates (both one minute average and five minute average) and the total convective energies released at each of the water application rates. Underwriters Laboratories, Inc. has constructed a calorimeter and water applicator to also conduct these commodity classification tests for warehouse commodities and for plastic pallets.

The Swedish National Testing and Research Institute has been utilizing the FMRC commodity classification test procedure to determine how the FMRC classifications compare to the Swedish classification system of four categories, L1, L2, L3, and L4. Persson (1993) has reported on the 1993 status of those tests, which include both the standard FMRC commodities and their European counterparts. Perhaps the ultimate outcome of this type of collaborative testing might eventually be a reconciliation of the US and European classifications on the basis of reproducible heat release rate data with uniform water application.

Small-array calorimetry tests have also been conducted to generalize the effects of various applied water densities on commodity heat release rates (Yu et al., 1994). Results are summarized in Section 5.7 in the context of sprinklered warehouse fire modeling.

5.4.4 LARGE ARRAY SPRINKLERED FIRE TESTS

The original plastics storage test program (Dean, 1975) included large-scale array sprinklered fire tests to confirm some of the commodity effects observed in the small array tests. For example, the observed differences between polystyrene cups and polyethylene cups was verified by conducting a pair of tests with a large palletized array of commodity stacked 4.6 m (15 ft) high under a 7.6 m (25 ft) ceiling outfitted with $\frac{1}{2}$ in orifice sprinklers at a design density of 0.30 gpm/ft^2 (12.2 L/min-m^2). The sprinklered fire with the polyethylene cups (Test SP-6) was significantly less severe than with the polystyrene cups (Test SP-7). This is evident from the large differences in water demand, the extent of fire damage in the array, and the duration of ceiling temperatures capable of damaging structural steel, as indicated by the data in Table 5.15. In fact, these differences have led to the designation of polyethylene as a Factory Mutual Group B plastic (NFPA 230 designates polyethylene as Group A plastic), while polystyrene is the prototype Group A plastic.

A pair of tests was conducted to compare the severity of fires in large arrays of foamed (expanded) polystyrene and high-density (unexpanded) polystyrene. The test with high-density polystyrene cups was the same one (Test SP-7) used for comparison with polyethylene cups. The equivalent test with polystyrene foamed meat trays developed more rapidly (the first sprinkler opened at 0:46 with the meat trays and at 1:37 with polystyrene cups), opened more sprinklers (74 versus 58), and produced damage throughout the entire array. However, high ceiling temperatures were sustained for a longer time with the polystyrene cups. The overall greater severity of this and other foamed plastic tests confirmed the small array test results. An example of a small array FPC

Table 5.15. Comparison of plastic commodities in large array fire tests (data from Dean, 1975)

Commodity (Fire Test #)	Total sprinklers opened	First sprinkler actuation time (min:sec)	Duration of $>1000\,°F$ ceiling gas temperatures	Percent fuel consumption
Polyethylene cups (SP-6)	13	3:06	1.8 minutes	7%
Polystyrene cups (SP-7)	58	1:37	13.1 minutes	55%

test with foamed plastic is shown in Plate 2. The water application rate neither suppressed nor controlled the cartoned polystyrene foamed meat tray fire in this four-tier high configuration. Thus there are special considerations and more stringent sprinkler protection requirements for expanded plastics in both NFPA 13 and CEN prEN 12845. However, the Fire Products Collector commodity classification tests have shown that at least some fire retardant foamed plastic formulations are equivalent to a Class IV commodity.

In view of the differences between plastic commodities and ordinary combustible (cellulosic) commodities, some large array tests were conducted to determine the severity of fires with mixed commodities. For example, a test was run to determine the effect of placing a plastic commodity in the bottom tier of a rack storage array with standard Class II commodity in the upper tiers. Results shown in Table 5.5 demonstrate that the single tier of plastic commodity caused 88 sprinklers to open whereas only 31 opened with Class II commodity throughout the array. Results with plastic in the upper tier were also significantly worse than the equivalent Class II commodity test. Therefore, the inclusion of only a limited quantity of Group A plastic commodity in a warehouse warrants sprinkler protection based on Group A plastic storage.

5.5 Sprinkler flow rate requirements

What are the sprinkler flow rate requirements for suppressing or controlling warehouse storage fires? In other words, what is the required flow rate per sprinkler and how many sprinklers are needed to control or suppress the fire? How are these requirements affected by sprinkler characteristics and by warehouse ceiling heights? The answers to these questions have traditionally been based on the results of large-scale sprinklered fire tests of the type described in Section 5.4.4, and shown in Plate 3.

Most of the large-scale fire tests have been conducted with an ignition source consisting of cellulose-cotton rolls soaked in 120 ml of gasoline or heptane and then placed in a polyethylene bag. The soaked rolls are usually placed in the lowest tier of storage, at the intersection of the transverse and longitudinal flue spaces, and then ignited with a torch. This produces a 1 m (3 ft) high flame on the exposed commodity. Placing the igniter in the flue space produces more rapid and repeatable storage involvement than when it is placed elsewhere. Comparison tests reported by Field (1985) indicate that other ignition locations eventually result in similar fire development in the flue space. A more important aspect of the igniter location is its position relative to the ceiling and in-rack sprinklers. Most tests have been conducted with the ignition site situated midway between the nearest sprinklers, usually centered between four sprinklers since that should result in the longest time delay to first sprinkler actuation. However, since it is now clear that the greatest challenge depends upon the sprinkler spray distribution and local density at the ignition site, many of the more recent tests have been conducted with other locations relative to the sprinklers.

Some of the key fire tests are described here, but the emphasis in the following discussion is on the considerations involved in generalizing the test results to develop general protection guidelines. The discussion is divided into the categories (1) ceiling spray sprinklers, (2) in-rack sprinklers, and (3) Early Suppression fast Response Sprinklers.

5.5.1 CEILING SPRAY SPRINKLERS

Required flow rates for conventional ceiling sprinkler systems are usually specified in terms of a required discharge density per unit floor area and an associated minimum demand area. These so-called area-density specifications are presented in the form of tables and charts, but could be represented for many warehouse storage applications in the form

$$Area = A_0 \exp(-a_1 \, Density) F_H(H) F_{sh}(h_s) F_{com} F_{SPT} F_{spr} F_{AW} F_{mis} F_{SF} \qquad [5.5.1]$$

where *Area* is the minimum water demand area (ft² or m²), *Density* is the water discharge density (gpm/sq-ft) or (mm/min), A_0, a_1 are empirical constants based on sprinklered fire tests, $F_H(H)$ is a function of ceiling-to-storage clearance height, H, $F_{sh}(h_s)$ is a function of storage height (h_s), F_{com} is the commodity classification factor, F_{SPT} is the sprinkler link temperature adjustment factor, F_{spr} is a sprinkler-specific factor to account for differences in sprinkler performance at the same water discharge density and link actuation temperature, F_{AW} is the aisle width adjustment factor, F_{mis} is the adjustment factor for miscellaneous effects such as double row versus multi-row racks, solid versus open racks, inrack sprinklers, and/or dry pipe sprinklers, and F_{SF} is the safety factor to account for other unanticipated effects that would decrease sprinkler effectiveness.

The basic premise in writing equation [5.5.1] is that there are no synergistic effects between the various parameters on the right hand side of the equation, i.e. all of the F factors are functions of one parameter only. Although this is not really a valid assumption (test data indicate that storage height, clearance, commodity, sprinkler temperature rating, and aisle width effects are, in fact, coupled to discharge density and should not be isolated as separate multiplying factors), there is not yet any more comprehensive method to account for synergisms.

The demand area versus discharge density empirical relationship for 6.1 m (20 ft) high double row rack storage of the standard Class II commodity (metal lined double triwall cartons) under a 9.1 m (30 ft) ceiling is shown in Figure 5.19. The data points in Figure 5.19 were measured in a series of large-scale rack storage fire tests in 1969–1970 as part of an extensive test program to generate data for the NFPA Rack Storage Standard (231C, 1998), which has now been incorporated into the general NFPA sprinkler standard (NFPA 13, 1999). A best-fit curve through the base line fire test data obtained with 74 °C (165 °F), 13 mm (1/2 in) orifice ($K = 5.6$ gpm/psi$^{1/2}$)

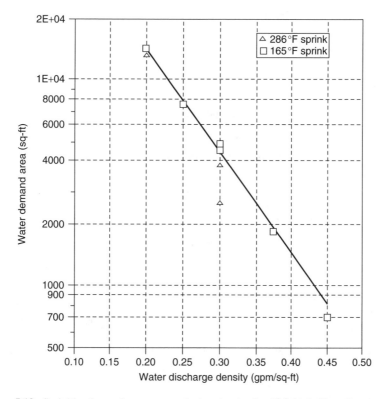

Figure 5.19. Sprinkler demand area versus design density for 20 ft high Class II commodity

sprinklers and 2.4 m (8 ft) aisle widths suggests that

$$A_0 = 1.37 \times 10^5 \, \text{ft}^2 (12.7 \times 10^3 \, \text{m}^2), \quad a_1 = 11.43 \, \text{ft}^2/\text{gpm} (465 \, \text{m}^2\text{-min/L})$$

for the base line data in Figure 5.19. Although these values fit almost the entire range of data in Figure 5.19, upper and lower limits of 558 m² and 186 m² (6000 and 2000 ft²) were established for specifying water discharge densities (NFPA 231C).

Some data for tests with 141 °C (286 °F) sprinklers are also indicated in Figure 5.19. Data from tests in which an excessive number of sprinklers opened are shown in Figure 5.19, but are not considered to be successful design basis fire control data. The only 141 °C (286 °F) sprinkler test data in Figure 5.19 used by the sponsoring Committee to develop design guidelines was the test with a density of 12.2 L/min-m² (0.30 gpm/ft²) that resulted in a 232 m² (2500 ft²) demand area. (The 12.2 L/min-m² (0.30 gpm/ft²) test that resulted in a 353 m² (3800 ft²) demand area was conducted ten years after the original rack storage test program.) The committee decided to adjust the 74 °C (165 °F) sprinkler data by drawing a parallel line through the 12.2 L/min-m² (0.30 gpm/ft²), 232 m² (2500 ft²) datum for the 74 °C (165 °F) sprinkler. Thus the factor F_{SPT} in NFPA 231C was assigned the following values:

$$F_{SPT} = \begin{cases} 1.0 \text{ for } 74\,°C \, (165\,°F) \text{ sprinklers} \\ 1.0 \text{ for } 100\,°C \, (212\,°F) \text{ sprinklers} \\ 2500/3800 = 0.60 \text{ for } 141\,°C \, (286\,°F) \text{ sprinklers} \end{cases}$$

More recent tests with Group A plastic commodity protected by 16 mm (0.64 in) orifice ($K = 11.4 \, \text{gpm/psi}^{1/2}$) sprinklers have shown that the preceding values for F_{SPT} are not universally applicable. Data from the pertinent two pairs of tests reported by Troup (1994, 1998) are shown in Table 5.16

Comparing the sprinkler demand areas in Tests 1 and 2 in Table 5.16,

$$\frac{F_{SPT} \text{ for 286F sprinklers}}{F_{SPT} \text{ for 165F sprinklers}} = \frac{9}{3} = 3.33$$

On the other hand, the corresponding demand areas in Tests 3 and 5 suggest

$$\frac{F_{SPT} \text{ for 286F sprinklers}}{F_{SPT} \text{ for 165F sprinklers}} = \frac{7}{13} = 0.54$$

Table 5.16. Effects of sprinkler link temperature on water demand area (data from Troup, 1994). Group A Plastic, ELO $K = 11.4 \, \text{gpm/psi}^{1/2}$ sprinklers, discharge density = 0.60 gpm/ft², 27 ft ceiling height

Test No.	1	2	3	4
Storage height (ft)	20	20	15.3	15.3
Ceiling to storage clearance (ft)	7	7	11.5	11.5
Ignition centered below (No. of sprinklers)	2	2	1	1
Sprinkler link temperature (°F)	165	286	286	165
Sprinkler demand area (ft²)	4 × 80 = 320	9 × 80 = 720	7 × 80 = 560	13 × 80 = 1040
Number of pallet loads consumed	3	9	6	5

WAREHOUSE STORAGE

Since the values of F_{SPT} are significantly different in the two pairs of test, we can only surmise that F_{SPT} is not only a function of link temperature, it seems to depend upon ceiling-to-storage clearance also. Other tests described by Goodfellow and Troup (1983) also demonstrate that lower temperature ratings either alone or in combination with lower Response Time Index (RTI) values can significantly reduce the water demand area for Large Drop sprinkler (0.64 in orifice) protection of 6.1 m (20 ft) high rack storage of polystyrene cups in compartmented cartons. The current interpretation of these findings is that higher temperature ratings can reduce the water demand for sprinklers that control the fire primarily by pre-wetting unburned commodity ahead of the spreading flame, but that lower temperature ratings reduce the demand area when the fire is at least partially suppressed by spray penetration through the fire plume. Unfortunately, the boundary between fire control by pre-wetting and fire suppression by penetration and extinguishment is not yet well defined even qualitatively. Modelling approaches to this important issue are described in Section 5.7.

Comparisons of test results in the original rack storage test program were used to obtain the following values for F_{com}, F_{AW}, and F_{mis}:

$$F_{com} = \begin{cases} 0.89 \text{ for Class I commodities} \\ 1.0 \text{ for Class II commodities} \\ 1.12 \text{ for Class III commodities} \\ 1.52 \text{ for Class IV commodities} \end{cases}$$

$$F_{AW} = \begin{cases} 1.0 \text{ for 8 ft (2.4 m) wide aisles} \\ F_{AW} = 1.18 \text{ for 4 ft (1.2 m) wide aisles} \end{cases}$$

$$F_{mis} = \begin{cases} 1.0 \text{ for double row rack wet pipe systems} \\ 1.3 \text{ for multiple row racks} \\ 1.3 \text{ for dry pipe systems} \end{cases}$$

The ceiling clearance function $F_H(H)$ has not been developed explicitly because only a limited number of comparison tests have been conducted with ceiling clearance as the only variable. These comparisons are between 3.05 m (10 ft) clearance and 0.94–1.2 m (3–4 ft) clearance, and NFPA 231C (now NFPA 13) provides a curve that can be represented as

$$F_H = 0.80 \text{ to } 0.88 \quad \text{for } 1.5 \text{ ft} < H < 4.5 \text{ ft}$$

$$F_H = 1.0 \quad \text{for } H > 4.5 \text{ ft}.$$

The storage height function, $F_{sh}(h_s)$, in NFPA 13 is specified as a factor for modifying the required design density rather than demand area. These storage height factors for required design density are shown in Figure 5.20a and 5.20b, for rack storage and for solid pile/palletized storage, respectively. F_{sh} is a steeper function of storage height for rack storage than it is for solid pile/palletized storage.

No explicit value for F_{SF} was adopted from the original rack storage test program. However, the tests were run with a constant discharge density (constant water pressure) rather than the declining density caused by the opening of additional sprinklers in a facility without any pressure controller. The discharge densities specified from the area-density relationships are minimum recommended densities at the hydraulically most remote sprinklers. Hence there is an implicit safety factor for facilities in which the initially opened sprinklers are at a higher pressure and discharge density than the hydraulically most remote sprinklers.

The dependence of water demand area on sprinkler deflector design for a given orifice size and discharge density was inadvertently demonstrated in a series of tests (Dean, 1980) intended

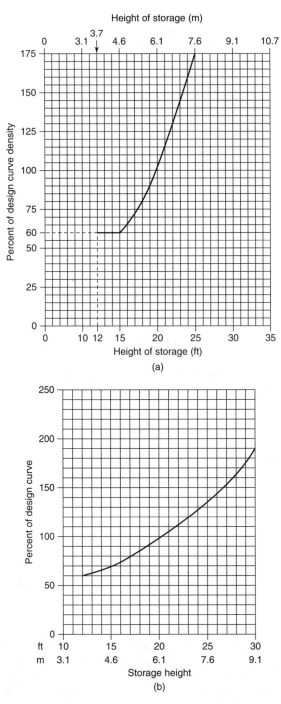

Figure 5.20. (a) Effect of rack storage height on required design density in NFPA 231C; (b) Effect of storage height on required design density in NFPA 231. Reprinted from (a) NFPA 231C®, *Rack Storage of Materials* and (b) NFPA 231®, *General Storage*. Copyright © 1998 National Fire Protection Association, Quincy, MA 02269. This reprinted material is not the complete and official position of the National Fire Protection Association, on the referenced subject which is represented only by the standard in its entirety

Table 5.17. Repeatability test series – data for 20 ft high rack storage of Class II commodity (from Dean, 1980)

Item	Test 69	Test 69R	Test 69R1	Test 69R2	Test 69R3
First sprinkler operation (min:sec)	3:00	2:16	2:24	2:56	2:46
Total operated sprinklers	25	38	36	31	24
Total measured water flow dm³/min (gpm)	2908 (787)	4259 (1125)	4012 (1060)	3388 (895)	3350 (885)
Actual discharge density mm/min (gpm/ft²)	12.6(0.31)	12.2(0.30)	12.0(0.29)	12.0(0.29)	15.0(0.37)
Max ceiling temp over ignition C (F)	638(1180)	843(1545)	932(1710)	663(1225)	805(1480)
Occurrence (min:sec)	3:45	13:29	11:54	15:24	12:11
Duration over 538 °C (1000 °F) (min)	1.4	13.7	15.4	5.8	13.8
East aisle jump (min:sec)	none	8:57	7:19	10:59	7:30
West aisle jump (min:sec)	none	14:08	12:30	none	none
Sprinkler brand	A	B	B	A	A
Sprinkler temperature	286	280	280	286	

to repeat a key test in the original rack storage test program. As indicated in Table 5.17, four repeat tests were conducted with 6.1 m (20 ft) high double row rack storage of metal lined double triwall commodity protected by 1/2 in orifice sprinklers discharging approximately 12.2 L/min-m² (0.30 gpm/ft²) from the 30 ft (9.14 m) ceiling. Three different 138–141 °C (280–286 °F) sprinklers were used and the variation in water demand area ranged from 232–353 m² (2500–3800 ft²). The more effective sprinkler was the one that generated more of a downward directed spray pattern that appeared to provide better penetration.

A given discharge density delivered from a large orifice sprinkler will be more effective than from a smaller orifice sprinkler. This effect was demonstrated by a pair of tests in the Plastics Storage test program (Dean, 1975); the test with a 17/32nd in orifice sprinkler produced a significantly lower water demand than the equivalent test with a 1/2 in orifice sprinkler.

The demonstrated advantage of the larger orifice sprinklers has spurred the recent development of several new large K factor sprinklers for warehouse storage applications. Table 5.18 lists the warehouse storage ceiling sprinklers currently manufactured in the US. The K factors range from 5.6 gpm/psi$^{1/2}$ to 25 gpm/psi$^{1/2}$, with the corresponding orifice sizes ranging from 13 mm to 24 mm. Most of the sprinklers have a maximum coverage area of 9.3 m² (100 ft²) per head for storage applications.

A recent unequivocal demonstration of sprinkler orifice size effects is the comparison tests between Extra Large Orifice (16.3 mm) sprinklers, and 17/32nd inch (13.5 mm) orifice sprinklers (Troup, 1998), and between the new K 25 (25.4 mm orifice) sprinklers and smaller orifice sprinklers (Troup and Vincent, 2001). Some results are shown in Figures 5.21a and b. Results for sprinkler demand area indicate that F_{spr} values for the ELO (K 11) sprinkler are in the range 0.50 to 0.67 compared to the smaller orifice sprinklers. The corresponding values for the K 25 sprinkler are 0.14 and 0.44. The other data shown in Figure 5.21 are the numbers of pallet loads burned in each test. In each case, the K 11 and K 25 sprinklers reduced the burn damage by at least 70% compared to K 8 or K 5.6 sprinklers!

The need to account for differences in sprinkler performance with the same discharge density has motivated inclusion of sprinkler K factor specifications in some of the NFPA sprinkler requirements for warehouse storage. For example, requirements for rack storage of plastic are given in

Table 5.18. Warehouse ceiling sprinklers

Sprinkler Type	Orifice diameter (mm)	Nominal K-factor[a] (gal/min-psi$^{1/2}$) [liter/min-kPa$^{1/2}$]	Minimum design pressure (psig)[a] [bar]	Minimum design flow per sprinkler (gpm) [lpm]	Maximum coverage area per sprinkler (ft^2) [m^2]
Half inch	12.7	5.6 [8.1]	7 [0.5]	15 [56]	100 [9.3]
17/32nd inch	13.5	8.0 [11.5]	7 [0.5]	21 [80]	100 [9.3]
Extra Large Orifice (ELO)	16.3	11.2 [16.1]	10 [0.7]	35 [134]	100 [9.3]
Large drop	16.3	11.2 [16.1]	25 [1.7]	56 [212]	130 [12.1]
ESFR K 14	17.8	14.0 [20.1]	50 [3.4]	99 [375]	100 [10.7]
ESFR K 17	19.5	16.8 [24.1]	35 [2.4]	99 [375]	100 [9.3]
ESFR K 25	23.6	25.0 [36]	15 [1.0]	98 [370]	100 [9.3]
K 17	19.5	16.8 [24.1]	Depends on Model Listing	Depends on Model Listing	100 [9.3]
K 25 extended coverage[b]	23.6	25.0 [36]	Depends on Model Listing	Model Listing	196 [18.2]

[a]Based on NFPA 13, 1999
[b]2001 Approved/ Listed in 2001

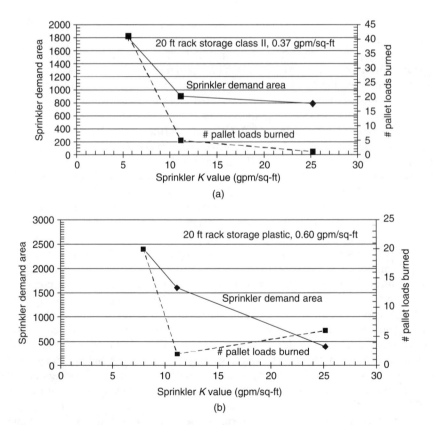

Figure 5.21. (a) Effect of K value on sprinkler performance for Class II rack storage; (b) Effect of K value on sprinkler performance for rack storage of plastic

terms of tabulated point values rather than area-density curves represented by equation (5.14). Table 5.18 summarizes the NFPA 13/231C requirements in terms of the alternative protection options for 20 ft high storage in a 25 ft high building, up to 25 ft high storage under a 30 ft high ceiling, and up to 35 ft storage under a 40 ft ceiling. In the case of the 20 ft high storage, use of the K factors of 11 or larger allows a 50% reduction in the water demand area, in accord with the previously cited test data. When storage heights exceed 20 feet, the sprinkler designer has the option of using either ESFR sprinklers or a combination of in-racks and conventional control mode ceiling sprinklers. Although the water supply requirements for ESFR sprinklers are specified in NFPA 13 in terms of a minimum discharge pressure and a minimum number of operating sprinklers, they have been converted to an equivalent water discharge density and demand area.

Since the new large orifice sprinklers are not yet manufactured in Europe, the CEN draft standard peEN 12845 does not recognize differences in performance among sprinklers. Ceiling sprinkler area-density tabulations are provided for various combinations of storage height and storage categories. For example, unexpanded plastic rack storage heights in the range 3.8–4.4 m (12.5–14.4 ft) require a discharge density of 30 mm/min (0.74 gpm/ft^2) over 300 m^2 (3200 ft^2). The corresponding NFPA 13/231C required density and demand area for up to 15 ft high plastic storage in racks (with as much as 10 ft clearance) are 0.60 gpm/ft^2 (24.4 mm/min) over 4000 ft^2. Storage heights greater 4.4 m require in-rack sprinklers in the CEN draft standard.

5.5.2 IN-RACK SPRINKLERS

In-rack sprinklers are usually conventional sprinklers equipped with a special water shield to prevent link wetting by water discharged from above. Figure 5.22a is a photograph of an upright in-rack sprinkler, and Figure 5.22b shows a pendent in-rack sprinkler equipped with a surrounding guard to prevent damage during commodity loading and unloading. In some cases the water shield is rigidly attached to the sprinkler, and in other cases separate shields/guards are installed over the sprinklers.

By situating these sprinklers within the rack itself, they are able to deliver water to commodity that would otherwise not receive sufficient water because of the combined effects of obstructions by upper tier racks and the difficulty of droplet penetration through the fire plume and ceiling jet. Hence in-rack sprinklers are often installed in high rack storage applications with high-challenge commodities. Table 5.19 lists several NFPA 13/231C required combinations of in-rack

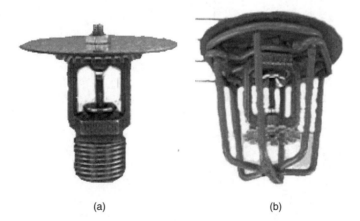

Figure 5.22. (a) Upright in-rack sprinkler; (b) Pendent in-rack sprinkler with guard (from Central/Tyco brochure)

Table 5.19. NFPA 13/231C sprinkler protection options for rack storage of Group A plastic commodities

Maximum storage height (ft)	Maximum ceiling height (ft)	Sprinkler K factor (gpm/psi$^{1/2}$)	Minimum discharge density (gpm/ft^2)	Minimum demand area (ft^2)	In-rack sprinklers required?	Restrictions
20	25	≤8.0	0.60	4000	No	Not for multiple row racks
		≤8.0	0.30	2000	Yes: One level in longitudinal flue	—
		11, 17, or 25	0.60	2000	No	Not for multiple row racks
25	30	≤11.0	0.45	2000	Yes: one level	—
		≤11.0	0.30	2000	Yes: Two levels	—
		17 (UL listings)	0.80	2000	No	FM does not recognize this application
		ESFR: K = 14, 17, or 25	1.0 ± 0.02^a	1200^b	No	Ceiling slope and obstruction restrictions
35	40	ESFR K = 14, 17, or 25	1.22 ± 0.02^a	1200^b	No	Ceiling restrictions. No expanded plastics
		≤11.0	0.45	2000	Yes: Multiple levels in longitudinal flue and on aisle face	—

^{a}Effective densities based on specified minimum pressure requirements and 100 ft^2 coverage per sprinkler
^{b}Minimum of 12 operating sprinklers with a maximum coverage area per sprinkler of 100 ft^2

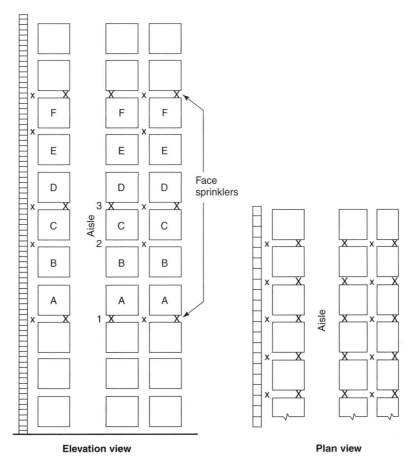

Figure 5.23. NFPA 13/231C in-rack sprinkler requirements for high rack storage of plastic (from NFPA 13). (1) Sprinklers labeled 1 shall be required where loads labeled A or B represent top of storage. (2) Sprinklers labeled 1 and 2 shall be required where loads labeled C represent top of storage. (3) Sprinklers labeled 1 and 3 shall be required where loads labeled D or E represent top of storage. (4) For storage higher than loads labeled F, the pattern for locating in-rack sprinklers as specified by Notes 2 and 3 shall be repeated. (5) X represents face and in-rack sprinklers. (6) Each square represents a storage cube that measures 4–5 ft (1.2–1.5 m) on a side. Actual load heights can vary from approximately 18 in to 10 ft (0.46–3.1 m). Therefore, there can be one load to six or seven loads between in-rack sprinklers that are spaced 10 ft (3.1 m) apart vertically

sprinklers and ceiling sprinklers for rack storage of Group A plastic at heights of 25–35 ft. Plastic commodity storage heights above 35 ft depend almost entirely on in-rack sprinklers. The CEN draft standard insists on in-rack sprinklers for plastic (Category 3) storage higher than 14.4 ft.

The parameters governing the specification of in-rack sprinklers are their design pressure (or flow rate), the design number of flowing sprinklers, their spacing horizontally and vertically (i.e. tier spacing), and their location relative to the longitudinal and transverse flue spaces and the aisle. Figure 5.23 shows the NFPA 13/231 required in-rack sprinkler locations for double row rack storage at heights greater than 25 ft. In multiple row rack applications, there are also requirements for solid barriers to restrict vertical flame spread to upper tiers. In both NFPA 13/231C and CEN prEn 12845 the minimum required flow rate per in-rack sprinkler is 30 gpm. NFPA 13/231C requires a minimum of seven flowing sprinklers on each of the top two levels, i.e. a total of 14 flowing sprinklers if multiple levels are required. CEN prEN 12845 requires a minimum of three flowing sprinklers per level for as many as three levels, i.e. a total of nine flowing sprinklers if three or more levels are required.

In-rack sprinklers are often a burden to warehouse operations because of the need to provide water filled piping to the racks, and the associated threat of inadvertent damage to the sprinklers and water damage to commodities during stock loading/unloading. The installed piping restricts possible re-configuring of rack design and location. The desire to provide effective sprinkler protection without requiring in-rack sprinklers has led to the development of larger orifice ceiling sprinklers including sprinklers capable of providing early suppression of high challenge rack storage fires.

5.5.3 EARLY SUPPRESSION FAST RESPONSE (ESFR) SPRINKLERS

ESFR sprinkler minimum water supply requirements are specified in NFPA 13 as a 12-sprinkler water demand. The 12 sprinklers correspond to two concentric rings of operating sprinklers when ignition is centered below four heads. The minimum required flow rate per sprinkler head depends on the storage and ceiling, as indicated in Table 5.19. The original ESFR sprinklers had a K factor of 14 gpm/(psig)$^{1/2}$ (144 L/min/(bar)$^{1/2}$), and had a minimum required operating pressure of 50 psig (340 kPa) to protect up to 25 ft high rack storage of Group A plastic commodity at a ceiling height up to 30 ft. The design basis for this requirement was a series of eleven large-scale tests conducted with polystyrene cups in cartons such that the results should be applicable to all unexpanded cartoned plastic as well as Class I–IV commodities (Yao, 1988). Storage heights varied from 4.3–7.3 m (14–24 ft) and ceiling clearance varied from 1.2–4.9 m (4–16 ft). Ignition location relative to the sprinklers was also varied. The maximum number of sprinklers opened was 11, and that occurred in a failure mode test (one plugged sprinkler).

New lift truck capabilities allow rack storage up to 10.7 m (35 ft), or 40 ft high in many new warehouses. This new storage height capability raised some questions about ESFR sprinkler effectiveness for cartoned storage up to 10.7 m (35 ft) in warehouses with ceilings up to 12.2 m (45 ft) high. The new ESFR $K = 25$ gpm/(psig)$^{1/2}$ sprinklers were developed by sprinkler manufacturers and certified by Factory Mutual and Underwriters Laboratories based on a combination of Actual Delivered Density (ADD) testing (Chan et al., 1994) and large-scale fire tests at these storage and ceiling heights. The minimum design pressure for the 40 ft high storage is 40 psig (corresponding to a required flow rate of 158 gpm) for one approved/listed $K = 25$ ESFR sprinkler and 50 psig (corresponding to 177 gpm) for another model.

These developments in ESFR sprinkler protection are applicable to most cartoned commodities. They also raise questions about their applicability to other commodities such as exposed plastics (both expanded and unexpanded) and special commodities such as rolled paper, tires, flammable liquids in small containers, etc. One way to compare the challenge of these commodities (in the context of ESFR sprinklers) to that of cartoned plastics is to conduct Required Delivered Density tests (Yu, 1989) in a manner analogous to the Commodity Classification tests as shown in Plate 2. In situations where the special commodities are deemed to represent a greater challenge than the cartoned plastics, special protection may be warranted. Large-scale sprinklered fire tests are conducted to determine the nature of any special sprinkler protection requirements or storage restrictions. Current ESFR sprinkler protection restrictions in NFPA 13/231C are exclusion of exposed, expanded plastic for storage heights up to 25 ft, and exclusion of any expanded plastics at higher storage heights. Open-top containers are also excluded from ESFR protection. The challenges and protection requirements for several special commodities are described in Chapter 6.

WAREHOUSE STORAGE

5.6 Sprinklered warehouse fire modeling

Although a generalized fire model does not currently exist for sprinklered warehouse fire, there has been progress toward that eventual goal. This section provides an overview of the relevant phenomena needed for such a model, and a summary of the analyses, correlations, and sub-models available for dealing with key phenomena.

5.6.1 CONCEPTUAL MODEL OVERVIEW

Figure 5.24 is a flow chart of the essential components needed for a sprinklered warehouse fire model. Input for the model would consist of (a) descriptions of the stored commodities

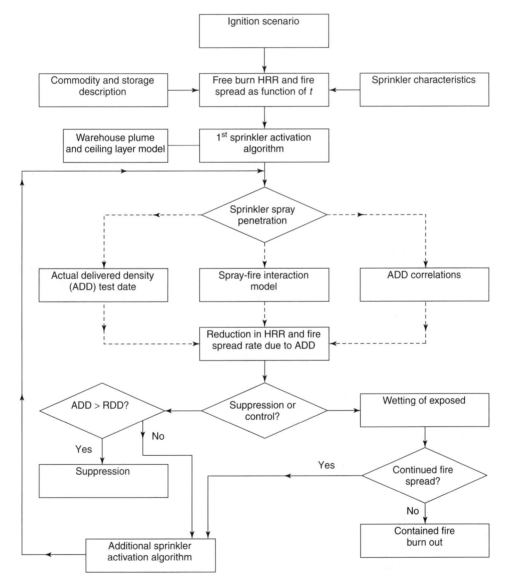

Figure 5.24. Sprinklered warehouse fire model flowchart

and storage/warehouse configuration, (b) the sprinkler characteristics and locations, and (c) the ignition location and initial fire size. Based on the ignition scenario and commodity/storage description, the free burn heat release rate and fire spread rate would be calculated as follows.

5.6.2 FREE BURN HEAT RELEASE RATES AND FLAME SPREAD RATES

There are two approaches to calculating free burn heat release rates and flame spread rates. The simplest approach is to use an empirical representation based on small array data correlations such as those described in Section 5.4.3. A more detailed and complex approach is to divide the exposed surface area of the storage array into numerous fuel surface elements with a given heat release rate per unit area. McGrattan *et al.* (2000) used a value of 600 kW/m² in their simulation of the prototype Group A plastic commodity. Each element starts burning when its surface temperature reaches the ignition temperature. McGrattan *et al.* calculated carton temperatures from local heat fluxes and a thermally thin representation and thermal parameter values described in Section 5.4.2. Alternatively, the surface temperature could be calculated from the following thermally thick representation Delichatsios (1993):

$$T_s - T_0 = \frac{1}{\sqrt{\pi k \rho c}} \int_0^t \frac{q''(\bar{t})}{\sqrt{t - \bar{t}}} d\bar{t} \qquad [5.6.1]$$

where q'' is the heat flux on the carton surface at time t, and the value of $k\rho c$ is obtained from the type of laboratory test data described in Section 5.4.2. Even if a primarily empirical prescription of the heat release history is used, equation [5.6.1] is useful for calculating the time of flame jump across the warehouse aisle as a result of radiative heating. In general, a numerical evaluation of equation [5.6.1] would be needed.

5.6.3 WAREHOUSE FIRE PLUMES AND CEILING JETS

A representation of the fire plume and ceiling jet is needed for the calculation of ceiling sprinkler actuation times. McGrattan *et al.* (2000) used a Computational Fluid Dynamic Large Eddy Simulation method with on the order of one million elemental control volumes for this analysis. However, the free burn plume and ceiling jet are amenable to relatively simple data correlation representations of the plume and ceiling jet illustrated in Figure 5.25. These correlations are qualitatively similar to the correlations described by Evans (1995), but are based on testing with warehouse storage arrays (Kung *et al.*, 1984).

The plume centerline temperature, T_c, and velocity, U_c, impinging on the ceiling is given by

$$T_c - T_0 = [Ag^{-1/3} c_p \rho_a T_a] Q_c^{2/3} (H - z_0)^{-5/3} \qquad [5.6.2]$$

$$U_c = 4.25(AQ_c)^{1/3}(H - z_0)^{-1/3} \qquad [5.6.3]$$

where $A = g/(c_p T_\infty \rho_\infty) = 0.0279 \, \text{m}^4 \, \text{kJ}^{-1} \, \text{s}^{-2}$, T_0 is the ambient temperature in the warehouse (°C), H is the ceiling clearance measured from the top of fuel array (m), Q_c is the fire convective heat release rate (kW) at time t, and Z_0 is the plume virtual origin, which is given by

$$z_0 = z_{00} + 0.095 Q_c^{2/5} (z_0 \text{ in m}, Q_c \text{ in kW}) \qquad [5.6.4]$$

where $z_{00} = -1.6$ m for two-tier rack storage and -2.4 m for three-tier and four-tier storage (Kung *et al.*, 1984).

Ceiling jet temperatures and velocity, U, correlations suggested by Kung *et al.* (1984) for warehouse storage fires are:

$$\Delta T / \Delta T_c = \exp\{-0.66[(r/b_c) - 1.5]^{1/2}\}, \quad \text{for } r/b_c > 1.5 \qquad [5.6.5]$$

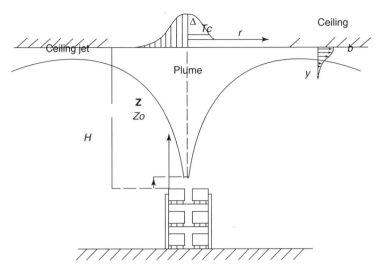

Figure 5.25. Warehouse fire plume and ceiling layer

and

$$U/U_c = \exp\{-0.44[(r/b_c) - 1.5]^{0.57}\}, \quad \text{for } r/b_c > 1.5 \quad [5.6.6]$$

where r is the radial distance from plume centerline to ceiling location, $\Delta T_c = T_c - T_0$, and b_c is the half-width of the plume at ceiling level.

The plume half-width (radius where velocity is one-half the centerline value) is given by (Kung et al., 1984)

$$b_c = 0.107(H - z_0)[1 + 0.106 Q_c^{2/3}(H - z_0)^{-5/3}]^{1/2} \quad [5.6.7]$$

The plume centerline temperature rise at ceiling level can be obtained by substituting $r = 0$ in equation [5.5.1]. The result is

$$\Delta t_c = (9.2 T_\infty/g)(A Q_c)^{2/3}(H - z_0)^{-5/3} \quad [5.6.8]$$

More general non-steady plume formulations have been developed by Delichatsios (1983) and by You (1989). Their results indicate that the preceding quasi-steady equations should be applicable for

$$\frac{(H - z_0)}{U_c} \left(\frac{1}{Q_c} \frac{dQ_c}{dt} \right) \gg 1$$

In particular, quasi-steady temperatures and velocities should be accurate to within 10% when the nondimensional parameter above is less than or equal to about 0.10. When this criterion is not satisfied, either a non-steady plume model is needed, or the quasi-steady correlations need to be adjusted to account for the gas travel delay time $(H - z_0)/U_c$, such that the instantaneous value of ceiling gas temperatures and velocities are based on the value of Q_c at $t - (H - z_0)/U_c$. The latter approach, which has been described by Evans (1995) for a t^2 fire development, can be called a Time Adjusted Quasi-Steady formulation.

One other special concern in warehouse fire ceiling jets is the location of ceiling sprinkler links so that they experience near-maximum temperatures and velocities. Measurements by Kung et al. (1984) indicate that the maximum temperatures and velocities occur at a depth of about

10 cm (3.9 in) below the ceiling, and that the temperature difference, ΔT_y, and velocity, U_y, at a distance y (cm) below the ceiling are given by

$$\Delta T_y/\Delta T = \exp\{-[(y - 10.2)^2/\delta_t^2]\} \quad [5.6.9]$$

and

$$U_y/U = \exp\{-[(d - 10.2)^2/\delta_v^2]\} \quad [5.6.10]$$

where

$$\delta_t = b_c[0.32 + s(r/b_c - 3.1)] \quad [5.6.11]$$

$$\delta_v = 0.67\delta_t \quad [5.6.12]$$

$$s = -0.0097 - 0.0013(H/1.32) + 0.0059(H/1.32)^2 \; (H \text{ in m}) \quad [5.6.13]$$

The temperature profile represented by equation [5.6.9] is shown schematically in Plate 3, and is plotted in Figure 5.20 for $H = 5$ m (15.4 ft) and two pairs of values of Q_c and r. The temperature rise in Figure 5.20 is reduced to one-half of its maximum value at a depth of 18–35 cm (7–14 in).

More recent warehouse fire test data obtained by Yu and Stavrianidis (1989) indicate that

$$\frac{\delta_T}{H - z_0} = 0.07 \quad \text{for} \quad \frac{r}{H - z_0} < 0.5, = 0.10 \quad \text{for} \quad \frac{r}{H - z_0} > 0.9$$

More general (and complex) correlations for δ_T and δ_v are given by Yu and Stavrianidis (1989).

5.6.4 SPRINKLER ACTUATION MODEL

The ceiling sprinkler link is subjected to convective heating by the hot ceiling jet flowing past it, and is cooled by heat conduction to the sprinkler frame and attached piping. The heat balance equation derived by Heskestad and Bill (1998) for the sprinkler link temperature, T_L, at time t is

$$\frac{dT_L}{dt} = \frac{\sqrt{u}}{RTI}(T_g - T_L) - \frac{C}{RTI}(T_L - T_f) \quad [5.6.14]$$

where T_g and u are the local ceiling jet temperature and velocity at the link position and depth beneath the ceiling, RTI is the sprinkler RTI measured in a plunge tunnel test, C is the link conduction parameter, and T_f is the sprinkler frame temperature.

In the general case of an arbitrary heat release rate history and significant radiant heating of the sprinkler link by the fire plume or flame, equation [5.6.14] requires numerical solutions via a computer. Two computer codes for this purpose have been developed by Yu (1992) at FMRC and by Sleights (1993) at WPI. Both codes account for conductive heat losses from the sprinkler link as well as the other effects described above. The Yu/FMRC code is based on proprietary data correlations and is not publicly available. The Sleights/FMRC code contains several alternative previously published ceiling jet correlations including the time adjusted quasi-steady (TAQS) correlations described here.

Sleight's compared calculated (using the TAQS correlations) and measured heat release rates at sprinkler activation for several dozen tests involving Class II and Group A plastic commodities with various storage heights. The overall difference for the numerous simulations shows the calculated sprinkler activation heat release rate on-average only 5.6% lower than the measured value. The root-mean-square difference was 19%. This agreement is encouraging in view of the complicated heat release rate histories and the various difficult-to-measure sprinkler parameters needed for calculations that include radiative and conductive heat transfer to/from the sprinkler link.

McGrattan et al. (2000) used equation [5.6.14] to determine multiple sprinkler actuation times in their Industrial Fire Simulator computer code. Local gas velocities and temperatures were determined from their CFD simulations, including the effects of water spray cooling of the plume and ceiling jet. Their calculated sprinkler actuation times agreed with large-scale data to within 5–10 seconds for the first four sprinklers surrounding the ignition site, but the second ring sprinklers actuated 15–30 seconds after the calculated times. Thus, the modeling of water spray effects is much more difficult and much less accurate than the modeling of the free burn plume and ceiling layer.

5.6.5 SPRAY-PLUME PENETRATION MODEL

The objective of a warehouse sprinkler spray-plume penetration model is to calculate the spray density that reaches the top of the storage array after passing through the flame and plume. This density is called the Actual Delivered Density (ADD), and depends upon both the spray characteristics and the size and location of the fire relative to the ceiling sprinklers. Chan et al. (1994) obtained the following ADD correlations for $\frac{1}{2}$ in orifice ($K = 5.6$ gpm/psi$^{1/2}$) and 17/32 in orifice ($K = 5.6$ gpm/psi$^{1/2}$) sprinklers 3 m above a fire plume simulator.

For a fire directly under one flowing sprinkler, the penetration, P, ratio of the ADD to the local spray density in the absence of any fire, is

$$P = C_1 \left(\frac{(\rho_w g)^{2/3}}{(\rho_a \mu_r)^{1/3}} \frac{D}{U_c} \right)^{0.75} \qquad [5.6.15]$$

where D is the sprinkler orifice diameter, and $\rho_a \mu_r$ evaluated at the plume maximum temperature $= 7.63 \times 10^{-6}$ kg^2/m^4-s, and $C_1 = 0.10 \pm 0.01$. When the fire plume was situated between either two sprinklers or four sprinklers, Chan et al. obtained the following correlation:

$$P = C_1' \left(\frac{g^{2/3}(\rho_w \sigma)^{1/3} D^2}{(\rho_a \mu_r)^{1/3} U_c Q_w^{2/3}} \right)^{C_2} \qquad [5.6.16]$$

where Q_w is the water flow rate, σ is the surface tension of water, and the empirical constants were found to have the following values:

For a fire situated between two sprinklers, $C' = 0.41 \pm 0.04$, and $C_2 = 0.85$.
For a fire situated between four sprinklers, $C' = 0.46 \pm 0.01$, and $C_2 = 1.0$.

In principle, the ADD can also be calculated from a theoretical model of the spray-plume interaction, if the spray drop size and velocity distribution is known from measurements of the type reported by Chan (1994). CFD model calculations have been conducted with varying degrees of success and different CFD codes by Alpert (1985), Bill (1993) and Nam (1996). Each calculation involved following several hundred individual drop trajectories and calculating the gas phase flow field at several thousand computational grid cells. The McGrattan et al. simulations in their Industrial Fire Simulator typically involved five to ten thousand drops. All the CFD computations are both computer and labor intensive, and are still more of a research project than a routine engineering computational tool. Like the ADD measurements, the calculations can be very useful for sprinkler design and development.

5.6.6 REDUCTION IN HEAT RELEASE DUE TO ACTUAL DELIVERED DENSITY

As with the free burn commodity fire tests, water application tests have been conducted both with laboratory scale materials and small storage arrays. The laboratory scale tests are conducted with different types of water applicators placed under a Cone Calorimeter or similar apparatus with an imposed radiant heat flux. Tewarson (1995) found that the reduction in heat release rate was linearly proportional to the applied water density, i.e.

$$\dot{Q}'' - \dot{Q}_0'' = -\kappa \dot{m}_w'' \qquad [5.6.17]$$

where $\dot{Q}_0''$ is the free burn heat release rate per unit surface area at the same imposed heat flux, $\dot{m}_w''$ is the water application rate per unit burning surface area, and the proportionality constant, κ, depends upon the material and the efficiency of water application including the possible presence of water film formation on the burning surface. Tewarson's data for cellulosic filter paper exposed to heat fluxes in the range 25–50 kW/m² are equivalent to a proportionality constant of about 9.0 kJ/g. This value is in good agreement with Tewarson's theoretical derivation which suggests, in the absence of water film formation,

$$\kappa = \varepsilon_w \frac{\Delta H_{ch}}{\Delta H_g} \Delta H_w \qquad [5.6.18]$$

where ε_w is the efficiency of water application, ΔH_{ch} is effective chemical heat of combustion of the material, ΔH_g is its heat of gasification, and ΔH_w is the heat of gasification of water, 2.6 kJ/g.

Hietanemi et al. (1999) reported on Cone Calorimeter experiments in which water spray application caused a reduction in heat release rate for only some of the materials tested. Specifically, the heat release rates were reduced for polypropylene and for nylon 66, but were increased for tetramethylthiuram monosulfide and for chlorobenzene. Hietanemi et al. suggest that the increase in heat release rate for the latter two materials might have been due to water spray deformation of the flame such that the flame surface area increased. Although there was no explanation as to why this effect would occur with only some materials, one might speculate that it could be associated with differences in char formation and flame structure. In any case, this effect would not be expected with a more gentle method of water application.

Yu et al. (1994) conducted Fire Products Collector tests to determine how water application would affect the heat release rate for burning Class II and Group A plastic commodities. Their observed reductions in heat release rate, $\dot{Q}$, satisfied the equation

$$\dot{Q} = \dot{Q}_0 e^{-k(t-t_0)} \qquad [5.6.19]$$

where $\dot{Q}_0$ is the heat release rate at the time of water application, t_0. The exponential factor k in equation [5.6.19] was correlated as

$$k = c_1 \dot{m}_w'' - c_2 \qquad [5.6.20]$$

$c_1 = 0.536$ and $c_2 = 0.0040$ for Class II commodity, when $0.006 < \dot{m}_w'' < 0.024$ kg/m²-s, $c_1 = 0.716$ and $c_2 = 0.0131$ for Group A plastic commodity when $0.012 < \dot{m}_w'' < 0.041$ kg/m²-s.

The water application density, $\dot{m}_w''$, used to obtain these values was calculated by assuming the water applied to the top of the burning array was distributed uniformly over all the exposed surfaces of the storage array. Yu et al. also presented a theoretical derivation for the exponential constant k that showed it should vary inversely as the depth of the fuel layer undergoing pyrolysis at the time of water application, i.e. fuels with deep pyrolysis layers would be less affected by water spray application than fuels with thin pyrolysis layers.

Hamins and McGrattan (1999) extended equation [5.6.19] to account for fires that slowly redevelop after an initial decrease in heat release rate at water application. They correlated data for boxes of Group A Plastic as

$$\dot{Q} = \dot{Q}_0 [e^{k_1(t-t_0)} + k_2(t-t_0)] \qquad [5.6.21]$$

where

$$k_1 = -M_w'' = -\dot{w}_w'' \frac{A}{U_w P} \quad [5.6.22]$$

and

$$k_2 = -0.020 \, M_w'' + 0.0015 \, s^{-1} \quad [5.6.23]$$

The new parameters in equation [5.6.22] are M_w'', the mass of water applied per unit exposed surface area, $\dot{w}_w''$, the delivered water flux density at the top of the commodity, A, the top surface area of a unit load of Group A plastic, U_w, the speed of a water film flowing down the side of the commodity (observed as 0.54 m/s), and P, the perimeter of a unit load of commodity. The data correlation coefficients associated with equations [5.6.22] and [5.6.23] were quite low, raising some doubt as to their general validity and accuracy beyond the actual test conditions.

5.6.7 FIRE CONTROL CRITERIA: CAN WETTED COMMODITY BE IGNITED?

Fire control is normally defined as limited fire spread due to the applied sprinkler spray wetting the exposed unburned fuel surfaces near the fire. This occurs both because the heat release rate and associated flame height is reduced, and because the wetted exposed surfaces are more difficult to ignite. Tewarson (1995) and Hamins et al. (1998) have presented data to show how the time to autoignition of increases with the mass of water on the surface of a material being tested under a Cone Calorimeter type apparatus. The Hamins et al. data for corrugated paper at two heat fluxes are shown in Figure 5.26. The theoretical lines in Figure 5.26 are based on the assumption that the ignition delay time is the time required to evaporate a mass of water, M_w'' per unit surface area in and on the sample, i.e.

$$t_i - t_{iq0} = \frac{M_w'' \Delta H_w}{q_{ex}''} \quad [5.6.24]$$

where t_{iq0} is the time to ignition at the imposed heat flux, q_{ex}'' in the absence of any water. The data are in relatively good agreement with equation (5.38).

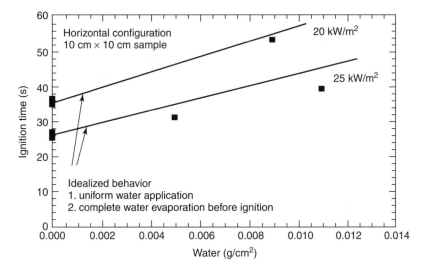

Figure 5.26. Time to ignition for wetted corrugated paper (from Hamins et al., 1999)

In the case where the water is being continuously applied to the fuel surface as it is exposed to the heat flux, the critical heat flux for ignition would be expected to increase by an amount needed to evaporate the water being applied. Thus,

$$q''_{crw} = q''_{cr} + \dot{m}''_w \Delta H_w \qquad [5.6.25]$$

where q''_{crw} and q''_{cr} are the critical heat fluxes for ignition with water and without water, respectively.

By comparing the calculated value of q''_{crw} from equation [5.6.25] to the value of radiant heat flux across the aisle of a burning storage array, it should be possible to determine whether or not the sprinkler discharge will prevent the flame from jumping the aisle.

5.6.8 FIRE SUPPRESSION CRITERIA

Tewarson (1995) measured the water flux application rate, $\dot{m}''_{ws}$, needed to extinguish polymer slab fires exposed to various heat fluxes, $\dot{q}''_{ex}$, in the FMRC Flammability Apparatus. He correlated his suppression data in the form

$$\dot{m}''_{ws} = c_1 \dot{q}_{ex} + c_2 \qquad [5.6.26]$$

Values of c_1 and c_2 determined for various horizontal and vertical oriented polymers are shown in Table 5.20.

The values of c_1 for vertical PMMA and polyoxymethylene surfaces are approximately equal to the inverse of $\Delta H_w = 2.6$ kJ/g, suggesting that all the water is being vaporized during suppression. The values of c_1 for the horizontal surfaces are approximately half ΔH_w^{-1}, suggesting that some of the water is accumulating on the surface and reducing the heat flux reaching the polymer surface.

Tewarson (1995) also calculated the value of $\dot{m}''_{ws}$ needed to immediately suppress asymptotically large burning surfaces by calculating the value of flame heat flux from such large flames. His value for PMMA is 26 g/m²-s, and for polystyrene it is 34 g/m²-s. Since these values correspond to delivered densities of only 0.034 gpm/ft² and 0.043 gpm/ft², it appears that most water sprinkler sprays are relatively inefficient in the sense that only a small fraction of the discharged water actually vaporizes on the burning fuel surfaces.

Yu et al. (1984) used equations [5.6.19] and [5.6.20] to determine the minimum delivered water fluxes required to achieve suppression of prototype Class II and Group A Plastic commodities four minutes after initiation of water discharge. Their calculated values were 6 g/m²-s for the prototype Class II commodity, and 17–20 g/m²-s for the prototype Group A plastic commodity. Their value for Group A plastic is approximately half the value calculated by Tewarson for immediate suppression of polystyrene. This is consistent with the four minute suppression criterion used by Yu et al. to determine their minimum required water flux rates.

A reality check on the required water fluxes needed to achieve suppression when applied to the top of the storage array can be obtained by recalling the values of Required Delivered

Table 5.20. Constants in correlation for minimum water flux for suppression (from Tewarson, 1995)

Polymer – orientation	c_1 (g/kJ)	c_2 (g/m²-s)
Polymethylmethacrylate – vertical	0.37	1.67
Polymethylmethacrylate – horizontal	0.22	1.56
Polyoxymethylene – vertical	0.42	1.97
Polyoxymethylene – horizontal	0.24	2.08
Polystyrene – horizontal	0.22	3.1

Density needed to achieve suppression of 3-tier, 4-tier, and 5-tier high rack storage of Group A plastic commodity. These RDD values are in the range 0.30–0.50 gpm/ft^2, when the water is discharged at times corresponding to ESFR sprinklers actuating on a 30 ft high warehouse ceiling. If the warehouse fire model does not explicitly model suppression in the interior of the storage array, these would be the values needed to assess the conditions needed for suppression. On the other hand, if a detailed model of the interior of the array is included in a model such as the NIST Industrial Fire Simulator, than equation [5.6.26] and Table 5.20 would be more appropriate criteria for fire suppression.

5.7 Cold storage warehouse fire protection

Cold storage warehouses present two special fire protection challenges. One challenge is the presence of foamed insulation wall and ceiling construction. The second challenge is the need for dry pipe sprinkler systems to protect large storage arrays of combustible items.

The May 3 1991 fire in a cold-storage warehouse complex in Madison, Wisconsin is indicative of the challenges posed by both the storage and insulation. As reported by Isner (1991), the fire originated in a 55 ft high building constructed with floor-to-ceiling storage racks. The roof assembly, which was supported by the rack structure, consisted of metal decking under foamed insulation covered by tar and gravel. Three separate storage areas, with storage temperatures ranging from $-23\,°C$ ($-10\,°F$) to $1\,°C$ ($+34\,°F$), had walls of foam insulation between metal sheathing. The palletized rack storage commodities consisted of 13 million pounds of butter, and an assortment of other refrigerated foods including meat, poultry, cheese, cranberries, and various vegetables. The storage area was protected by a dry-pipe ceiling sprinkler system designed to discharge 0.15 gpm/sq-ft. There were no in-rack sprinklers.

When Madison firefighters responded to the alarm at the warehouse, they found the ceiling sprinklers operating over 'an apparent pile of rubble engulfed in yellow flames' in one of the freezer storage areas. There was also a separate roof fire, which spread from an initial area of about 150 ft^2, to involve most of the 260 × 170 ft roof in that section of the building. Concern for potential rack storage collapse, which in fact occurred 47 minutes after arrival, limited manual firefighting to hosestreams discharged from outside the storage area. Plate 4 is a photograph of the melting butter and flames after the collapse of the rack structure and an exterior wall. The fire eventually spread to two other storage buildings and caused a loss in excess of $100 million. The most important deficiency in the Madison, WI warehouse was the absence of dry-pipe in-rack sprinklers, and the inadequate ceiling sprinkler discharge density.

Design of a dry-pipe sprinkler system for a cold-storage warehouse presents several challenges besides installing in-rack sprinklers. The design numbers of flowing in-rack sprinklers and ceiling sprinklers must be increased to account for the delay time between sprinkler link actuation and the arrival of water at the open sprinklers. NFPA 13 allows a delay time of as long as 60 seconds, which can cause a rack storage fire to increase by a factor of two-to-four beyond the size at which a wet pipe system would begin discharging water. There are also serious reliability issues associated with the dry pipe valves, the normally dry piping in the refrigerated area, and the dry-pipe sprinkler heads. The dry pipe reliability issue is the mechanical complexity of many dry pipe valves that render them vulnerable to hang-up when not regularly maintained and tested. The reliability issues associated with the dry piping and sprinkler heads are their propensity to corrosive deterioration and to be occluded by condensation and freezing of moisture in the air-filled piping. The former issue can be mitigated by periodic sample testing of sprinkler heads (UL testing has indicated that more than half the heads did not discharge water at the minimum sprinkler operating pressure), and replacement of obstructed heads. The latter issue can be mitigated by using dry air with a dew-point substantially lower than the storage temperature.

The March 1992 Missouri cold-storage warehouse fire started when grease residue on an insulated wall was ignited during hot cutting operations. Workers used portable extinguishers on the fire, but it continued to spread up the wall and ceiling, which were insulated with steel-faced expanded polystyrene. There was no automatic sprinkler protection in the 25 ft high warehouse, which was fully loaded with palletized frozen foods. The fire eventually involved an 180,000 ft^2 section of the warehouse between two firewalls (*NFPA Journal*, November/December 1993). The bursting of an ammonia refrigeration line during the fire jeopardized the manual firefighting operations. The loss was estimated to be $100,000.

The most tragic cold-storage warehouse fire occurred in November 1999 in Worcester, Massachusetts. After being shutdown for several years, the warehouse was occupied by a pair of homeless people who inadvertently started a fire by knocking over a burning candle. Unfortunately, the homeless couple left the warehouse without reporting the fire. When Worcester firefighters responded, they entered the warehouse in search of any occupants. A sudden, rapid increase in fire size caused visibility to deteriorate while six firefighters were disoriented and trapped inside. The firefighters perished in the unsprinklered warehouse, which was insulated with an assortment of highly combustible wall/ceiling insulation.

A 1997–1998 series of fire tests in the FM 25 ft corner test facility has provided valuable information on the contribution of different types of insulation to cold-storage room fires. Insulation panels included steel-faced expanded polystyrene in thickness varying from 102–254 mm (4–10 in), and polyisocyanurate insulated steel deck roofing. The thick expanded polystyrene panels presented the greatest challenge, and required discharge densities well in excess of 0.2 gpm/ft^2, even when wet pipe systems were used. One effective cold room sprinkler configuration described in the 1998 FMRC Update article (Vol. 12, No. 2) consisted of dry pendant heads at the bottom of short dry pipe extensions from a wet pipe system installed in a nearby wet system in a heated area of the warehouse. Readers are advised to consult FM Data Sheets and Approval Guides for updated specifications of recommended discharge densities and areas, as well as for listings of approved insulation assemblies not requiring this level of protection.

References

Alpert, R., Numerical Modeling of the Interaction Between Automatic sprinkler Sprays and Fire Plumes, *Fire Safety Journal*, **9**, 157–163, 1985.

ASTM E 162, 'Test Method for Surface Flammability of Materials Using a Radiant Energy Source,' American Society for Testing and Materials, 1983.

ASTM E 906, 'Standard Test Method for Heat and Visible Smoke Release Rates for Materials and Products,' American Society for Testing and Materials, 1987.

ASTM E 1321, 'Standard Test Method for Determining the Material Ignition and Flame Spread Properties,' American Society for Testing and Materials, 1990.

ASTM E 1354, 'Standard Test Method for Heat and Visible Smoke Release for Materials and Products Using an Oxygen Consumption Calorimeter,' American Society for Testing and Materials, 1994.

Babrauskas, V., Burning Rates, Section 3, Chapter 1, *The SFPE Handbook of Fire Protection Engineering*, SFPE, NFPA, 1995.

'Before the Fire: Fire Prevention Strategies for Storage Occupancies', National Fire Protection Association Ad Hoc Task Force, 1988.

Bill, R.G., Numerical Simulations of Actual Delivered Density Measurements, *Fire Safety Journal*, **17**, 227–240, 1993.

Buckley, J.L., 'Stored Commodity Test Program: Part III – Commodity Classification,' FMRC JI 0N0R4.RU/ 0N1J8.RU, Prepared for the Society of the Plastics Industry, April 1988.

CEN, 'Automatic Sprinkler Systems, Design and Installation and Maintenance,' prEN 12845 (Draft), European Committee for Standardization, April 2001.

Chan, T.-S., Measurements of Water Density and Drop Size Distributions of Selected ESFR Sprinklers, *J. Fire Protection Engineering*, **6**, 79–87, 1994.

Chan, T.-S., Kung, H.C., Yu, H.-Z. and Brown, W., Experimental Study of Actual Delivered Density for Rack-Storage Fires, *Proceedings Fourth Intl. Symp. on Fire Safety Science*, pp. 913–924, 1994.

Chicarello, P.J. Troup, J.M.A. and Dean, R.K., 'Large-Scale Fire Test Evaluation of Early Suppression Fast Response Sprinkler Applications,' Prepared for the National Fire Protection Research Foundation, FMRC JI 0N1E7.RR, 1986.

Dean, R., 'Stored Plastics Test Program,' FMRC JI 20269, June 1975.

Dean, R.K., 'Investigation of Conditions Potentially Affecting Rack Storage Fire Severities,' Factory Mutual Research Corporation J.I. 0E0J1.RR, October 1980.

Dean, R., 'Stored Commodity Test Program: Part II – Large-Scale Tests of Cartoned Polyurethane Foam,' FMRC JI 0N0J7.RR, March 1987.

Delichatsios, M.A., A Scientific Analysis of Stored Plastic Fire Tests, *Fire Science and Technology*, **3**, 73–103, 1983.

Delichatsios, M. and Chen, Y., Asymptotic, Approximate, and Numerical Solutions for the Heatup and Pyrolysis of Materials Including Reradiation Losses, *Combustion and Flame*, **92**, 292–307, 1993.

Evans, D., Ceiling Jet Flows, *SFPE Handbook of Fire Protection Engineering*, Chapter 2-4, SFPE, NFPA, 1995.

Factory Mutual Loss Prevention Data Sheet 2-2, 'Early Suppression Fast Response Sprinklers,' May 1988.

Factory Mutual Engineering Corporation Loss Prevention Data Sheet 8-0S, 'Commodity Classification,' 1979.

Factory Mutual Data Sheet 8-9, 'Storage of Plastics and Elastomers,' 1981.

Factory Mutual Loss Prevention Data Sheet 8-33, 'Carousel Storage and Retrieval Systems,' 1998.

Factory Mutual Loss Prevention Data Sheet 8-34, 'Protection for Automatic Storage and Retrieval Systems,' 1998.

FMRC Update, 'Advances in Commodity Classification, A Progress Report,' Vol 4 No 1, Factory Mutual Research Corporation, 1990.

'Flammability Test Method/Requirements for Packaging Materials,' Underwriters Laboratories Report UL – USNNC154/86NK26091, Prepared for Wright-Patterson Air Force Base, 1988.

Field, P. and Murrell, J., The Fire Hazard and Protection of Bin Storage, *Fire Surveyor*, **17**(6), 5–15, December 1988.

Goodfellow, D.G. and Troup, J.M.A., 'Large-Scale Fire Tests to Study Sprinkler Sensitivity,' FMRC JI 0H4R7.RR, 1983.

Hamins, A. and McGrattan, K., 'Reduced-Scale Experiments to Characterize the Suppression of Rack-Storage Commodity Fires,' NISTIR 6439, November 1999.

Hietaniemi, J., Kallonen, R. and Mikkola, E., Burning Characteristics of Selected Substances: Influence of Suppression with Water, *Fire and Materials*, **23**, 149–169, 1999.

Heskestad, G. and Delichatsios, M.A., The Initial Convective Flow in Fire, *Proceedings Seventeenth Intl. Combustion Symposium*, pp. 1113–1122, 1978.

Heskestad, G., 'Fire Plumes,' Chapter 2-2, *SFPE Handbook of Fire Protection Engineering*, SFPE, NFPA, 1995.

Heskestad, G. and Bill, R., Quantification of Thermal Responsiveness of Automatic Sprinklers Including Conduction Effects, *Fire Safety Journal*, **14**, 113–125, 1988.

Ingason, H., 'Rack Storage Fires,' Safetynet Seminar at www.safetynet.de/Activities/35.htm, February 2001.

Isner, M., $100 million Fire Destroys Warehouses, *NFPA Journal*, 37–41, November/December 1991.

Khan, M. 'Evaluation of Fire Behavior of Packaging Materials,' FMRC RC87-TP-7, presented at Defense Fire Protection Association Symposium, 1987.

Kung, H.C., Spaulding, R.D. and You, H-Z, 'Response of Sprinkler Links to Rack Storage Fires,' Factory Mutual Research Corporation J.I. 0G2E7.RA (2), November 1984.

Lee, J.L., 'Early Suppression Fast Response (ESFR) Program Phase 1: Determination of Required Delivered Density (RDD) in Rack Storage Fires of Plastic Commodity,' FMRC JI 0J0J5.RA, 1984.

Lee, J.L., 'The Effect of Different Storage Configurations on Required Delivered Density (RDD),' Factory Mutual Research Corporation J.I. 0M0E2.RA, April 1986.

Lee, J.L., 'Stored Commodity Fire Test Program Part 1: Fire Products Collector Tests,' FMRC JI 0N0R4.RU/0N1J8.RU, Prepared for The Society of the Plastics Industry, July 1987.

McGrattan, K., Hamins, A. and Stroup, D., 'Sprinkler, Smoke & Heat Vent, Draft Curtain Interaction – Large Scale Experiments and Model Development,' NISTIR 6196, September 1998.

McGrattan, K., Hamins, A. and Forney, G., Modeling of Sprinkler, Vent, and Draft Curtain Interaction, *Proceedings Sixth International Symposium*, Intl Assn for Fire Safety Science, 2000.

NFPA 13, 'Automatic Sprinkler Systems,' National Fire Protection Association, 1999.

NFPA 230, 'Fire Protection of Storage,' National Fire Protection Association, 1999.

NFPA 231, 'Standard for General Storage,' National Fire Protection Association, 1987.

NFPA 231C, 'Rack Storage,' National Fire Protection Association, 1987.

Persson, H. Sprinkler Protection of Warehouse- A New Method for Classification of Commodities, *INTERFLAM 93*, pp. 489–497, 1993.

Quintiere, J.G. and Harkleroad, M.T., New Concepts for Measuring Flame Spread Properties, in *Fire Safety, Science and Engineering*, ASTM STP 882, T.Z. Harmathy, Ed., American Society for Testing and Materials, 1985.

Quintiere, J.G., Surface Flame Spread, Section 2/Chapter 14, *SFPE Handbook of Fire Protection Engineering*, SFPE, NFPA, 1995.

Sleights, J.E., A Sprinkler Response Computer Program for Warehouse Storage Fires, MS Thesis, Worcester Polytechnic Institute, December 1993.

Spaulding, R.D., 'Evaluation of Polyethylene and Polyethylene Terepthalate Commodities Using the Fire Products Collector,' FMRC JI 0P0J2.RA, May 1988.

Tewarson, A., 'Experimental Evaluation of Flammability Parameters of Polymeric Materials,' *Flame Retardant Polymeric Materials*, vol 3, Plenum Press, 1982.

Tewarson, A., Generation of Heat and Chemical Compounds in Fires, Section 3/Chapter 4, *SFPE Handbook of Fire Protection Engineering*, SFPE, NFPA, 1995.

Troup, J.M.A., 'Large-Scale Fire Tests of Rack Stored Group A Plastics in Retail Operation Scenarios Protected by Extra Large Orifice (ELO) Sprinklers,' FMRC Report J.I. 0X1R0.RR, prepared for Group A Plastics Committee, November 1994.

Troup, J.M.A., Extra Large Orifice (ELO) Sprinklers: An Overview of Full-Scale Fire Test Performance, *J. of Fire Protection Engineering*, **9**, 27–39, 1998.

Troup, J.M.A. and Vincent, B., Fire Test Performance Evaluation of K-Factor 25 Control-Mode (Density/Area) Extended-Coverage Sprinklers for Storage Occupancies, *NFPA World Safety Congress & Exposition*, May 2001.

Ward, R.P., Survey of Large Fires in Sprinklered and Non-Sprinklered Warehouses and Storage Areas, *Fire Prevention*, March 1985.

Yao, C., Early Suppression Fast Response Sprinkler Systems, *Chemical Engineering Progress*, 38–43, September 1988.

You, H.-Z., 'Transient Plume Influence in Measurement of Effective Convective Heats of Rack Storage Fires,' FMRC JI 0N1J0.RA(2), 1989.

Yu, H.-Z., 'RDD Test Protocol for ESFR Sprinkler Applications,' FMRC J.I. 0N1J0.RA, Factory Mutual Research Corp, 1989.

Yu, H.-Z., 'A Sprinkler Response Prediction Computer Program for Warehouse Applications,' FMRC J.I. 0R2E1.RA, Factory Mutual Research Corp, 1992.

Yu, H.Z., Lee, J.L. and Kung, H.C., Suppression of Rack Storage Fires by Water, *Proceedings of the Fourth International Symposium on Fire Safety Science*, pp. 901–912, 1994.

6 STORAGE OF SPECIAL COMMODITIES AND BULK MATERIALS

Several commodities are packaged or stored in ways that accentuate their flammability characteristics and introduce special fire protection considerations beyond those of ordinary warehouse storage. Roll paper, nonwoven fabric roles, and rubber tires, are three such commodities discussed in this chapter. Other special commodities discussed here are aerosol products and solid oxidizers. Bulk storage of unpackaged materials such as powders, coal, wood chips, and grain is another special storage consideration addressed in this chapter.

6.1 Roll paper

6.1.1 COMMODITY DESCRIPTION

Weight per unit sheet area, surface texture, and the possible presence of special coatings are the primary material factors affecting the flammability of roll paper. The NFPA standard (formerly NFPA 231 F, now in NFPA 230) and Factory Mutual standard (FM Data Sheet 8-21) for roll paper use the categorization scheme based on paper-weight per unit area shown in Table 6.1.

All papers with a gauzy texture, regardless of weight, are classified as lightweight paper in the FM Standard and as tissue paper in the NFPA standard. Plastic coatings or laminates are sometimes applied to paper to provide liquid or corrosion resistance for products such as milk cartons and special wrappings. The presence of these coatings can significantly influence paper flammability and extinguishability as described in Section 6.1.3. Roll wrappings covering the rolls also influence (and in most cases actually determine) the paper classifications in the FM and NFPA standards. In particular, FM Data Sheet 8-21 specifies that paper of any weight can be treated as heavyweight paper if it is banded and completely wrapped with either a single layer of paper weighing a minimum of $200 \, g/m^2$ ($40 \, lb/1000 \, ft^2$), or two layers of paper weighing at least $100 \, g/m^2$ ($20 \, lb/1000 \, ft^2$).

The most common mode of roll paper storage is on-end in stacks or columns. On-end stacking allows the outer layers of the rolls to unwind (exfoliate) as they burn. This promotes rapid flame spread by shedding charred/wet outer layers and continually exposing dry surface area to the flames. Exfoliation can be mitigated in many cases by restraining the rolls with steel circumferential bands or by minimizing the spacing between adjacent rolls. However, spacing is often determined by clearance requirements of clamp lift trucks used to transport the rolls. The NFPA and Factory Mutual standards have categorized on-end storage arrays as closed, standard, or open arrays depending on roll spacing, as shown in Figure 6.1. Since air access for combustion

Table 6.1. Fire protection classes of roll paper

Grade/class	Basis weight		Examples
	(lb/1,000 sq-ft)	(g/sq-m)	
Heavyweight	>20	>98	Kraft, Linerboard
Mediumweight	10–20	49–98	Newsprint, Bond
Lightweight	<10	<49	Tissue, cigarette

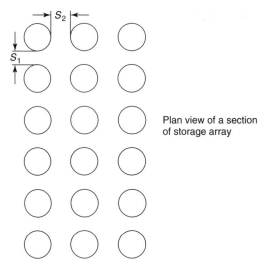

Plan view of a section of storage array

	FM D.S. 8-21		NFPA 231F	
Array type	S_1 (cm)	S_2 (cm)	S_1 (cm)	S_2 (cm)
Closed	< 10	< 10	< 2.5	< 5
Standard	0	10	< 2.5	> 5
Open	> 10	> 10	> 2.5	> 5

Figure 6.1. Roll spacing in closed, standard, and open arrays

can be another important factor in large storage arrays, the roll paper fire hazard is lowest in the closed array, and greatest in the open array, irrespective of the exfoliation characteristics of the roll.

On-side storage prevents exfoliation and rapid vertical flame spread and therefore represents a significantly reduced hazard compared to on-end storage. Fire protection requirements in the FM and NFPA standards are reduced accordingly. However, horizontal storage on axial rods with separation between rolls is considered to be tantamount to on-end storage for the determination of protection requirements.

6.1.2 LOSS EXPERIENCE

Incident reports summarized in the FM standard provide insight into some of the special protection requirements associated with on-end storage of roll paper. One of these considerations is the potential for extensive water damage and water-borne soot damage when drains become clogged with clumps of wet paper. The problem is exacerbated when accumulated water causes the stacks to topple. Trenches, skids, and pallets are recommended as mitigating measures. Water impermeable wraps may also help prevent this type of damage.

One large-loss fire in 1985 demonstrated the loss potential of roll paper with special coatings capable of self-heating and re-igniting after apparent extinguishment. The potential for re-ignition has been accounted for in the recommended duration of water supplies for sprinkler systems and hose streams.

Roll paper stacks eventually topple over during a fire. The sudden collapse of roll paper stacks was responsible for firefighter fatalities in at least one warehouse fire. Several paper companies now prohibit manual firefighting in stacked roll paper fires.

6.1.3 ROLL PAPER FIRE TESTS

Fire products collector calorimetry tests

Heat release rates for various grades and sizes of roll paper have been measured in the Factory Mutual Research Fire Products Collector. The test configuration used for the 1.07–1.14 m (42–45 in) diameter roll tests is illustrated in Figure 6.2. Four 2.7 m (105 in) high stacks with a 15 cm (6 in) separation between stacks were used in each test. As in the commodity classification tests described in Chapter 5, one free burn and several water application tests were conducted for each of the papers.

Free burn heat release rates for roll papers with basis weights in the range 17–49 g/m² (3.5–10 lb/1000 ft²) were similar. The convective heat release rate increases with time as

$$Q_c = 6.24 \times 10^{-10} (t - t_0)^{8.83} \quad \text{for } t - t_0 > 30 \text{ s} \qquad [6.1.1]$$

where Q_c = convective heat release rate (kW), and $t - t_0$ = time from self-sustained burning (s).

The exponent (8.83) in equation [6.1.1] is much greater than the exponents for the cartoned warehouse commodities described in Chapter 5. Thus, lightweight and medium weight roll paper fires develop faster than almost any other warehouse commodity (the only known exceptions besides flammable liquids being exposed vertical polystyrene foam insulation and one of the nonwoven rolls discussed in Section 6.2).

Although the various lightweight and medium weight roll papers seem to have the same fire growth rate, the peak heat release rates depend on paper basis weight. The light tissue (3.5 lb per 1000 ft² basis weight) peak heat release rate is about 60% greater than the peak values for the 8–10 lb/1000 ft² papers.

Fire Products Collector tests were also conducted with 20–24 cm (8–9 in) diameter rolls of heavyweight papers with different gaps between the rolls, and with different coatings and surface finishes. The lowest heat release rates were measured with kraft (heavyweight) paper with small gaps (3 in) between the rolls. Polyethylene coatings on both the smooth paper and the crepe paper caused the maximum heat release rate to increase substantially. The fires also developed more rapidly in the coated papers.

Water application tests were also conducted with the roll papers listed in Table 6.1. Results for the 42 in diameter newsprint rolls at 0.14 and 0.27 gpm/ft² are shown in Figure 6.5 along with the free burn heat release rate curve. Both delivered water densities suppressed the fire, with

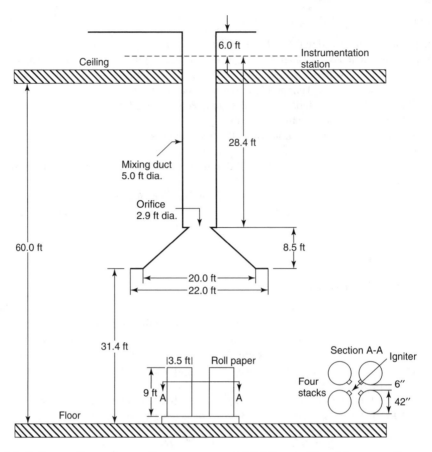

Figure 6.2. Roll paper fire products collector test setup. © 2002 Factory Mutual Insurance Company, with permission

0.27 gpm/ft^2 producing more rapid suppression than 0.14 gpm/ft^2. Peak convective heat release rates as a function of water application rate for the two tissues and for newsprint are plotted in Figure 6.6. Tissue data in Figure 6.6 indicate that it is considerably more difficult to suppress than newsprint.

Water application tests for the 1.07 m (42 in) diameter newsprint rolls showed that delivered densities of 0.14 gpm/ft^2 and 0.27 gpm/ft^2 (5.7 and 11 l/min-m^2) provided fire suppression, with suppression occurring more rapidly at the higher delivered density. Tissue roll fires were considerably more difficult to suppress than the newsprint fires. Sprinkler discharge densities need to be substantially greater than the delivered water densities because of the water losses in the fire plume and flame.

Sprinklered fire tests

The American Paper Institute has sponsored an extensive series of large-scale roll paper sprinklered fire tests at the Factory Mutual Test Center. A typical storage configuration is shown in Figure 6.3. The tests were designed to determine the effectiveness of ceiling sprinklers in controlling fires in (6.1 m and 7.6 m (20 and 25 ft) high stacks of roll paper.

STORAGE OF SPECIAL COMMODITIES AND BULK MATERIALS

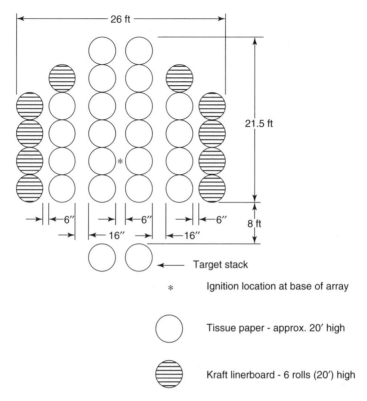

Figure 6.3. API roll paper test setup. Reprinted with permission from NFPA 231F *Storage of Roll Paper*. Copyright © 1996 National Fire Protection Association, Quincy, MA 02269. This reprinted material is not the complete and official position of the National Fire Protection Association, on the referenced subject which is represented only by the standard in its entirety

Table 6.2. Comparison of sprinklered fire test results for linerboard and tissue paper with 0.29 gpm/ft²

Test #	A3	B1
Paper type	Kraft Linerboard	Tissue
Stack height	20′	21′–10″
Clearance to ceiling	10′	8′–2″
Sprinklers opened	20	88
Sprinkler demand area	2000 ft²	8800 ft²
Maximum gas temp	1550 °F	1680 °F
Target ignited?	No	Yes

The tests clearly showed that protection effectiveness is strongly dependent upon the type of paper. For example, two tests conducted with identical sprinkler protection and similar storage arrays demonstrate how much more difficult it is to protect tissue paper than kraft linerboard (heavyweight). Both tests had 17/32" orifice, 280 F sprinklers with 10 × 10 ft spacing and 14 psig constant pressure (0.29 gpm/ft²). Unbanded rolls arranged in the FM standard array were employed in both tests with stack heights and test results as shown in Table 6.2.

The excessive water demand and extent of fire spread in the tissue paper test compared to the kraft linerboard test demonstrated that significantly higher water discharge densities are needed to control a roll tissue fire than a linerboard fire. Similar comparisons between test results from kraft paper tests and newsprint tests established a difference in protection requirements for mediumweight and heavyweight grades of paper. Furthermore, a test with kraft linerboard wrapping around the newsprint demonstrated that this type of wrapping (and closed array storage with a maximum flue space of 2 in) reduces the flammability of newsprint to that of linerboard for purposes of sprinkler protection specification.

Test results for the tissue paper tests are summarized in Table 6.3. All but one of the tests were conducted with $K = 7.8$ gpm/psi$^{1/2}$ sprinklers. The most successful combination of storage height and sprinkler protection occurred in Test B5. The ceiling height (25') and storage-to-ceiling clearance (5'-2") were lowest in Test B5 and the sprinkler discharge density (declining from 0.92

Table 6.3. Summary of roll paper tissue tests (data from NFPA 231F)

Test number	B1[a]	B2	B3	B4	B5[c]	B6[c]
Test date	10/4/79	7/23/80	7/30/80	10/15/80	7/28/82	8/5/82
Paper type	Tissue	Tissue	Tissue	Tissue	Tissue	Tissue
Stack height (ft-in.)	21-10	20-0	21-8	18-6	19-10	25-3
Paper banded	No	No	No	No	No	No
Paper wrapped	No	No	No	No	No	No
Fuel array	Std	Std	Std	Std	Std	Std
Clearance to ceiling (ft-in.)	8-2	10-0	8-4	11-6	5-2	4-9
Clearance to sprinltlen (ft-in.)	7-7	9-5	7-9	10-9	4-7	4-2
Sprinkler orifice (in.)	17/32	17/32	17/32	0.64	17/32	17/32
Sprinkler temp. rating (F)	280	280	280	280	280	280
Sprinkler spacing (ft × ft)	10 × 10	10 × 10	10 × 10	10 × 10	10 × 10	10 × 10
Water pressure (psi)	14[b]	60	95	50	138 Initial 102 Final	138 Initial 88 Final
Moisture content of paper (%)	9.3	9.3	10.2	6.0	8.2	9.2
Fint sprinkler operation (min:sec)	0:43	0:32	0:38	0:31	0:28	0:22
Total sprinklen open	88	55	26	64	17	29
Final flow (gpm)	2575[b]	1992	1993	4907	1363	2156
Sprinkler demand area (ft')	8800	3300	2600	6400	1700	2900
Avg. discharge density (gpm/ft')	0.29[b]	0.60	0.77	–	0.92 Initial 0.80 Final	0.96 Initial 0.74 Final
Max. one min. avg. gas temp. over ignition (F)	1680[b]	1463	1634	1519	[d]	[e]
Duration of high temp. within acceptable limits	No	Yes	Yes	Marginal	Yes	Yes
Max. one min. avg. fire plume gas velocity over ignition (ft/sec)	–	40.7	50.2	47.8	–	–
Target ignited	Yes	Yes	No	No	No	Briefly
Extent of fire damage within acceptable limits	No	No	Marginal	Marginal	Yes	Marginal
Test duration (min)	17.4	20	20	25.5	45	45

[a] Phase I Test
[b] Pressure Increased to 50 psi at 10 min
[c] Phase 111 Tests Decaying Pressure
[d] Max. Steel Temp. Over Ignition 341 °F
[e] Max. Steel Temp. Over Ignition 132 °F

to $0.80\,\text{gpm/ft}^2$ as the 17 sprinklers opened) was higher than in Tests B1 to B4. Test B6 used the same high discharge density, but results were only marginally successful because of the 25 ft high storage stacks, compared to the 20 ft high stacks in Test B5.

The lack of success in the tissue tests with 0.30 and $0.60\,\text{gpm/ft}^2$ discharge densities (Tests B1 and B2) are consistent with the Fire Products Collector data (which were obtained several years after the sprinkler tests), which showed high heat release rates for 9 ft high storage even with a *delivered* density of $0.51\,\text{gpm/ft}^2$. The relative success with $0.80-0.90\,\text{gpm/ft}^2$ and 5 ft clearance demonstrates that even roll tissue fires can be controlled with ceiling sprinklers producing an Actual Delivered Density in the range $0.50-0.80\,\text{gpm/ft}^2$ to the top of the rolls.

6.1.4 ROLL PAPER PROTECTION REQUIREMENTS

NFPA 231 F (now in NFPA 13) and FM Data Sheet 8-21 provide sprinkler protection minimum requirements based on the large-scale test results, including some more recent tests not described in Section 6.1.3. Although most of the specifications are for high storage heights, there are also specifications for storage heights comparable to those used in the Fire Products Collector testing. In the case of 10 ft high open array storage of medium weight paper, both standards call for $0.30\,\text{gpm/ft}^2$ over a $2000\,\text{ft}^2$ demand area. Based on the Fire Products Collector data for newsprint, this density should provide rapid control or suppression assuming that at least half the discharge density penetrates the 5 MW fire plume.

In the case of storage heights of at least 20 ft, the FM specifications for $K = 7.8\,\text{gpm/psi}^{1/2}$ sprinklers are slightly more conservative (higher water densities and/or demand areas in some cases) than the NFPA 13 specifications. Figure 6.4 shows the discharge density versus demand area specifications for 20 ft high storage of tissue and medium weight paper (unbanded, standard array) under a 25 ft high ceiling. The decaying density data from tissue Test B5 (Chicarello and Troup, 1982) are also plotted in Figure 6.4 to illustrate the relationship between the test data and FM and NFPA specifications. The FM specification is a two-point demand, while the NFPA 13 specification is only a single point specification ($0.76\,\text{gpm/ft}^2$ over $2500\,\text{ft}^2$).

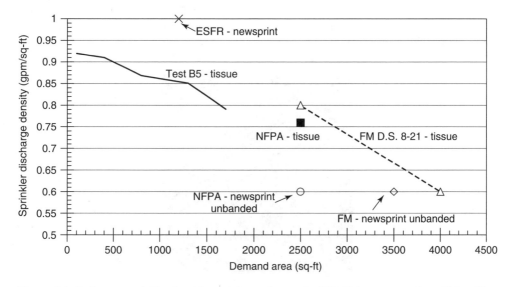

Figure 6.4. Roll paper sprinkler densities and demand areas for 20 ft high storage under a 25 ft ceiling

FM and NFPA also specify ESFR sprinkler protection requirements for medium weight and heavy weight roll paper. The requirement for 20 ft storage under a 25 ft high ceiling is 12 ESFR sprinklers operating at a minimum flow rate of 100 gpm over 100 ft^2 coverage per sprinkler. The corresponding ESFR discharge density of 1.0 gpm/ft^2 over 1200 ft^2 is also indicated in Figure 6.4. The required ESFR discharge rate increases as the storage height increases or the ceiling height increases over 30 ft. There are no specifications for ESFR protection of tissue paper (unless the tissue paper is wrapped to qualify for medium weight protection).

The NFPA standard indicates that high expansion foam would provide adequate protection, but the FM offers no such suggestion. Both standards allow significantly reduced densities for on-side storage than for on-end storage.

6.2 Nonwoven roll goods

6.2.1 COMMODITY DESCRIPTION

A nonwoven material is defined as a fabric-like matting of natural or synthetic fibers held together by chemical, mechanical, or thermal bonding. The most common fibers are polypropylene, polyester, rayon, polyethylene, cellulose/wood pulp, cotton, and nylon. Chemical binding is achieved with polymers such as acrylics or vinyl acetate copolymers. Mechanical binding is achieved by entangling the fibers with needlepunches, air jets, or water jets. Thermal bonding is applicable to thermoplastic fibers that can be melted and solidified while entwined. Figure 6.5 is a micrograph showing the polyethylene fibers in a thermally bonded nonwoven. The nonwoven trade association's *Nonwoven Fabrics Handbook* provides an extensive description of nonwoven, materials, manufacturing processes, and current applications.

Nonwoven fabrics are made and stored on hollow core roles similar to those for roll paper. Roll sizes and storage modes are similar to roll paper for most fibers. The exceptions are the soft, high-loft nonwovens which often are stored on racks or in pyramid piles on floor.

Nonwoven fabrics are manufactured by paper companies, chemical companies, and textile companies, as well as some independent companies devoted to nonwovens. They are used in a wide variety of products including baby diapers, feminine hygiene products, medical and hospital disposables, industrial and household wipes, mailing envelopes, blanket insulation, wearing apparel linings, carpet backing, house wraps and roofing material, air conditioning filters, and interior automotive trim.

Figure 6.5. Micrograph of a polyethylene nonwoven fabric construction (from www.Tyvek.com)

STORAGE OF SPECIAL COMMODITIES AND BULK MATERIALS

Nonwoven materials are capable of rapid flame spread and high heat release rates because their fibrous construction allows air access and relatively low thermal inertia (as represented by the product of thermal conductivity, density, and specific heat). Storage as on-end rolls promotes outer layer unwinding and exposure of unburned layers as in roll paper. Thermoplastic nonwovens pose special flame spread problems because of their propensity to melt when they burn.

6.2.2 LOSS EXPERIENCE

Until recently, nonwoven roll goods have not been present in sufficient quantity to be responsible for many fires. The fires that have been reported to date have primarily occurred in cartoned storage of relatively small rolls. Dense smoke in these fires hampered manual firefighting efforts. One fire involving 15 ft high palletized and rack storage of cartoned polyester and rayon rolls damaged the roof deck and opened 37 ceiling sprinklers (FM loss 72-2970).

6.2.3 FIRE TESTS

Fire products collector tests

Fire Products collector tests have recently been conducted with both on floor storage and on rack storage of nonwoven rolls. The on-floor storage configuration was identical to that of 42 in roll paper (see Figure 6.1). The rack storage configuration consisted of a 2 pallet wide, 2 pallet deep, 2 tier high configuration identical to that used in the Fire Products Collector commodity classification tests described in Chapter 5. Both free burn and water application tests were conducted with both storage configurations.

Nonwovens tested under the Fire Products Collector during a Factory Mutual Research project are listed in Table 6.4. Rolls of thermally bonded (spunbond) polypropylene with basis weights ranging from $20\,g/m^2$ (4.1 lb per 1000 ft^2) to $60\,g/m^2$ (12 lb/1000 ft^2) were tested, as well as a polypropylene-wood pulp blend (65% wood pulp) with a basis weight of $190\,g/m^2$.

The Roll Loft Factor (RLF) values listed in Table 6.3 are defined as follows:

$$RLF = \frac{\pi L(D_0^2 - D_i^2) \sum_i \rho_i m_i}{4(W_0 - W_c)} - 1$$

$$RLF = \frac{V_{roll} \dfrac{M_{fuel}}{V_{fuel}}}{M_{fuel}} - 1$$

Table 6.4. Nonwoven and roll papers tested used in FMRC fire products collector tests

Fiber	Binder	Basis weight (g/m^2)	Roll loft factor	Heat of combustion (kJ/g)
Polypropylene	Spunbond	20–26	3–4	45
Polypropylene	Spunbond	35	2.8	45
Polypropylene	Spunbond	57–60	2.5	45
Polypropylene	Spunbond	25	3.7	45
Cellulose-65% Polypropylene-35%	Spunbond	190	16.6	28
Polyester	Acrylic (15%)	85	94.2	23
Rayon	Acrylic (30%)	22	4.0	18.8
Tissue	–	17	4.7	16
Tissue	–	38	3.8	16
Newsprint	–	49	3.8	19

Where L = roll length, D_0 = roll outer diameter, D_i = roll inner (core) diameter, W_0 = roll weight, W_c = core weight, ρ_i = bulk density of the ith component, and, m_i = mass fraction of the ith component.

Since the roll consists of fuel plus air,

$$RLF = \frac{(V_{fuel} + V_{air})}{V_{fuel}} - 1 = \frac{V_{air}}{V_{fuel}}$$

In other words, RLF is equal to the air/fuel ratio of the roll. Rolls with high values of RLF (nominally greater than 25 per Factory Mutual Data Sheet 8-23) are termed *high-loft nonwovens*. They are typically used as batting and fiberfill insulation. Since they have a high air/fuel ratio, they are expected to burn efficiently and completely.

The on-floor free burn heat release rate curves for several nonwovens are shown in Figure 6.5, along with the three roll papers listed in Table 6.4. The curve labeled Coform in Figure 6.5 refers to the cellulose-polypropylene blend with a Roll Loft Factor of 17. Its free burn heat release rate history is very similar to the curves for newsprint and tissue, i.e. rapid fire development with the convective heat release rate increasing to 4 MW about 12 seconds after ignition. The lightweight (22 g/m^2) rayon-acrylic nonwoven had an even more rapid fire-growth rate, reaching 4 MW in about six seconds. The on-floor tests with the polypropylene nonwoven showed a much slower fire development accompanied by unraveling of the outer layers of the roll, and polypropylene dripping before burning.

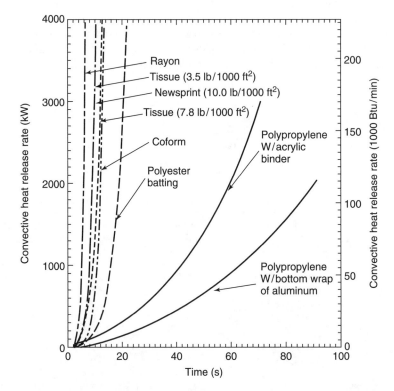

Figure 6.6. Paper and nonwoven roll heat release rates. © 2002 Factory Mutual Insurance Company, with permission

Plate 1 Fire following dust explosion in textile factory. (Photo by Jim Patten, ©*The Eagle-Tribune*, Lawrence, Massachusetts, USA)

Plate 2 Small-array of cartoned expanded plastic burning under the FM Fire Products Collector

Plate 3 Rack storage sprinklered fire test. ©2002 Factory Mutual Insurance Company, with permission

Plate 4 Burning pallet loads of butter in Madison, WI cold storage warehouse fire. Reprinted with permission from NFPA *Journal November/December 1991*, Copyright ©1991 National Fire Protection Association, Quincy, MA 02269. This reprinted material is not the complete and official position of the National Fire Protection Association, on the referenced subject which is represented only by the standard in its entirety

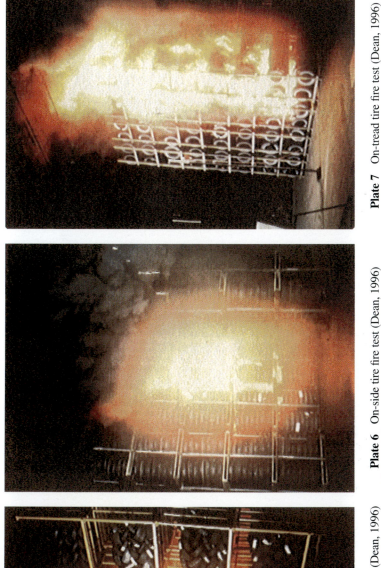

Plate 5 Interlaced tire storage fire test (Dean, 1996) **Plate 6** On-side tire fire test (Dean, 1996) **Plate 7** On-tread tire fire test (Dean, 1996)

Plates 5, 6 and 7 ©2002 Factory Mutual Insurance Company, with permission

Plate 8 Flareup during water spray application to burning fatty acid

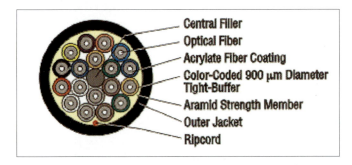

Plate 9 Example of a fiber optic cable cross-section

Plate 10 Various stages of Ostankino tower fire. From Reuters

STORAGE OF SPECIAL COMMODITIES AND BULK MATERIALS

The water application test results for the polypropylene-wood pulp were also similar to the data for the roll papers tested, with only small reductions in peak heat release rate at the discharge densities used. This is in contrast to the polypropylene spunbond, which did experience melt burning but at a rate that was easily controlled with water delivered densities as low as $0.14 \, \text{gpm/ft}^2$.

Polypropylene melting was a far greater problem in the on-rack tests. The different basis weight polypropylene spunbonds required five to six minutes to become fully involved. However, the peak convective heat release rates for the polypropylene spunbonds were much larger than those of any of the roll papers (it exceeded the 10 MW capacity of the Fire Products Collector) with and without water application.

Sprinklered fire tests

Sprinklered fire tests have confirmed that nonwoven rolls are extremely difficult to protect effectively. Rack storage tests with high-loft polyester batting protected by $K = 11 \, \text{gpm/psi}^{1/2}$ sprinklers discharging $1.0 \, \text{gpm/ft}^2$ resulted in complete burn out of the intermediate-scale array. Thus ceiling sprinklers need to be supplemented by in-rack sprinklers. Rack storage tests with in-rack sprinklers in the longitudinal flue and rack face (aisle) were able to provide effective protection for polypropylene nonwovens.

Nonwoven roll on-floor storage is even more of a challenge than rack storage because of the need to rely on ceiling sprinklers only. The author is aware of only one on-floor nonwoven roll test in which the fire was controlled. That particular test involved 21 ft high storage of a polyester-rayon 50/50 blend with a low roll loft factor. ESFR sprinklers discharging 100 gpm for each 100 ft^2 of sprinkler coverage did limit fire spread; but 12 sprinklers opened. Since this is more than were opened in any of the rack storage tests with the prototype Group A plastic commodity under a 30 ft high ceiling, the installation guidelines based on early suppression of the standard plastic commodity tests are not applicable.

The uncontrolled fire tests with on-floor storage involved a 50% polyester – 50% rayon nonwoven, and the following nonwovens listed in Table 6.4: spunbound polypropylene, polyester-acrylic, and rayon-acrylic. Water flow rates per sprinkler were in the range 60–105 gpm, corresponding to discharge densities of 0.60–1.05 gpm/ft^2. Two tests were conducted with $K = 7.8 \, \text{gpm/(psig)}^{0.5}$ sprinklers, and three tests were conducted with large drop sprinklers ($K = 11 \, \text{gpm/(psig)}^{0.5}$). Storage heights varied from 15–20 ft high under a 30 ft ceiling.

6.2.4 SPRINKLER PROTECTION REQUIREMENTS FOR NONWOVENS

Guidelines for sprinkler protection of nonwoven roll storage are provided in FM Data Sheet 8-23. In the case of on-floor storage, the lack of success with the 10–15 ft clearances in the fire tests, motivated recommendations for false ceilings in order to avoid these clearances above nonwoven stacks in a warehouse. Even with these low clearances, large demand areas (as much as 5000 ft^2) are recommended for on-floor storage.

Rack storage protection specifications in FM Data Sheet 8-23 include options for either using combined flue and face in-rack sprinklers, or a mid-height barrier in conjunction with face sprinklers. Adjustments to these recommendations are expected as additional test data and loss experience is acquired.

6.3 Rubber tire storage

Rubber tire and synthetic rubber storage present a special challenge in warehouse fire protection because the deep-seated fires produce copious volumes of smoke and are particularly difficult to extinguish. Flaming within the hollow torroidal inner surface of the tire is often shielded from the sprinkler spray, and the steel belts in the tire remain hot enough to cause re-ignition after an initial apparent suppression. Automatic sprinklers of suitable water discharge density can and will control tire storage fires, but, there is a tendency for the tires to re-ignite if the sprinkler water flow is shut off too early or if airflow into the storage array is allowed to increase.

The NFPA 231D standard for storage of rubber tires (now incorporated into NFPA 13 and NFPA 230) describes four stages of a typical sprinklered tire storage fire. The characteristics of the four stages, called the incipient, active, critical, and overhaul stages are listed in Table 6.5. Manual firefighting efforts are usually futile and dangerous during the active stage in which the sprinklers are trying to gain control of the fire.

The three most commonly used tire storage configurations are: on-side (horizontal) storage (Plate 6), on-tread (vertical) storage (Plate 7), and laced (oblique angle with alternating rows) storage (Plate 5). The tires are often loaded into open frame portable racks as illustrated in the photographs. Storage heights can range from a few feet for on-floor storage to over 25 ft for rack storage. On-tread tire storage usually represents a greater fire protection challenge than on-side storage because burning can continue within the horizontal flue formed by the wheel holes, and propagate extensively while being shielded from the sprinkler spray. However the pallets often used in on-side storage also shield the smaller vertical flues in on-side piles. Interlaced storage has proven to be the most challenging sprinkler protection storage mode because it seems to allow fire redevelopment even more than the other storage configurations.

Rubber tire protection guidelines originally stem from several series of sprinklered fire tests at the Factory Mutual Test Center and from a 1971 test series in France (Cleremont-Ferrand, 1973). The French tests involved 20 ft high caged tire portable rack storage in an aircraft hangar with an arched roof. The tests demonstrated that a sprinkler discharge density of $0.55\,\text{gpm/ft}^2$ could control the fire providing there was ample (almost unlimited) water supply for at least 34 sprinklers. The French tests also provided an excellent demonstration of the advantages of high expansion foam used in conjunction with ceiling sprinklers. Protection guidelines (NFPA 231D and FM Data Sheet 8-3) for open portable rack storage now provide the option of using a high

Table 6.5. Stages of a sprinklered tire warehouse fire

Stage	Initiating event	Duration (min)	Visibility	Smoke color	Recommended firefighting
Incipient	Ignition	2–5	Decreasing	Black	Portable extinguisher; hose line; remove burning tires from storage array.
Active	Sprinkler Actuation	60–90	Virtually None	Turning from black to gray	Preparation only; *No ventilation*.
Critical	Apparent Extinguishment; Diminished Smoke	?	Improving	Grey while controlled	Slowly ventilate with sprinklers still flowing
Overhaul	Smoke Clearing	24 × 60	Clear	None	Small hose streams while sprinklers still flowing; remove burned tires outside warehouse. Shut off sprinklers and monitor fire area for 24 hours.

ceiling sprinkler design density (0.60 gpm/ft² over 5000 ft² *and* 0.90 gpm/ft² over 3000 ft² for up to 25 ft high on-side storage) without high expansion foam or a much lower density (usually 50% reduction) with foam.

A 1970 Factory Mutual tire fire test inadvertently demonstrated the relationship between building ventilation and sprinkler control. Sprinkler control of 18 ft high on-tread storage was established with a discharge density of 0.60 gpm/ft² until the test building was ventilated to reduce the heavy smoke concentration. When the building was ventilated, an intense fire began to spread through the array and could not be controlled even with 95 flowing sprinklers (50 opened after the building was ventilated) over an area of 4750 ft². Sprinkler control was eventually achieved by reducing the building ventilation.

Figure 6.7 illustrates the tendency of tire fires to redevelop even when subjected to significant water spray densities. The heat release rates plotted in Figure 6.7 were obtained from 20 ft high Required Delivered Density tests conducted with a delivered density of 0.55 gpm/ft² under the Fire Products Collector. The prototype Group A plastic commodity is suppressed with this density as is evident from its heat release rate curve in Figure 6.7. On the other hand, the on-tread tires showed a very different behavior. The fire remained in an incipient stage for the first four minutes after ignition. Sudden fire development at four minutes triggered the actuation of the water spray to the top of the tire array at that time. The water did immediately reduce the fire intensity and maintained the heat release rate under 30,000 btu/min (780 kW) for the next 3 1/2 minutes, but the heat release rate eventually grew back up to 150,000 btu/min (2600 kW) when the test had to be terminated.

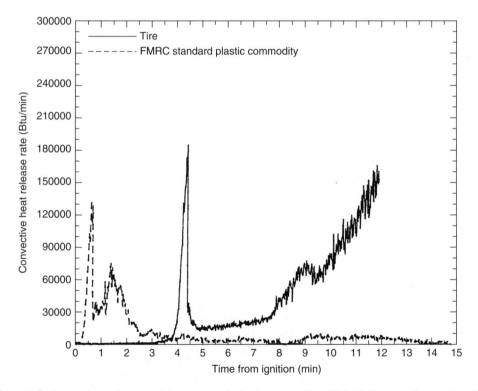

Figure 6.7. Comparison of on-tread tire array and plastic commodity (20 ft high) heat release rates with 0.55 gpm/ft² delivered density. © 2002 Factory Mutual Insurance Company, with permission

As with many other commodities, recent tests have demonstrated the advantages of larger orifice sprinklers for protecting high-piled rubber tires. ESFR sprinklers and Large Drop sprinklers have successfully protected 25 ft high tire storage in a series of fire tests conducted by FM (Dean, 1996) and shown in Plates 5–7. The Large Drop sprinklers discharging 100 gpm over each 100 ft^2 of coverage (density of 1.0 gpm/ft^2) protected both on-side and on-tread storage. The ESFR sprinkler, which had previously been shown to protect 25 ft high on-tread storage at a water flow rate of 100 gpm per sprinkler, needed a flow rate of 125 gpm per sprinkler to protect 25 ft high interlaced storage. The interlaced storage test opened 14 ESFR sprinklers, and the NFPA and FM standards require designing for 20 flowing ESFR sprinklers since this is a fire control application rather than an early suppression application.

One special consideration in dealing with the sprinkler and hosestream runoff from a tire storage fire is the presence of oil generated during the pyrolysis of the tires. Based on reports of the oil recovered from large fires of scrap tires, there have been between 0.1 and 1.0 liter of oil produced for every tire burned. This oil should be separated and recovered from the runoff before allowing it to reach groundwater or nearby lakes or rivers. Fire-fighting guidelines for outdoor scrap tire fires are contained in Appendix C of NFPA 231D, which is based on the report of the International Association of Fire Chiefs (1995).

6.4 Aerosol products

Aerosol products often require special warehouse fire protection because of their liquefied gas (propellant) and flammable liquid contents, which are released suddenly upon can rupture. Can rupture produces a fireball, possible can rocketing (depending on the can failure mode), and a residual pool fire. Flammability classifications and associated fire protection guidelines are discussed following a brief description of generic aerosol products and historic warehouse fires involving aerosols.

6.4.1 PRODUCT DESCRIPTION

A typical aerosol product consists of propellant, concentrate, solvent, and additives in a small, pressurized metal container. Room temperature product pressures are usually in the range 30–80 psig (210–550 kPa) and container hydrostatic strengths are in the range 200–300 psig (1400–2100 kPa). Container volumetric capacities are in the range 5–27 fluid ounces (150–800 ml). Single piece aluminum cans account for about 8.5% of the US aerosol market, and about 30% of the European market (Johnsen, 1982, p. 75).

The most common aerosol propellants in the US since 1978 have been various blends of isobutane and propane. The blend is usually specified as A-x, where A stands for Aerosol Grade and x is the blend vapor pressure at 70 °F. Thus A-31 would consist of 100% isobutane and would have a room temperature vapor pressure of 31 psig, while A-110 would represent 100% propane with its 110 psig vapor pressure at 70 °F. If a vapor pressure below 31 psig is desired at 70 °F, n-butane (17 psig at 70 °F) is blended with isobutane.

In 1975, saturated chlorofluorocarbon propellants (generic formula $C_wCl_yF_z$) were used in approximately half the 2.7 billion aerosol units produced in the US. Since 1978, these saturated chlorofluorocarbons have been banned from use in non-essential aerosol products in many countries, because of their threat to the ozone layer. Hydrocarbon propellants replaced the CFCs for most products. However, hydrofluorocarbons (e.g. P-152a, CH_3CHF_2), carbon dioxide, and dimethyl ether (CH_3OCH_3) have been used to a limited extent in aerosol products. Since the flammability, water miscibility, and volatility of these substitute propellants vary widely, they can have a significant influence on the overall flammability of a particular aerosol product.

According to *The Aerosol Handbook*, the following aerosol products have annual productions in the US of over 100 million units: paints and coatings, antiperspirants and deodorants, hair sprays, automotive products (e.g. carburetor cleaner), foods (e.g. whipped cream), shaving creams, waxes and polishes, insecticides, air fresheners, and laundry aids. These myriad applications suggest that aerosol products are present in warehouses serving several different industries including paints, automotive parts, supermarkets, and general merchandise. Indeed, each type of warehouse has experienced at least one major fire involving aerosol product storage as described in Section 6.4.2.

6.4.2 AEROSOL WAREHOUSE FIRES

Since the 1978 ban on chlorofluorocarbon aerosol propellants in the US, there have been several multi-million dollar warehouse fires involving aerosol products. The list of 12 Very Large Loss warehouse fires in Table 5.1 includes four fires where aerosol storage was a significant factor. The role of aerosols in these four fires is reviewed here.

The 1979 Supermarkets General fire in Edison, New Jersey involved a supermarket distribution center with 290,000 ft^2 rack storage of merchandise including the following aerosol products: shaving cream, hair spray, deodorant, and petroleum based insecticide. Storage was 18–23 ft high and protected with ceiling sprinklers with a design density of 0.30 gpm/ft^2.

Employees noticed a small fire near cardboard cartons of aerosol product. As the cans began exploding, employees evacuated and called the fire department. The sprinkler system could not cope with the intense fire and an exterior wall collapsed soon after the fire department arrived. The warehouse was totally destroyed and resulted in approximately $35 million in property damage in 1979 dollars. This fire motivated the first Factory Mutual aerosol storage fire tests summarized in Appendix B of NFPA 30B.

The 1982 K Mart distribution center fire in Falls Township, Pennsylvania involved 1.2 million ft^2 of storage subdivided by fire walls into four sections as described in Appendix B. There was both palletized and rack storage of aerosols and many other products to a maximum height of 15 ft. Ceiling sprinkler protection was designed to deliver 0.40 gpm/ft^2 over the hydraulically most remote 3000 ft^2, using 286 °F, 17/32nd in orifice sprinklers on the 30 ft ceiling.

The K Mart fire apparently began when a carton of carburetor cleaner aerosol cans fell from a forklift truck. The cans reportedly ruptured or were punctured and the released vapor/liquid probably was ignited by sparks from the electric powered lift truck. According to the NFPA Journal report (reference B3.1), there were 40–50 pallet loads of carburetor cleaner (1056 cans per pallet load) in the immediate vicinity of the forklift. Inventory records indicate there were about 580,000 cans of petroleum based aerosol product and about 480,000 cans of alcohol-based aerosols in the distribution center. Of course, there were myriad other combustible commodities including 102,000 gallons of non-pressurized high flash point liquids, rubber tires, cans of butane lighter fluid, and 14-ounce propane cylinders.

Eyewitnesses report that the K Mart fire developed rapidly with rocketing cans, fireballs, and thick black smoke. Despite the operating sprinklers, flame and smoke penetrated the roof five to ten minutes after ignition. Fire eventually spread to the other three quadrants of the distribution center partly because of a lack of fire doors on some wall openings and partly due to the structural collapse of the firewalls as described in Appendix B. The loss was in excess of $100 million.

The 1985 MTM Partnership warehouse fire in Port Elizabeth New Jersey involved a 530,000 ft^2 24 ft high public warehouse subdivided into five sections separated by concrete block walls. One of the five sections was leased to a company that manufactured petroleum based aerosol products including paints and carburetor cleaners. The aerosols were stored in double row racks to a height of 17 ft. Commodities in the other four sections included drums of high flash point liquids, furniture, sporting goods, and motorcycles. Ceiling sprinkler protection provided 0.30 gpm/ft^2 over the most remote 4500 ft^2 using 165 °F 1/2-inch orifice sprinklers.

The fire originated in the section containing the aerosols. The exact cause is not clear but is believed to be of electrical origin associated possibly with a conduit ground fault or a power surge. There are no eyewitness accounts of the early stages of the fire (that section of the building was not occupied when the fire started), but there are many later observations of aerosol cans rocketing from the building and fireballs rising '300 ft into the night sky.' Some of the rocketed paint cans are reported to have landed over a thousand feet from the building. The fire walls did collapse during the fire (possibly because they were tied to a steel ceiling beam that extended the length of the building) and the warehouse was a total loss estimated at $150 million.

The 1987 Sherwin-Williams warehouse fire in Dayton, Ohio destroyed a 180,000 ft^2 automotive paint distribution center. An estimated 1.5 million gallons of paint, coatings, and solvents were stored in the warehouse. The fire started when un-pressurized metal containers of solvent dropped from a lift truck, and the pool of liquid was ignited by the truck. Besides the flammable solvent in metal containers, storage in the vicinity of fire origin included flammable liquids in plastic containers and aerosol paint cans. The aerosol paint was eventually observed to rocket up through the damaged roof, but it is not clear whether the aerosol contribution was a major factor in the overpowering of the ceiling sprinkler system and the $50 million loss of the entire warehouse.

Although these four losses represent some of the most spectacular warehouse fires involving aerosol products, there have been many others including several in Europe and Africa. One fire started when aerosol cans were stored too close to steam radiators such that the cans burst from overheating. Brief accounts of some of the fires, which occurred prior to 1982, can be found in *The Aerosol Handbook*.

6.4.3 AEROSOL PRODUCT FORMULATION EFFECTS

The first clear evidence of how product formulation affects aerosol storage fire severity came from a series of one pallet high sprinklered fire tests under a 30 ft ceiling at the Factory Mutual Test Center in 1980. Three different simple formulations were used, all with 35% hydrocarbon propellant. The remaining 65% was water in the first formulation, isopropanol in the second formulation, and toluene in the third formulation. The water-based formulation was effectively protected with 0.30 gpm/ft^2 ceiling sprinklers despite the occurrence of several can ruptures and accompanying fireballs. A test with the alcohol-based formulation was significantly more severe but still was controlled in the sense that less than half the cans were damaged. The equivalent test with the toluene formulation went rapidly out of control and had to be aborted. In fact, a single pallet load of the toluene formulation opened 36 sprinkler heads and resulted in nearly all the roughly 1000 cans being ruptured.

After this preliminary series of tests with simple formulations, the aerosol industry became involved in the testing of more realistic simulations of commercial product formulations. A series of two pallet load tests, one pallet high, was conducted under a 20 ft ceiling equipped with 1/2 in orifice sprinklers discharging at 30 gpm per head, i.e. at 0.30 gpm/ft^2. Four different oil-based formulations were utilized with hydrocarbon propellant and varying amounts of water. The severity of the fire, as measured in terms of sprinkler water demand, ceiling air temperatures, target ignition, and percent product damage, increased with increasing proportion of petroleum liquid in the product. Factory Mutual Research and the aerosol industry interpreted the differences in fire severity as a guide to establish boundaries between low level (Level 1), medium level (Level 2), and high level (Level 3) aerosol storage hazard. In particular, a maximum of 20% hydrocarbon (besides the propellant) was allowed in a Level 1 product, and a maximum of 55% petroleum liquid was allowed in a Level 2 hydrocarbon-water-additive formulation. Maximum amounts of alcohol were also specified for Level 1 and for Level 2 products. One key assumption made in generalizing these boundaries was that the amount of propellant in the formulation did not affect its classification. Subsequent tests conducted with formulations containing a large proportion of propellant demonstrated that the propellant can indeed have a significant affect on product protectability. As a result of these tests, the generic classification boundaries were revised to account for effects of large concentrations of flammable propellant in the product formulation.

The current Factory Mutual and NFPA 30B generic aerosol product classification boundaries are based entirely on the overall heat of combustion, ΔH_c, of the can contents, i.e.

$$\Delta H_c = \sum_i m_i \Delta H_{ci} \qquad [6.4.1]$$

where m_i is the mass fraction of component i in the product-propellant formulation, and ΔH_{ci} is the chemical heat of combustion for component i. Heats of combustion for many common aerosol formulation components are listed in Appendix A of NFPA 30B. The current generic boundaries between Level 1, Level 2, and Level 3 aerosols are:

Level 1 $\Delta H_c < 20\,\text{kJ/g}$
Level 2 $20\,\text{kJ/g} < \Delta H_c < 30\,\text{kJ/g}$
Level 1 $\Delta H_c > 30\,\text{kJ/g}$

Based on these boundaries, aerosol manufacturers, NFPA 30B requires aerosol manufacturers to label their aerosol product cartons to identify the Level, and allow warehouse operators to determine where they should be stored and how they should be protected.

As an example, the laundry pre-wash formulation that was considered the upper bound Level 2 product in the original aerosol classifications has a calculated heat of combustion of about 24 kJ/g, as shown in Table 6.6. The value of 24 kJ/g for the product heat of combustion represents the middle of the range for a Level 2 aerosol, rather than an upper bound.

The simplified classification based on equation (6.4.1) ignores product properties such as vapor pressure, water miscibility, and can strength, all of which have a significant influence on the sprinkler protection challenge. Some of these factors are accounted for in a single-can Aerosol Flammability Test (Rizzo, 1986; Zalosh, 1986), but the most definitive method of classifying aerosol products is to conduct a 12-pallet load sprinklered fire test. The test uses Large Drop Sprinklers installed on a 25 ft high ceiling and discharging 80 gpm per sprinkler (0.80 gpm/ft^2), and results are compared to data obtained with benchmark Level 1, Level 2, and Level 3 products. The protection provides effective control of Level 1 products, marginal control of Level 2 products, and is inadequate for Level 3 products (Zalosh, 1986).

The American classification system described above has not been adopted in other countries. In the United Kingdom, aerosol products are classified as flammable or nonflammable depending on the weight percent of flammable ingredients in the formulation. The demarcation is 45 wt% flammable ingredients with flash points below 100 °C. Flammable aerosol products according to

Table 6.6. Aerosol product heat of combustion calculation

Component	Weight fraction	ΔH_{ci}	$m_i \Delta H_{ci}$ (kJ/g)
A-40 propellant isobutane-propane blend	0.07	43	3.0
Petroleum distillate	0.50	41	20.5
Water	0.41	0	0
Additives	0.02	<40	<0.8
Total			23.5–24.3

this definition are subject to warehouse storage restrictions in the UK. One important difference between US and European packaging practices is that cardboard cartons are used predominantly in the US, while polyethylene stretch wrap without any carton is usually used in the UK.

6.4.4 SPRINKLER PROTECTION GUIDELINES

Level 1 aerosol products in the US are treated as an ordinary warehouse Class III commodity, per NFPA 30B. Rack storage protection for Level 3 products with more than 5 ft clearance to the ceiling requires 0.60 gpm/ft^2 at the ceiling over at least 1500 ft^2 (demand area depends upon clearance) and two lines of in-rack sprinklers at every tier. Alternatively, Early Suppression Fast Response sprinklers provide aerosol warehousers with an option that eliminates the need for in-rack sprinklers for Level 2 and Level 3 aerosols at rack storage heights up to at least 15 ft. The water discharge per ESFR sprinkler is 121 gpm for Level 3 products and 100 gpm for Level 2 products. As with most other ESFR applications, the water supply has to provide for a minimum of 12 flowing sprinklers, and there are restrictions on the allowable ceiling slope and near-ceiling obstructions.

NFPA 30B also describes protection requirements for limited quantities of aerosol products in general purpose warehouses. The limitations on allowable amounts of Level 2 and Level 3 products depend on whether the general-purpose warehouse has sprinkler protection meeting the requirements for Level 2 products, or for Level 3 products. There are also requirements for segregating these products from other storage in the warehouse.

6.5 Solid oxidizers

NFPA 430 (2000) defines an oxidizer as "any material that readily yields oxygen or other oxidizing gas, or that readily reacts to promote or initiate combustion of combustible materials." They are a special hazard because they can significantly increase the burning rate and flame temperature of many combustible materials, and they can react exothermally with many materials to cause spontaneous ignition. In addition, the other oxidizing gas referred to in the preceding definition is often chlorine, which poses toxic exposure issues.

Oxidizing swimming pool chemicals (sanitizers) have received a great deal of attention in recent years because of their involvement in several large-loss storage fires. The two most commonly used chlorinating sanitizers are calcium hypochlorite (Ca_2ClOH) and trichloroisocyanuric acid ($C_3Cl_3N_3O_3$). Both react exothermally when wet and/or when mixed with combustible liquids. The June 1988 fire in Springfield, Massachusetts apparently started when a leaking roof allowed rain to enter fiberboard barrels of trichloroisocyanuric acid tablets. According to the Springfield Fire Department reports provided by James Controvich (Director, Springfield Office of Emergency Preparedness), an early fire in a processing machine and ductwork was successfully suppressed by soda ash (sodium carbonate), but a later fire in another location had to be suppressed with copious amounts of water. Chlorinated decomposition and combustion products eventually caused the evacuation of 60 000 residents.

More recent fires have occurred in building supply stores that sell swimming pool chemicals as granules, tablets, or sticks in high-density polyethylene containers of the type shown in Figure 6.8. A typical fire scenario in these facilities involves inadvertent mixing with either combustible materials or incompatible swimming pool supplies stored in close proximity to the containers of chlorinated chemicals. Figure 6.9 shows the smoke plume from one such fire in a home building supply store in Georgia in 1996. Although the fire is reported to be of undetermined origin, it did start in a rack of the chlorinated swimming pool chemicals. According to the NFPA incident report, there was no barrier between the oxidizers and incompatible chemicals. The

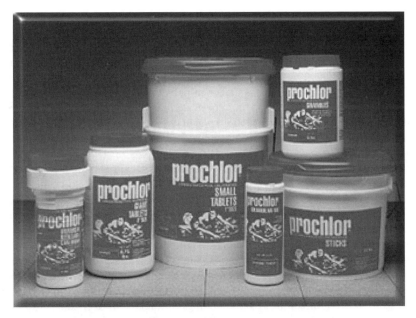

Figure 6.8. Packaged swimming pool chemical containing trichloroisocyanurate (from http://home.twcny.rr.com/dsone/sunspot/sunspotpage1.html)

Figure 6.9. Swimming pool chemical initiated fire at building supplies warehouse. Reprinted with permission from NFPA *Journal January/February*. Copyright © 1998 National Fire Protection Association, Quincy, MA 02269. This reprinted material is not the complete and official position of the National Fire Protection Association, on the referenced subject which is represented only by the standard in its entirety

ceiling sprinkler system was not designed to discharge densities and supply demand areas now called for in NFPA 430. NFPA reported the loss to be $9 million.

National Fire Protection Research Foundation sponsored fire tests at Underwriters Laboratories have demonstrated differences in the fire growth rate and intensity of various chlorinated oxidizers. Tests under the U.L. Fire Products Collector were conducted with two 2/3rd pallet loads of trichloroisocyanuric acid tablets and granules in polyethylene containers placed in corrugated fiberboard cartons protected by water spray nozzles. According to Nugent (1998), the fire heat release rates measured with the chlorinated isocyanurates were comparable to those of a Class 4 warehouse storage commodity. However, the smoke is both toxic and corrosive because it contains chlorine and hydrogen chloride. Furthermore, water spray application apparently produced nitrogen trichloride, which rapidly decomposed in a series of loud and rapid popping bursts.

The U.L. Fire Products Collector tests also included calcium hypochlorite in granular form in polyethylene bottles and in surlin bags. Decomposition of calcium hypochlorite in these tests produced sufficient oxygen to cause dramatically high heat release rates that were beyond the capacity of the 10 MW Fire Products Collector. Approximately 658 kg (1450 pounds) of calcium hypochlorite were completely consumed within 1 to 2 minutes after the fire penetrated the cartons! Moreover, the rapid formation of oxygen caused the surlin bags and polyethylene containers to burst and propel burning packaging material and undecomposed hypochlorite throughout the test room. Thus, there is potential for very rapid fire spread beyond the area of origin. After the calorimetry tests, rack storage fire tests with in-rack sprinklers were conducted to provide the basis for the protection contained in the new edition of NFPA 430.

NFPA 430 divides oxidizers into four classes, with Class 1 being the lowest fire protection challenge, and Class 4 requiring the most restrictions and protection. The qualitative descriptions of the four classes, and some examples in each class, are listed in Table 6.7.

The qualitative distinctions between the four oxidizer classes require the NFPA Technical Committee on Hazardous Chemicals to make subjective interpretations of any available test data or loss experience when placing specific chemicals in each category. Based on these interpretations, trichloroisocyanuric acid is designated as a Class 1 Oxidizer, and calcium hypochlorite is a Class 2 Oxidizer at concentrations less than 50% by weight, and a Class 3 at concentration greater than 50%.

Ammonium perchlorate is a Class 4 oxidizer because it has been demonstrated to undergo explosive decomposition such as occurred upon fire exposure in 1987 in a devastating incident at the PEPCON manufacturing facility in Henderson, Nevada. The ammonium perchlorate being stored in large tote containers detonated and produced blast damage with a TNT equivalent of greater than 500,000 lb TNT (Minizewski, 1992). Since the PEPCON plant was one of only two plants in the US that manufactured ammonium perchlorate for use in solid rockets, its destruction caused a serious delay in US rocket launches that year.

Table 6.7. Oxidizer classes in NFPA 430

Oxidizer class	Primary hazard (per NFPA 430)	Examples
1	Causes slight increase in burning rate	Inorganic nitrates and nitrites
2	Causes moderate increase in burning rate or spontaneous ignition	Calcium chlorate, calcium chlorite, sodium perchlorate
3	Causes severe increase in burning rate or undergoes vigorous decomposition	Calcium hypochlorite, sodium chlorite
4	Undergoes explosive reaction upon exposure to heat or shock	Ammonium perchlorate

Although several of the oxidizers decompose exothermally when mixed with water, it is generally believed that copious water spray application is the most effective method of extinguishment because the water acts as a heat sink when applied in large quantities. NFPA 430 provides sprinkler protection guidelines for storage of Class 2 and Class 3 oxidizers in warehouses and in retail stores such as the building supply stores that have experienced the swimming pool chemical fires. The warehouse storage guidelines for Class 2 oxidizers specify a relatively modest ceiling sprinkler density of $0.35\,\text{gpm/ft}^2$ ($14\,\text{mm/min}$) over $3750\,\text{ft}^2$ for palletized storage up to 12 ft high, but call for in-rack sprinklers in every tier of rack storage. The requirements for retail storage include installation of both horizontal barriers in every tier, and vertical barriers at all rack uprights in the area of oxidizer storage. These barriers are intended to confine the fire and allow the in-rack sprinkler discharge to achieve fire control. They are also intended to prevent mixing of incompatible materials (such as combustible liquids) with the oxidizers.

6.6 Bulk storage

6.6.1 GENERAL DESCRIPTION

Many industrial materials are stored in bulk either in outdoor piles, or in silos, bunkers, or bins. Some powders, pellets, and chips are stored in large bags or bulk containers in warehouses. In addition, many manufacturing processes include curing of bulk material at elevated temperatures and controlled humidities. The large quantities in bulk storage, and the possible increased temperature when placed in storage, raise some special fire protection issues in addition to the considerations discussed for other commodities and materials. These special issues include the prevention and detection of spontaneous combustion, prevention of explosions in silos and bunkers, and effective firefighting methods for large stockpiles and silos with deep-seated smoldering fires.

Spontaneous combustion in bulk storage begins either with a slow exothermal chemical or biochemical reaction (oxidation, microbiological respiration or decay, auto-decomposition, or polymerization, for example), or by heat released due to moisture condensation or absorption. As these processes continue, the temperature in the interior of the stockpile gradually increases above ambient. The magnitude of the temperature rise is governed by the rate of heat liberation in comparison to the rate of heat dissipated to the surrounding air. If the rate of heat dissipation is not sufficient, the temperature may eventually increase to the point where combustion reactions begin. The combustion often occurs as smoldering, but can transit to flaming combustion if there is a sudden increase in the rate of oxygen transport to the combustion zone.

Table 6.8 lists several bulk storage materials that have been involved in fire incidents attributed to spontaneous ignition. The materials are grouped by what is believed to be the primary initial heat generation mechanism in most incidents. Often there are multiple mechanisms, such as biological heating from ambient temperature to about $70\,°\text{C}$, followed by oxidation to the ignition temperature. Similarly, in the case of moisture induced heating, it is often difficult to differentiate between the effects of moisture and other heating modes. Back (1981) has provided an excellent review of moisture effects in hygroscopic materials, and some guidance on ways to prevent or mitigate these effects.

Particle size, bulk density, moisture level, and the presence of contaminants or additives can have a major effect on whether or not a particular material will undergo self-heating by one or more of these mechanisms. For example, the presence of absorbed oil or asphalt on many porous materials can dramatically increase their propensity for self-ignition. Therefore, it is important to have spontaneous heating/ignition test data for the particular bulk material at the conditions of storage.

Table 6.8. Examples of bulk materials prone to spontaneous heating

Microbiological heating	Oxidative heating	Moisture induced heating	Other chemical processes
Bagasse (sugar can residue)	Activated Carbon	Chlorinated Oxidizers	Monomers
Compost	Coal (Low-Rank)	Calcium Oxide (Lime)	Nitrocellulose
Grains	Cotton	Cotton Bales	Peroxides
Hay (moist)	Foam Rubber	Dry Paper Rolls	such as lauroyl peroxide,
Mulch (moist)	Metal Filings and Powders	Insulating Boards (dried) (lingo-cellulosic)	and benzoyl peroxide.
Pecans	Particle Board	Potassium Phosphide	
Sewage Sludge	Peat	Wool Bales	
Soya Beans	Sawdust		
Walnuts	Wood Chips		

6.6.2 SPONTANEOUS IGNITION TESTING

A variety of laboratory tests are available to characterize the spontaneous heating/ignition characteristics of combustible materials. If the material is homogeneous and can be characterized with samples on the order of a few milligrams, the traditional thremochemical analytical test methods of Differential Thermal Analysis (DTA), Thermogravimetric Analysis (TGA), and Differential Scanning Calorimetry (DSC) can be employed. DTA provides the exotherm initiation temperature or activation energy, TGA measures the mass decomposition involved in the exotherm, and DSC provides the heat of reaction. There are several detailed descriptions of these methods in textbooks such as Wendlandt (1974) and in review papers.

If the material is inhomogeneous, and/or is readily available in sample sizes in the range one to 100 g, larger scale uniform heating test methods can be used to determine the temperature at which spontaneous heating/ignition begins. Table 6.9 is a tabulation of the standardized tests of this type. Most of the tests entail placing the sample in an instrumented wire mesh basket or other container, and then placing the sample (in some cases alongside an inert sample) in an oven, furnace, or special apparatus to slowly heat the sample while its temperature remains near-uniform. The tests continue for either a designated duration, or until the exotherm is initiated. In several tests the sample is in the form of a powder or otherwise pulverized to fit in the basket. The small particle size facilitates air access to the interior of the sample so that oxidation or combustion can proceed throughout the sample.

6.6.3 SPONTANEOUS IGNITION THEORY

Spontaneous ignition theory is intended to utilize the spontaneous heating test data to determine the relationship between the size of the storage pile and the critical temperatures at which the low-level heat generating processes escalate to spontaneous combustion. Classical spontaneous ignition theory (sometimes called thermal explosion theory because of its initial application to unstable explosives and propellants) is based on the simplifying assumption that there is one exothermal chemical reaction as opposed to the two-stage self-heating process described in Section 6.6.1. This reaction is assumed to generate heat at a rate governed by an Arrhenius reaction temperature dependence. The approach is to first model a steady-state heat transfer process in which the heat generated by this reaction is balanced by the heat transferred at the surface of the pile, and then to use the model to determine the conditions under which a steady-state solution is not possible because the surface heat transfer rate is not adequate. Comprehensive descriptions of this approach are provided by Bowes (1984), Drysdale (1998), and Beever (1995); only a brief summary is presented here.

Table 6.9. Standard tests for spontaneous heating/ignition

Test designation (reference)	Test material	Sample size	Heating apparatus	Test duration (hr)
ASTM E771	Liquid or Particulate Solid	≥ 10 ml	Insulated Vessel with air supplied to sample well	24
ASTM D3523	Liquid or Particulate Solid	10 g	Open top double chamber for test sample and inert sample	4–72
BAM[a] SADT	Liquid or Solid	400 ml	0.5 liter dewar in over	168
VDI[b] 2263	Powder	8 ml	Oven with preheated air flow	?
Bu Mines Adiabatic Heating Oven (RI 8473)	Coal Dust	100 g	Oven with preheated air flow	Varies
JRIIS SIT[c] (Kotoyori, 1988)	Liquid or Solid	0.5–3 g	Open or closed cell adiabatic heating apparatus	Varies

[a] Bundesandalt fur Materialprufung (Berlin), Self-Accelerating Decomposition Temperature
[b] Verein Deutscher Ingenieure
[c] Japanese Research Institute of Industrial Safety Self-Ignition Test

Using the approximation that the temperature rise above ambient is small in comparison to the ratio of the activation energy to the universal gas constant, the differential equation governing the steady-state heat transfer with Arrhenius rate internal heat generation is (Drysdale, 1998, p. 199; Bowes, 1982, p. 27):

$$\nabla^2 \vartheta = -\delta e^\theta$$

$$\delta = \frac{E \rho Q A r^2}{R T_a^2 \lambda} e^{-E/RT_a} \quad [6.6.1]$$

where θ is a nondimensional temperature rise, $\vartheta = \frac{(E(T-T_a))}{RT_a^2}$, and δ is the Frank–Kamenetskii parameter, which represents the ratio of the heat generation rate at T_a to the heat conduction rate from the center of the material. The other parameters in equation [6.6.1] and the Frank–Kamenetskii parameter are as follows: E is the reaction activation energy (J/mol), A is the pre-exponential factor in the Arrhenius reaction rate equation (s^{-1}), R is the universal gas constant = 8.314 J/mol-K, ρ is the material bulk density (kg/m^3), Q is the heat of reaction (J/kg), r is characteristic length of the storage pile (m), λ is the material thermal conductivity (W/m-K), and T_a is the material storage temperature (K).

The boundary conditions associated with equation [6.6.1] are

$$\frac{d\vartheta}{dz} = 0 \quad \text{at } z = 0 \text{ (center of storage pile)}$$

$$-\frac{d\vartheta}{dz} = \alpha \theta_s \quad \text{at } z = r \text{ (surface of storage pile)} \quad [6.6.2]$$

where h is the natural convection heat transfer coefficient at the pile surface, and α is the Biot number.

Unstable storage conditions, as predicted by this theory, correspond to situations in which the value of δ is larger than a critical value denoted by δ_c. The value of δ_c depends upon the geometry

Table 6.10. Critical values of Frank–Kamenetskii parameter for different shape storage piles (compiled from values in Bowes, 1984)

Pile geometry	Dimension	δ_c
Plane slab	Height $2r \ll$ width and length	0.88
Rectangular box	Height $2r$, width $2w$, length $2l$	$= 0.88(1 + (r/w)^2 + (r/l)^2)$
Cube	$2r \times 2r \times 2r$	2.52
Cylinder	Diameter $\ll$ Height	2.00
Cylinder	Diameter/Height $= 2r/2l$	$\delta_c(r) = 2.0 + 0.84(r/l)^2$
Cone[a]	$D/H = 1$	9.1 ± 0.1
Diameter $= D$	$D/H = 2$	3.2
Height $= H$	$D/H = 4$	1.7

[a] δ_c for cones is based on the characteristic dimension $H/2$

of the storage pile and the value of α. In most spontaneous heating applications with piles of at least several meters characteristic dimension, the Biot number is sufficiently large (greater than about 20) for the asymptotic limit to be applicable. Values of δ_c for different geometry storage piles are listed in Table 6.10.

By taking logarithms of the defining equation above for the Frank–Kamenetski parameter, we can obtain

$$\ln \frac{\delta_c T_a^2}{r^2} = M - \frac{P}{T_a} \quad [6.6.3]$$

where
$$M = \ln \frac{EQ\rho A}{R\lambda}$$
and $P = E/R$.

Writing the equation in this form provides a framework for correlating and extrapolating the results of the spontaneous heating initiation tests in which the critical values of storage temperature, T_a, are measured for a laboratory sample of dimension r. If the tests are repeated with other size samples and results are plotted in the form $\ln(\delta_c T_a^2/r^2)$ versus $1/T_a$, the slope of the best-fit line through the data should correspond to the value of material parameter P, and the extrapolated intercept at $1/T_a = 0$ should correspond to the material parameter M. Results will depend upon the units for T_a and r used to plot the data and obtain the best-fit.

Data correlations using the preceding technique have been conducted by researchers to determine the values of M and P for several materials susceptible to spontaneous heating/ignition. Results are given in Table 6.11.

Using the tabulated values of M and P, the preceding equation can be used to calculate the value of r for a maximum safe storage pile at an ambient temperature T_a, or the maximum safe material curing temperature T_a for a given size curing slab. Solving equation [6.6.3] for the critical radius, r_c, for spontaneous ignition,

$$r_c = T_a \sqrt{\delta_c e^{-(M-P/T_a)}} \quad [6.6.4]$$

As an example, consider the case of the chemically activated carbon with the M and P values listed in Table 6.11, which is to be stored in a silo with a height/diameter ratio of 1, such that $\delta_c = 2.84$. Using those values in equation [6.6.4], we can determine the relationship between the silo radius, r_c, and the ambient storage temperature, T_a, for the critical self-ignition conditions. Results are shown in Figure 6.10 for the curve labeled Bowes, $E = 97\,\text{kJ/mol}$ (the value of E

Table 6.11. Spontaneous heating material constants

Material	M	P	Reference
Wood-Fibre Insulating Board	34.55[a]	12,145[a]	Bowes (1984)
Chemically Activated Carbon	35.903[a]	11,670[a]	Bowes (1984)
Animal Feedstuff (Distillery Byproduct)	39.88	8404	Beever (1995)
Milk Powder	41.84	9497	Beever (1995)

[a] The Bowes values for M and P are based on r in mm in equation [6.6.3]

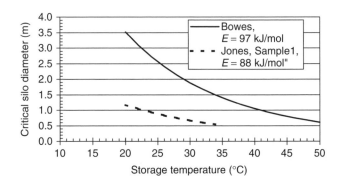

Figure 6.10. Critical silo diameter for chemically activated carbon

corresponding to P). They indicate that the critical silo radius for a storage temperature of 20 °C is 3.5 m, and it gets progressively smaller as the storage temperature increases.

The other curve in Figure 6.10 is calculated from the definition of δ and the parameter values obtained by Jones (1998) for his activated carbon Sample # 1. Jones used a different laboratory test method and data analysis approach. Using a laboratory micro-calorimeter, Jones measured the heat generated from one size sample (120 mg) at various storage temperatures. A curve fit to the data allowed Jones to calculate both the activation energy ($E = 88$ kJ/mol), and the pre-exponential factor ($A = 5 \times 10^5$ s^{-1}). He estimated the following other property values for activated carbon: $Q = 25$ kJ/g, $\rho = 360$ kg/m^3, $\lambda = 0.055$ W/(m°K). These values were used in the following equation for r_c versus T_a:

$$r_c = T_a \sqrt{\frac{\delta_c \lambda R}{Q E \sigma A e^{-E/RT_a}}} \qquad [6.6.5]$$

to obtain the lower curve in shown in Figure 6.10. For a given storage temperature, the most sensitive parameter in equation [6.6.5] is the reaction activation energy, E. The lower activation energy obtained by Jones for his sample produced correspondingly smaller critical silo radii for a given storage temperature. These differences demonstrate the importance of obtaining site-specific samples and test data, rather than rely on generic material data to determine safe storage conditions. Jones (1998, 1999) also reported on results for 10 other activated carbon samples with activation energies ranging from 88 kJ/mol to 113 kJ/mol. In a later paper he also reported results for samples of sawdust, bituminous coal, and lignite.

Even if the material is stored at temperatures and pile sizes that could lead to self-ignition, the time required for the material to ignite may be exceedingly long. Correlations for the time-to-ignition are given in Bowes (1984) and Beaver (1995). In one example calculation provided by Beaver for animal feedstuff storage, the storage conditions were found to be supercritical, but the calculated time-to-ignition (180 days) was significantly longer than the summer storage period.

6.6.4 DETECTION AND SUPPRESSION OF BULK STORAGE FIRES

Automatic detection of smoldering within outdoor stockpiles entails use of thermal monitoring devices. There are two options: monitoring the pile surface with an infrared system, or installing some type of temperature monitor (usually thermisters) within the pile interior. A decision as to which option to use and where to monitor should be based on knowledge of the size and location of the smoldering region. This can be ascertained by viewing calculated results of Handa et al. (1983) simulating the spontaneous heating and eventual ignition of a large coal stockpile.

Figure 6.11 shows the calculated isotherms reported by Handa et al. for a 22 m high trapezoidal pile with a 12 m wide top surface and a 64 m wide base. Isotherms slowly develop in the pile as oxidation occurs at a rate dependent on oxygen diffusion into the pile interior from the top and side surfaces. The maximum temperature after 30 days is 40 °C, only about 8 °C above the ambient temperature. The highest temperatures at this time occur near the upper portion of the sloping side of the pile. After 40 days, the maximum temperature of 55 °C occurs well within the pile, with slightly lower temperatures along the upper surface. After 45 days, the maximum temperature of 80 °C occurs near the junction of the sloping and upper surface. Ignition occurs at 47 days just beneath the outer edge of the upper surface of the pile. Surface temperatures in that area are in the range 50 °C to 100 °C.

Based on the isotherms in Figure 6.11, the most promising regions for thermal detection would be the outer edge of the upper surface of the pile, and the interior near the junction of the sloping side and upper surface. If thermal imaging is used to search for signs of spontaneous heating, the imaging device should be situated sufficiently high to see the upper surface, which is 22 m high in this case. Thus the thermal monitor would need to be installed on a tall structure or column. Alternatively, optimally situated pre-installed temperature monitors would have to anticipate the location of the outer surfaces of the stockpile.

Once an incipient or burning fire is detected, the usual practice is to remove the burning material in order to cool it and curtail heating of the rest of the stockpile. Recommendations to this effect are offered in NFPA 850 (1996) and other guidelines developed for outdoor unenclosed storage of coal and wood chips. If the fire has burrowed into the interior of a very large fire before suppression efforts begin, it is common for these manual suppression efforts to take weeks or months before the fire is completely extinguished.

Smoldering fires in bunkers and silos provide several other options for detection and suppression. Detection is often achieved with methane and CO detectors as recommended in NFPA 850. A CO concentration of 1.25 percent should alert plant operators to take action according to NFPA 850. It would also be prudent to install oxygen concentration monitors per the recommendations of Tuomisaari et al. (1998), because the oxygen monitors provide a means of knowing if and when suppression has been achieved. Locating and monitoring hot spots on the silo/bunker wall (with an infrared camera) is another means of fire detection and confirmation of fire suppression.

Effective suppression methods for bunker and silo fires are different than those for enclosed stockpiles because of the threat of dust explosions and because of the opportunity to use gaseous suppression agents. Van Wingerden and Alfert (1994) have described silo fire incidents in which explosions have occurred because attempts to remove the burning particulate prior to suppression actually generated combustible dust clouds within the silo. Water spray discharge within the silo can also generate combustible dust clouds, and water accumulation in the silo can result in

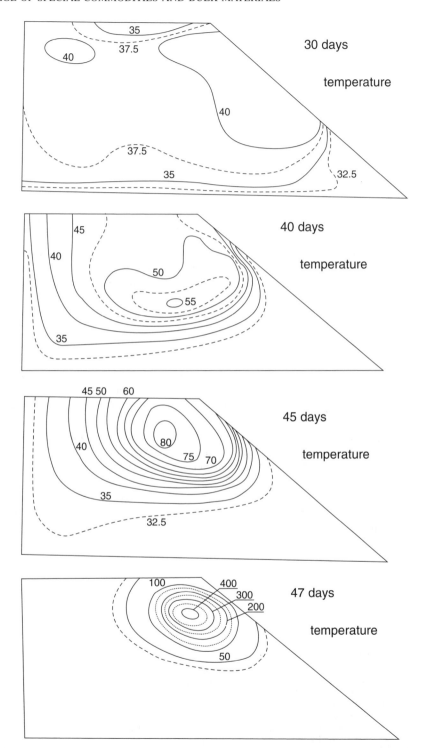

Figure 6.11. Calculated isotherms during spontaneous heating in coal pile (based on Handa et al., 1983)

sufficiently high hydrostatic pressures to damage the silo structure. Therefore water discharge into the silo, bunker is not recommended. However, low expansion foam has been used successfully both in VTT laboratory tests (Tuomisaari et al., 1998) and in several incidents.

The most success in extinguishing silo/bunker fires has been achieved with carbon dioxide, and to a lesser extent with nitrogen. For example, Tuomisaari et al. conducted approximately 50 laboratory tests with carbon dioxide or nitrogen applied to barrels of smoldering wood chips or peat. The gaseous agents were applied to the top of the barrel in some tests and to the bottom in other tests. Better results (i.e. earlier suppressions) were achieved with bottom application because the gases flowed up through the smoldering material and gradually extinguished the fire. These differences were accentuated in tests with long pre-burns prior to agent discharge, because those tests generated the largest burning regions in the barrel.

Tuomisaari et al. varied the amount of CO_2 used and the rate at which was discharged in their smoldering wood chip tests. The most complete suppression was achieved with the largest amount used: 1.5 kg-CO_2 per m^3 of barrel volume. At room temperature, this corresponds to 0.82 m^3-CO_2 per m^3 of barrel volume. If the carbon dioxide displaces the air in the silo, this volume ratio would reduce the oxygen concentration to 4%. However, it is more likely that the carbon dioxide mixes with the air, such that an air-CO_2 mixture is vented, and the remaining oxygen concentration becomes 12%. This is further reduced by the combustion process, which was observed to reduce oxygen concentrations to 2–5% in the VTT tests. The most rapid suppression was achieved with the largest CO_2 application rate, but the amount of CO_2 required to achieve suppression was larger with large discharge rates than with smaller discharge rates.

NFPA 850 also has some guidance on the amounts of carbon dioxide required to suppress smoldering coal fires in silos and bunkers. Appendix A of NFPA 850 cites experience at one utility that suggests the available amount of carbon dioxide for fighting a coal silo fire should be 3 m^3-CO_2 per m^3 of silo. This is more than three times the amount recommended by Tuomisaari et al. based on their laboratory tests. The additional CO_2 is presumably needed because of additional leakage and more non-uniformities in the large silos.

Patience is required in allowing the silo/bunker fire to be fully suppressed before attempting to remove the coal or other bulk combustible. Suppression was achieved within a few hours in the 200 liter (53 gallon) barrel tests at VTT, but often requires several days in silos with volumes on the order of 850 m^3 (30,000 ft^3). During that time, it is very helpful and reassuring to have oxygen and CO readings from detectors installed at the top of the silo. Absent those readings, firefighting personnel have to rely on smoke emissions and silo wall temperatures as evidence of fire suppression progress.

References

Association of the Nonwoven Fabric Industry (INDA), *Nonwoven Fabrics Handbook*, Cary, North Carolina, 1999.
ASTM D 3523, 'Standard Test Method for Spontaneous Heating Values of Liquids and Solids (Differential Mackey Test),' ASTM, 1980.
ASTM E 771, 'Standard Test Method for Spontaneous Heating Tendency of Materials,' ASTM, 1981.
Beever, P., Self-Heating and Spontaneous Combustion, *SFPE Handbook of Fire Protection Engineering*, 2nd Edition, Section 2, Chapter 12, SFPE, NFPA, 1995.
Back, E.L., Auto-Ignition in Hygroscopic, Organic Materials – Especially Forest Products – as Initiated by Moisture Absorption from the Ambient Atmosphere, *Fire Safety Journal*, **4**, 185–196, 1981/82.
Bowes, P.C., *Self-Heating: Evaluating and Controlling the Hazards*, Elsevier, 1984.
Chicarello, P. and Troup, J., 'Decaying Pressure Fire Tests in Roll Tissue Paper Storage,' FMRC Report J.I. 0H3R2.RR prepared for the American Paper Institute, September 1982.
Cleremont-Ferrand, 'Fire Extinguishment Tests on Tire Storage,' November 1971 (English Translation by Centre National de Prevencion et de Protection, 1973).
Dean, R., 'Protection Investigation of 25-ft High Rubber Tire Storage,' FMRC Report J.I. 0Z2R6.RR for the Rubber Manufacturers Association, July 1996.

Drysdale, D., *Fire Dynamics*, 2nd Edition, John Wiley & Sons Ltd, 1998.
Factory Mutual Loss Prevention Data Sheet 8-21, 'Roll Paper Storage,' FM Global, 1991, Revised October 1995.
Factory Mutual Loss Prevention Data Sheet 8-3, 'Tire Storage,' FM Global, 2001.
Factory Mutual Loss Prevention Data Sheet 8-23, 'Rolled Nonwoven Fabric Storage,' FM Global, 1991, revised 1993.
Handa, T., Morita, M., Sugawa, O., Ishii, T. and Hayashi, K., Computer Simulation of the Spontaneous Combustion of Coal Storage, *Fire Science and Technology*, **3**, 13–23, 1983.
International Association of Fire Chiefs, 'Guidelines for the Prevention and Management of Scrap Tire Fires,' 1995.
Johnsen, M., *The Aerosol Handbook*, 2nd Edition, Wayne Dorland Company, Mendham, NJ, 1982.
Jones, J.C., Recent Developments and Improvements in Test Methods for Propensity Towards Spontaneous Heating, *Fire and Materials*, **23**, 239–243, 1999.
Jones, J.C., Towards an Alternative Criterion for Shipping Safety of Activated Carbons, *Journal of Loss Prevention in the Process Industries*, **11**, 407–411, 1998.
Jones, J.C., Towards an Alternative Criterion for Shipping Safety of Activated Carbons. Part 2. Further Data Analysis and Extension to Other Substrates, *Journal of Loss Prevention in the Process Industries*, **12**, 183–185, 1998.
Kotoyori, T., Critical Ignition Temperatures of Chemical Substances, *Journal of Loss Prevention in the Process Industries*, **2**, 1989.
Kuchta, J.M., Rowe, V.R. and Burgess, D.S., 'Spontaneous Combustion Susceptibility of U.S. Coals,' Bureau of Mines RI 8474, 1980.
Miniszewski, K., PEPCON Explosion Incident, *Journal of Fire Protection Engineering*, 1992.
NFPA, 'Bulk Retail Store Fire,' NFPA Fire Investigation Report, 1996.
NFPA 13, 'Standard for Sprinkler System Installation,' National Fire Protection Association, 2000.
NFPA 30B, 'Manufacture and Storage of Aerosol Products,' National Fire Protection Association, 1998.
NFPA 230, 'Standard for Fire Protection of Storage,' National Fire Protection Association, 1999.
NFPA 231D, 'Standard for Storage of Rubber Tires,' National Fire Protection Association, 1998.
NFPA 231F, 'Standard for Storage of Roll Paper,' National Fire Protection Association, 1996.
NFPA 420, 'Code for the Storage of Liquid and Solid Oxidizers,' National Fire Protection Association, 2000.
NFPA 850, 'Recommended Practice for Fire Protection for Electric Generating Plants and High Voltage Direct Current Converter Stations,' National Fire Protection Association, 1996.
Nugent, D., Oxidizing Pool Chemicals, *NFPA Journal*, **92**, July/August 1998.
Nugent, D., Summer Safety: Pool Chemical Regulations, *NFPA Journal*, **94**, May/June 2000.
Rizzo, J., Aerosol Flammability Test, *Fire Journal*, 1986.
Tuomisaari, M., Baroudi, D. and Latva, R., 'Extinguishing Smouldering Fires in Silos,' Brandforsk Project 745-961, VTT Publication 339, Technical Research Centre of Finland, 1998.
VDI Richtline 2262, Untersuchungsmethoden zur Ermittlung von sicherheits-technischen Kenngrossen von Stauben, VDI-Verlag GmbH, Dusseldorf, 1988.
Van Wingerden, K. and Alfert, F., 'Detection and Suppression of Smoldering Fires in Industrial Plants,' Christian Michelsen Research Report CMR-94-F25068, 1994.
Wendlandt, W.W., *Thermal Methods of Analysis*, John Wiley & sons Ltd, 1974.
Zalosh, R., 'Aerosol Product Flammability Classification Test Program,' FMRC J.I. 0M1R4.RU, July 1986.

7 FLAMMABLE LIQUID IGNITABILITY AND EXTINGUISHABILITY

Flammable liquids pose special problems because they are readily ignited, burn with high heat release rates, spread flame rapidly and extensively, and are difficult to extinguish. This chapter provides quantitative data on ignitability and extinguishability and pool fire and spray fire phenomenology.

7.1 Incident data

Table 7.1 shows the reported occurrence of various flammable liquid fire ignition sources at Factory Mutual insured facilities for the ten year period from 1975 through 1984. Except for the relatively large number of spontaneous ignitions and the relatively few incendiary ignitions, the distribution is similar to that shown in Figure 1.2 of Chapter 1 for manufacturing facilities in general.

Electrical ignitions sources are most frequent and are associated with both electrical equipment and electrostatic discharges. Electrical equipment ignitions are usually due to sparking, arcing, or shorting. Some examples from the loss reports are: (1) arcing caused by solvent leakage onto an extension cord and subsequent deterioration of the electrical insulation; (2) accidental pulling of energized electrical cable out of a control box such that an arc is generated; (3) an electrical short in the starter motor of a forklift truck; and (4) arcing in an electrical junction box knocking off the unfastened cover of the box and burning a hole in an adjacent hydraulic fluid line. Electrostatic ignitions typically occur in paint spraying operations, in liquid transfer piping, and in mixing and filtering operations. Examples are cited in Sections 7.3 and 7.5.

According to the incident data in Table 7.1, spontaneous ignition is the second most frequent ignition source for flammable liquid fires. These incidents are associated with liquids that are prone to spontaneous heating at ambient temperatures (for example, linseed oil), from leakage of hot liquid onto pipe insulation, or by release of liquid at a temperature above its autoignition temperature. An example of the latter type occurred on a resin reactor with a sample port that was heated to assure liquid withdrawal of the slurry reactants; the sample port was inadvertently heated above the resin autoignition temperature.

Hot surfaces represent the third most prevalent ignition source for flammable liquid fires according to the data in Table 7.1. Typical examples are spilled liquids ignited by the exhaust pipe of a gas- fired lift truck or by the hot exhaust duct of a furnace or some other fuel-fired equipment. Radiant heaters and induction heaters adjacent to ruptured hydraulic oil lines have also been identified as the hot surface responsible for several ignitions.

Table 7.1. Flammable liquid fire ignition sources

Source	Incidents	%
Electrical	559	25.6
Spontaneous	317	15.5
Hot surface	278	12.7
Burner flame	204	9.4
Overheating	198	9.1
Incendiarism	160	7.3
Cutting/welding	137	6.3
Friction	120	5.5
Miscellaneous sparks	95	4.4
Smoking	69	3.2
Exposure	30	1.4
Molten material	12	0.6
Total	2137	100.0

Burner flames are common ignition sources for heat transfer fluids accidentally released within or in the vicinity of process heaters. Damage in these fires occurs when flame flashes back to the fluid release site. Criteria for flashback and sustained liquid burning are discussed in the following sections.

7.2 Ignitability temperatures

7.2.1 FLASH POINTS AND FIRE POINTS

The most often cited measure of a flammable liquid's ignitability is its flash point. Flash point is defined experimentally as the lowest temperature (at atmospheric pressure) at which the vapor-air mixture near the surface of a liquid sample can be ignited under specified test conditions. Flash points are termed either closed cup or open cup flash points depending on the nature of the test apparatus.

There are at least four different closed cup flash point test methods and two different open cup flash point test methods described in ASTM Standards. ASTM E 502-84 summarizes the various methods and provides guidance on the selection of an appropriate closed cup method depending on liquid viscosity and expected flash point range. NFPA 30 prescribes similar guidance for closed cup test methods. The basic difference between the methods is the rate of sample heating and whether or not the liquid is stirred during the test. The ignition source in all three NFPA 30 specified closed cup methods (the Tag method, the Pensky–Martens method, and the Setaflash method) is a small flame directed into the vapor space of the closed cup tester at specified intervals.

Closed cup and open cup flash points for a number of common flammable liquids are listed in Table 7.2. The open cup flash points are typically 3–9 °C (6–16 °F) higher than the corresponding closed cup flash points. However, there are certain liquids (e.g. hexane) which produce lower open cup flash points than closed cup flash points. NFPA 497 (1997) has a much more extensive list of closed cup flash points. Hshieh (1997) has developed an empirical correlation relating closed cup flash point to liquid boiling point for general organic compounds including silicone compounds.

Closed cup flash points are the primary basis for classifying flammable liquids in fire protection codes and standards. NFPA 30, for example, has the following classification scheme:

Table 7.2. Flammable liquid ignitability properties

Liquid	Closed cup flash point (F)[a]	open cup flash point (F)	fire point (F)	Boiling point (F)[a]	Autoignition temperature (F)[a]
Acetone	0	15	–	133	869
Acrylonitrile	23	32	–	171	898
Benzene	12	–	–	176	1040
n-Butanol	84	110	110[c]	243	689
Carbon Disulfide	−22	–	–	115	194
Cyclohexane	−4	–	–	179	473
n-Decane	115	131	142[c]	345	410
Ethanol	55	71	–	173	685
Ethyl Ether	−49	–	–	95	320
Ethylene Glycol	232	240	–	387	752
Fuel Oil – No. 2	255[b]	–	264[b]	–	500[b]
Fuel Oil – No. 6	295[b]	–	351[b]	–	629[b]
Gasoline[e]	−46	–	–	Varies	536
Glycerol	320	350	–	554	698
Heptane-n	25	30	–	198	536
Hexane-n	−7	−14	–	156	437
Isopropanol	53	60	–	181	750
Jet A Fuel[d]	100–150	–	–	335–570	>435
JP-4 Jet Fuel[d]	−10 → +30	–	–	140–455	>445
Kerosene	>100	–	–	304–574	410
Methanol	54	60	–	148	725
Methyl Ethyl Ketone	28	34	–	175	960
Methyl Salicylate	214	–	–	432	850
Mineral Spirits	104	–	–	300	473
Motor Oil	420[b]	–	435[b]	–	690[b]
Naptha, V.M.& P	28	–	–	211–320	450
Phenol (Carb. Acid)	175	185	–	358	817
Styrene	90	100	–	293	914
Toluene	40	45	–	231	1026
Turpentine	95	–	–	300	488
Vinyl Acetate	18	30	–	163	801
Xylene-o	63	75	–	291	867

[a] Closed cup flashpoint, boiling point, and ignition temperature data are from NFPA 325M-1977 unless otherwise noted
[b] Data from Modak, EPRI NP-1731, 1981
[c] Data from Glassman and Dryer (1980)
[d] Jet A and Jp-4 data from AGARD-AR-132-Volume 2
[e] Data from NFPA 497-1997

NFPA Class	Flash point range	
IA	<73 °F	Boiling Point <100 °F
IB	<73 °F	Boiling Point >100 °F
IC	>73 °F	<100 °F
II	>100 °F	<140 °F
IIIA	>140 °F	<200 °F
IIIB	>200 °F	

Thus, Class I liquids can readily be ignited at ordinary (IA and IB) or warm (IC) room temperatures, while Class II liquids can readily be ignited in extremely warm environments such as might occur on a hot summer day in an unventilated enclosure. Class III liquids require some type of process or exposure fire preheating for ignition in a pool configuration.

The United Kingdom utilizes the following classification scheme:

U.K. Classification	Flash Point Range
Highly Flammable Liquid	<32 °C (90 °F)
Flammable Liquid	32–60 °C (90–140 °F)
Combustible Liquid	>60 °C (140 °F)

A theoretical flash point can be defined as the liquid temperature which produces a lean flammability limit vapor-air mixture over the liquid surface. This theoretical flash point can be calculated from equilibrium vapor pressure-temperature data and lower flammable limit data. Drysdale (1985) has provided sample calculations for pure liquids and for flammable liquid mixtures satisfying Raoult's Law. Glassman and Dryer (1981) and Thorne (1977) have shown several examples in which the theoretical flash point for most liquids is within 2 °C (3.6 °F) of their closed cup flash points. Thorne (1977) has also shown how flash points of mixtures of flammable and nonflammable liquids can be calculated from vapor-air-inert gas flammability diagrams.

Mixtures of flammable and nonflammable components can produce some unexpected flash points. If the nonflammable component is more volatile than the flammable component, the nonflammable component vapor can inert the vapor space of a closed cup and prevent any flash flame. The same mixture may exhibit a flame when tested in an open cup or re-tested in a closed cup after some of the liquid has been allowed to evaporate. On the other hand, there have been some reports (e.g. Wray, 1984) of volatile nonflammable chlorinated hydrocarbons that have actually lowered the flash point of a hydrocarbon solvent when added to it in small proportions. In view of these unanticipated data, both open cup and closed cup flash points are needed to properly evaluate the ignitability hazard for liquid mixture spill scenarios.

Glassman and Dryer (1981) did some interesting experiments to investigate the effects of experimental conditions on measured open cup flash points for various liquids. They used a sustained pilot flame ignition source at various heights above the surface of the pure liquid samples. Their data demonstrate that the open cup flash point determined in this manner decreases as the flame ignition source approaches closer to the liquid surface. Thus it is possible to ignite vapors at temperatures below those measured in both the open cup and the closed cup methods which place the pilot flame at a specified height (6.3 mm or 1/8 in the ASTM D-1310 open cup) above the liquid surface. When Glassman and Dryer repeated the tests with a spark ignition source, significantly higher liquid temperatures were required for ignition. Thus the sustained flame ignition source must have locally heated the liquid surface.

If the open cup flash point test is repeated at higher liquid temperatures, eventually the liquid sample will continue to burn after the pilot flame is removed. The lowest temperature at which sustained burning occurs in the open cup apparatus is called the fire point of the liquid. Some fire points are listed in Table 7.2. They are 5–31 °C (9–56 °F) higher than the corresponding liquid closed cup flashpoint; the median difference is 11 °C (20 °F).

Neither the closed cup not the open cup flash point can be relied upon to provide safe lower bound liquid temperatures in the following situations:

- Enclosure pressures below one atmosphere.
- Energetic spark ignition sources.
- Chemically unstable liquids or liquids in containers that can undergo slow reactions so as to produce volatile or pyrolytic products.

Examples of these situations and data for aircraft fuels are presented in DOT/FAA/AR-98/26; examples and data for asphalts and residual fuel oils are reported by Zalosh (1997).

7.2.2 AUTOIGNITION TEMPERATURES

Autoignition is the spontaneous ignition of a material in the absence of an imposed ignition source such as a pilot flame or spark. The autoignition temperature of a liquid is the minimum temperature at which autoignition occurs under specified test conditions. Liquid autoignition data are usually obtained by raising the temperature of a container to some measured value, pouring the liquid sample into the container, and observing if and when the liquid ignites. The measured autoignition temperature is known to depend on the container size and material. ASTM E-78 specifies that the test be conducted in a 500 ml (16.9 fl oz) borosilicate flask suspended in a furnace.

Table 7.2 contains autoignition data obtained, for the most part, from NFPA 325M-1977. In all cases, the autoignition temperature is well above the liquid boiling point. Thus, autoignition actually refers to the spontaneous ignition of the vapor or vapor-air mixture in the heated test apparatus.

If a flammable liquid is exposed to a surface heated to the liquid autoignition temperature, the liquid may or may not ignite, depending on the surface size, material, and orientation. Small heated surfaces must be considerably hotter than the liquid autoignition temperature and must be in contact with the liquid for a relatively long time before the liquid will ignite. For example, Modak (1981) demonstrated the difficulty of igniting various high flash point liquids with simulated welding spatter. A steel rod with a 1.3×0.3 cm (0.5×0.12 in) cross-section was heated to its melting point (1600 °F or 871 °C) while suspended 0.5 m (1.6 ft) above an open container holding fuel oil or motor oil. The molten steel dripped into the open container for about one minute. Five oils were tested with autoignition temperatures in the range 500–690 °F (260–365 °C). Only one of the five oils ignited (the oil with the highest viscosity), and it self-extinguished after 10 seconds.

Ignition tests with a much larger heated surface were conducted by Holmstedt and Persson (1986) using hydraulic fluids with autoignition temperatures in the range 350–545 °C (662–1013 °F). The fluids were sprayed onto a 65×75 cm (25.5×29.5 in) hot steel plate. Ignition occurred when the hot plate temperature was approximately equal to the liquid autoignition temperature. Thus, autoignition temperatures are an appropriate measure of ignitability for spray impingement onto large heated surfaces.

The case of liquids dripping onto heated surfaces is much more complicated, with ignition temperatures depending on ventilation conditions as well as surface size and fluid properties. A review of this type of data for jet aircraft fuels with spontaneous ignition temperatures in the range 212–280 °C (414–536 °F) concludes that these fuels should not be allowed to impinge upon surfaces at temperatures above 240 °C unless a complete assessment of the particular situation and data is conducted (AGARD-AR-132-Vol 2).

7.2.3 TIME TO REACH FIRE POINT

Suppose a spill of a Class II or Class III flammable liquid is exposed to the flame radiation from an adjacent fire as depicted in Figure 7.1. The incident radiant flux is partially absorbed at the liquid surface and partially transmitted to the underlying liquid and the floor. As the liquid surface is heated, there is convective heat transfer to the air (and vapor) above the surface, and eventually there may be re-radiation from the surface. The net heat transfer to the surface is therefore somewhat less than the incident radiant flux.

We wish to estimate the time required for the spilled liquid to be heated from the ambient temperature to its flash point or fire point. This problem has been studied theoretically and experimentally by Modak (1981) and by Putorti (1994). The one-dimensional heat transfer equation for

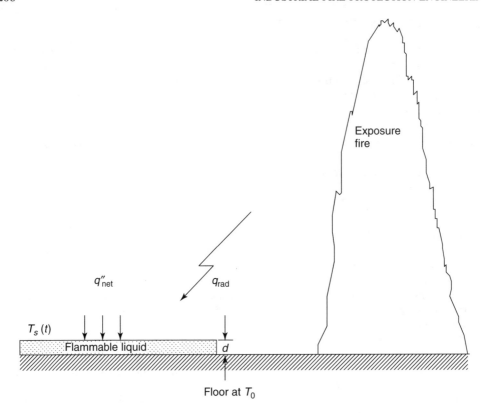

Figure 7.1. High fire point liquid spill heated by exposure fire

a liquid subjected to a radiant heat flux such that a fraction, γ, of the incident flux is transmitted through the free surface of the liquid is

$$k\frac{\partial^2 T}{\partial y^2} + \gamma \dot{q}''(1-r)\beta e^{-\beta y} = \rho c_p \frac{\partial T}{\partial t} \quad [7.2.1]$$

with the free surface boundary condition

$$\text{at } y = 0, \; -k\frac{\partial T}{\partial y}\bigg|_{y=0} = (1-r)\dot{q}'' - h_c(T_s - T_0) - \varepsilon\sigma(T_s^4 - T_0^4)$$

where T_s is the liquid surface temperature (C), T_0 is the initial floor temperature (C), q'' is the net heat flux to the liquid surface (W/m²), k is the liquid thermal conductivity (W/m-°K), ρ is the liquid mass density (kg/m³), c_p is the liquid specific heat (J/kg-°K), γ is the liquid surface transmissivity, such that $(1 - \gamma - r)$ is the surface emissivity or absorptivity, β is the absorption coefficient per unit depth (m^{-1}), h_c is the convective heat transfer coefficient, and r is the surface reflectivity.

In the case of confined deep spills, there is a period of time for which liquid depths are sufficiently large for the liquid to absorb all the incident heat flux without transmitting any heat to the floor. Modak has shown that this thick layer approximation is valid when $d \ll \sqrt{\alpha t}$ where d is the liquid layer thickness (cm), α is the liquid thermal diffusivity (cm²/s), and t is the time of exposure (s) to flame radiation. Modak's solution to equation [7.2.1] for the deep liquid case, for which $r = h_c = 0$, is:

$$T_s - T_0 = 2q'' \sqrt{\frac{t}{\pi k \rho c}} - q'' \left(\frac{\gamma}{\beta k}\right) [1 - e^{-\beta^2 \alpha t} \text{erfc}(\beta\sqrt{\alpha t})] \qquad [7.2.2]$$

The first term on the right hand side of equation [7.2.2] is the solution to the heat conduction equation for a semi-infinite solid or for a liquid surface with an emissivity equal to one. According to Modak, an emissivity of one (zero transmissivity) might be applicable to dirty or contaminated liquids. However, the transmittance for most clean hydrocarbon liquids is best approximated with $\gamma = 0.55$ and $\beta = 48 \text{ m}^{-1}$. Modak also suggests the following property values for hydrocarbon liquids in the temperature range 100–300 °C (212–572 °F):

$$\alpha = 6.6 \times 10^{-8} \text{ m}^2/\text{s}, \quad k\rho c_p = 7.5 \times 10^5 \text{ W}^2\text{-s/m}^4\text{-K}^2$$

While Putorti uses an average value between the initial temperature and the fire point. These values for SAE 30 and SAE 50 are

$$\alpha = 7.3 \times 10^{-8} \text{ m}^2/\text{s}, \quad k\rho c_p = 2.7 \times 10^5 \text{ W}^2\text{-s/m}^4\text{-K}^2$$

Figure 7.2 shows the calculated and measured liquid surface temperature rise for a 120 mm (4.7 in) deep spill of motor oil exposed to a radiant heat flux of 13.8 kW/m² (1.2 Btu/sec-ft²). The

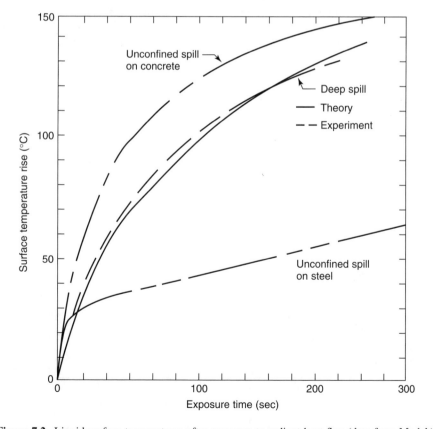

Figure 7.2. Liquid surface temperatures after exposure to radiant heat flux (data from Modak)

calculated temperature rise, which is virtually identical to the measured curve, is based on equation [7.2.2] with a net heat flux of 10 kW/m^2 (0.9 Btu/sec-ft^2). The other two curves in Figure 7.2 are for unconfined spills, 0.75 mm (0.03 in) deep, on a concrete floor and on a steel floor. The unconfined spill on concrete produced a more rapid temperature rise than the deep spill, but the unconfined spill on steel resulted in a relatively slow temperature rise because of the high thermal conductivity of steel.

Modak and Putorti both compared their calculated times to ignition (time at which T_s is equal to the fire point, T_{fire}) to measurements for confined liquid layers. Putorti's calculations are based on the assumptions $\gamma = 0$ and $h_c = 10 \text{ W/m}^2\text{-C}$ ($0.03 \text{ Btu/min-ft}^2\text{-F}$). In that case, the relationship between ignition time and incident heat flux should be as indicated in Figure 7.3, i.e. the function of ignition time on the ordinate in Figure 7.3 should equal the incident heat flux. Putorti's data for SAE 30 and SAE 50 motor oil (fire points of 282 °C or 540 °F and 291 °C or 556 °F, respectively) are consistent with the calculations for heat fluxes less than about 25 kW/m^2 (2.2 Btu/sec-ft^2) and $r = 0.45$.

Alternatively, Modak's solution (equation [7.2.2]) can be solved by iteration for the time to ignition. Modak compared his solution to his ignition time data for #2 fuel oil ($T_{fire} = 129°C$ or 264 °F) and a phosphate ester hydraulic fluid with a fire point of 313 °C (595 °F). He obtained good agreement when he assumed the net heat flux to the liquid surface was equal to 65% of the incident heat flux. This can be rationalized as being equivalent to the net heat flux corresponding to $r = 0.35$ or to $r = 0$ and $h_c = 10 \text{ W/m}^2\text{-C}$ ($0.03 \text{ Btu/min-ft}^2\text{-F}$) and surface re-radiation with a surface emissivity of 0.45. More recently, Hshieh and Julien (1998) have obtained cone calorimeter time-to-ignition data for a variety of silicone fluids, and found that the ignition times varied as the incident heat flux to a power that varied from -1.33 to -2.84. If Putorti's theory were applicable to these fluids, the exponent should be -2.0. The fact that it varies around this value indicates

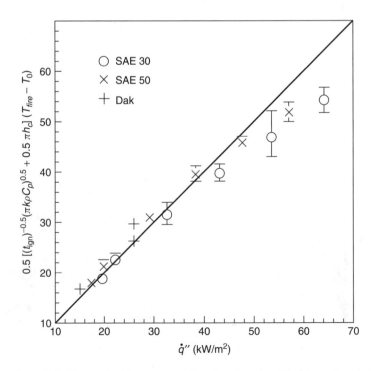

Figure 7.3. Time to ignition versus radiant heat flux (modified from Putorti)

FLAMMABLE LIQUID IGNITABILITY AND EXTINGUISHABILITY 209

that the theory presented here should be considered an approximation that does not necessarily apply to all fluids. The silicone fluids are particularly difficult to describe theoretically because they have widely varying thermal properties.

7.3 Electrostatic ignitability

Electrostatic ignition of a flammable liquid usually occurs through the sequence of: (1) electrostatic charge generation in/by a flowing fluid; (2) charge accumulation in the liquid and/or on a container or pipe or other object in the container; and (3) electrostatic discharge in the form of either an energetic spark or brush discharge. A phenomenological description of these processes is presented here along with a discussion of their applicability to flammable liquid ignition scenarios.

Ionized particles, present to some extent in all liquids, are responsible for electrostatic charge generation. The required concentration of ions is extremely low. Bustin and Dukek (1983) estimate that only one ion per 1.6×10^{12} neutral molecules of hydrocarbon would suffice to generate the highest charge levels measured in industrial equipment. Trace amounts of electrochemically active impurities, as little as 1 g (0.035 oz) per 1000 tons (909 kg) of liquid, are often the source of these high charge levels in petroleum products (Klinkenberg and van der Minne, 1958).

Figure 7.4 provides a conceptual illustration of how ion separation leads to electrostatic charging in the case of liquid flow through a pipe. Ions of one polarity (negative in Figure 7.4) are selectively adsorbed onto the liquid-wall interface, leaving the liquid with a layer of opposite charge (positive in Figure 7.4) a short distance away from the wall. The thickness of this electrical double layer, δ, is (Klinkenberg and van der Minne)

$$\delta \approx \sqrt{D\tau}$$

where $\tau = \varepsilon\varepsilon_0/\kappa$, D is the molecular diffusivity of the ions (cm²/s), ε_0 is the dielectric constant for a vacuum (8.854×10^{-12} C/V-m), ε is the dielectric constant of the liquid relative to ε_0, and κ is the electrical conductivity of the liquid (ohm-m)$^{-1}$.

The characteristic time, τ, is an important fluid property both for electrostatic charge generation and decay. Values of τ and of electrical conductivity and dielectric constant are listed in Table 7.3 for a number of flammable liquids and for water. Using the values of τ for water (1.8×10^{-4} sec) and a midrange value for gasoline (20 sec), and an estimated value of 2×10^{-5} cm²/s for the

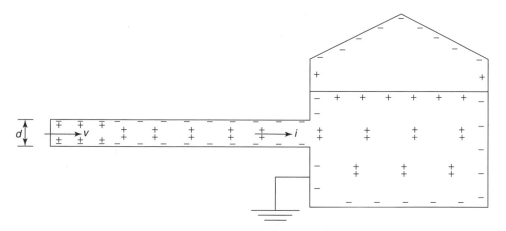

Figure 7.4. Streaming current and charge accumulation due to ion entrainment from electrical double layer

Table 7.3. Electrostatic properties of common liquids

Liquid	Dielectric constant[a]	Conductivity (ohm-cm)$^{-1}$[b]	Relaxation time (s)
Acetone	21	6×10^{-8} ?	3.2×10^{-5}
Acetonitrile	38	7×10^{-6}	4.7×10^{-5}
Benzene[c]	2.3	7.6×10^{-8}	2.6×10^{-6}
n-Butanol	17	1.0×10^{-8}	1.5×10^{-8}
Ethanol	24	1.3×10^{-9}	1.6×10^{-3}
Ethyl Acetate	6.0	1.0×10^{-9}	5.3×10^{-4}
Ethyl Ether	4.3	$<4 \times 10^{-13}$	>1
Ethylene Glycol	38	$<1 \times 10^{-13}$	>34
Gasoline (no additive)	2	10^{-13}–10^{-15}	2–200
Glycerol	43	3×10^{-7}	1.4×10^{-5}
Hexane	1.9	1×10^{-16}	1700
Jet Fuel (no additive)[d]	2.1	1.4×10^{-14}–4×10^{-14}	~9
Jet Fuel (with additive)[d]	2.1	1.5×10^{-12}–5×10^{-12}	~0.10
Kerosene (no additive)	2	~5×10^{-15}	35
Methanol	33	4.4×10^{-7}	6.7×10^{-6}
iso-Propanol	20	3.5×10^{-6}	5.1×10^{-7}
Toluene	2.4	$<1 \times 10^{-14}$	>21
Water	79	4×10^{-8}	1.8×10^{-4}

[a] Dielectric constants are 25 °C data from the *Handbook of Analytical Chemistry*
[b] Conductivity data are from *Generation and Control of Static Electricity*, The Sherwin Williams Company, 1988
[c] Benzene conductivity can vary by several orders of magnitude
[d] Jet fuel data at 20 °C from *Handbook of Aviation Fuel Properties*, CRC Report No. 530, 1983

molecular diffusivity (Klinkenberg and van der Minne, 1958, p. 42) the calculated values of δ for water and for gasoline are 0.6 micrometers (0.02 mil) and 200 micrometers (7.8 mil), respectively.

When the liquid is flowing, its ions are swept away from the fluid-solid interface and entrained into the flow. The resulting flow of ions is called the streaming current, i_s, usually expressed in microamperes. When the flowing liquid enters a receiving container or tank, as shown in Figure 7.4, it has a net (positive) charge which induces a separation of charge on the container/tank walls.

The magnitude of the streaming current, i_s, developed during pipe flow depends on the fluid velocity, v, the pipe diameter, d, the wall material and surface condition, and the fluid electrokinetic properties. The following general equation for i_s is applicable to long pipes in which an equilibrium charge has been developed:

$$i_s = K_e d^m v^n \qquad [7.3.1]$$

where K_e is the so-called *electrification parameter*, with dimensional units of microamps-meter$^{-(m+n)}$-sec^n, when d is in meters, v is in meter/sec, and i_s is in microamps. Various values for the exponents m and n and for K_e have been suggested by different researchers as follows:

Reference	K_e	m	n	Derivation
Gibson and Lloyd (1970)	0.6–1.4	1.6	2.4	Empirical
Klinkenberg and van der Minne (1958)	$K_e(\kappa)$	0.75	1.75	Theoretical
Koszman and Gavis (1962)	$K_e(\kappa)$	0.87	1.87	Theoretical
Mancini (Schön) (1987)	3.75	2	2	Empirical

The different values obtained for the exponents may be due in part to different levels of turbulence in the various experiments (the Klinkenberg and van der Minne exponents are based on fully developed turbulent flow) as well as possible contamination problems. The value of K_e depends upon liquid electrical conductivity and dielectric constant in a manner that is not amenable to any simple theory. Experiments conducted by Klinkenberg and van der Minne with doped liquids of varying electrical conductivity indicate that the maximum value of K_e occurs at a conductivity of 25–100 (ohm-cm)$^{-1}$. The value of K_e in this conductivity range is 0.1 to $0.5\,\mu\text{Am}^{-2.5}\text{s}^{1.75}$ based on m and n values of 0.75 and 1.75, respectively. The wide variation in values for K_e, m, and n suggest that calculations using equation [7.3.1] should be considered accurate to within an order of magnitude.

The electric charge per unit volume of liquid is called the charge density, σ, usually expressed in units of microcoulomb per cubic meter ($\mu\text{C/m}^3$). In the case of pipe flow, σ is the ratio of electron flow rate to volumetric flow rate, i.e.

$$\sigma = i_s/(v\pi d^2/4) \qquad [7.3.2]$$

Substituting from equation [7.3.1]

$$\sigma = 4K_e d^{(m-2)} v^{(n-1)}/\pi \qquad [7.3.3]$$

If the Mancini (originally Schön) values of m and n are used in equation [7.3.3], the charge density is seen to be independent of pipe diameter and directly proportional to velocity. Current guidelines in the paint and coatings industry (Sherwin-Williams Company, 1988) recommend a maximum velocity of 4.6 m/s (15 ft/s) in piping systems in order to limit static charge generation. This would correspond to a maximum charge density of $22\,\mu\text{C/m}^3$ based on the Mancini values of K_e, m, and n.

Peak charge densities corresponding to the peak streaming currents (1 μA) measured by Klinkenberg and van der Minne for large flow rates (1 m^3/min) of kerosene through pipes and hoses are of the order of $60\,\mu\text{C/m}^3$. Significantly higher charge densities (on the order of $1000\,\mu\text{C/m}^3$) and streaming currents have been measured when jet fuel and similar liquids flowed through filter media at high flow rates (Bustin and Dukek). The enhanced charge generation capability of filters is apparently due to the large surface area of filter material in contact with the flowing liquid.

One overt manifestation of the charge generation propensity of filter material occurred in a 1983 fire at an ink and varnish manufacturing plant. A plant employee was withdrawing a sample of ethyl acetate (closed cup flash point of 26 °F or -3 C, conductivity of 10^{-9} (ohm-cm)$^{-1}$) from an ungrounded transfer pipe. When he opened a valve to get a sample in a five-gallon pail, the ethyl acetate sprayed out because of the inadvertent presence of air in the line along with the ethyl acetate. The employee closed the valve and placed some cheesecloth over the open end of the pipe to prevent further spraying. As soon as the valve was reopened, a spark ignited the ethyl acetate as it passed through the cheesecloth. Louvar et al. (1994) describe other electrostatic ignition incidents involving liquid flow through a filter and tank filling.

The preceding incident also illustrates how charge generation phenomena can be intensified when the liquid has a second phase, which in this case was air that entered the ethyl acetate line because of a low liquid level at the ethyl acetate storage tank. The second phase in many other cases is water intrusion into a gasoline or kerosene supply. Charges are generated at the phase-to-phase interface, particularly when the second phase is present in the form of suspended droplets or sinks down through the lighter liquid, as would be the case with water entering a hydrocarbon storage tank.

When a charged liquid exits from a filter (or some other efficient charge generator) and enters a pipe, there is a competition between charge dissipation and charge accumulation in the pipe. In this case, equation [7.3.1] must be modified to account for charge variation along the pipe length.

The streaming current in this more general case is (Klinkenberg and van der Minne, p. 60)

$$i_s = K_e d^m v^n (1 - e^{-z/v\tau}) + i_0 e^{-z/v\tau} \qquad [7.3.4]$$

where z is the distance along the pipe from its entrance, and i_0 is the initial current entering the pipe.

Equation [7.3.4] reduces to equation [7.3.1] when $z \gg v\tau$, i.e. for long pipe lengths.

Bustin and Dukek have conducted bench scale tests of charge decay downstream of a micropore filter. They found that the exponential decay represented by the last term in equation [7.3.4] is observed with liquids that have conductivities greater than 10^{-14} (ohm-cm)$^{-1}$. However, hydrocarbon liquids with smaller conductivities (corresponding to values of τ greater than about 20 s) dissipate charges more rapidly than would be expected on the basis of $e^{-t/\tau}$ decay. These low conductivity liquids develop an effective conductivity which is proportional to the charge density. As a result, the charge relaxation observed with these low conductivity liquids satisfied the equation

$$\sigma = \sigma_0/(1 + \mu\sigma_0 t/\varepsilon\varepsilon_0) \qquad [7.3.5]$$

where σ is the charge density (μC/m^3) at time t (s), σ_0 is the initial charge density (μC/m^3) after filter, μ is the ion mobility (m^2/volt-sec), ε_0 is the dielectric permittivity for vacuum (8.85 × 10^{-6} μC/V-m), and ε is the relative dielectric constant of liquid.

The Bustin and Dukek measurements indicated that the effective value of ion mobility is about 10^{-8} m^2/V-sec. Using this value, the time required for an initial charge density of 500 μC/m^3 to decay 90% can be calculated for a typical hydrocarbon such as kerosene. Solving equation [7.3.5] for t and letting $\sigma_0/\sigma = 10$, yields $t = (9)(2) (8.85 \times 10^{-6})/(500 \times 10^{-8}) = 32$ seconds. The corresponding 90% decay time calculated on the basis of exponential decay and a time constant of 35 seconds for kerosene (from Table 7.3) is 80 seconds, which is significantly longer than the observed decay time.

Figure 7.5 illustrates how the preceding charge generation and dissipation processes might occur during the filling of a tank truck. The liquid is electrostatically charged during pumping,

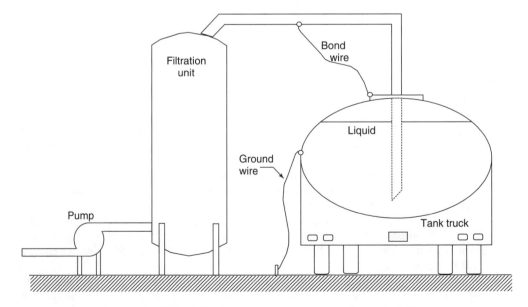

Figure 7.5. Electrostatic charge generation and decay during tank filling

pipe flow, and filtration. This charge generation in the flowing liquid causes a streaming current as indicated in the graph at the top of the page. After exiting the filtration unit, the charge and associated streaming current are reduced via current flow to the grounded transfer pipe. However, the liquid retains some charge upon entering the tank truck. The retained charge is gradually dissipated via current flow to the inside tank wall. The associated voltage distribution in the tank truck at the completion of filling and at a characteristic time τ later are sketched in the graph alongside the tank truck. If the voltage difference between the liquid free surface and the tank wall is sufficiently great, an electrostatic spark can be generated in the vapor space of the tank. If the liquid is at a temperature above its closed cup flash point such that a flammable vapor-air mixture exists in the tank, the spark or brush discharge (shown in Figure 7.5) can trigger an explosion and subsequent liquid fire.

The electric charge developed in a tank such as the tank truck in Figure 7.5 can be estimated if the liquid charge generation and relaxation processes are known. If the liquid charge is generated from the streaming current, i_s, in the fill pipe and filter, and if charge decay is due to constant conductivity current flow to the tank wall, the charge balance equation is

$$\frac{dQ}{dt} = i_s - \frac{Q}{\tau} \qquad [7.3.6]$$

where Q is the charge in the tank at time t (microcoulomb), i_s is the streaming current entering tank (microampere), τ is the liquid charge relaxation time constant (sec), and t is the time from the onset of tank filling (sec).

The solution to equation [7.3.6] for zero initial charge in the tank, and constant i_s, is

$$Q = i_s \tau \{1 - e^{-t/\tau}\} \qquad [7.3.7]$$

The charge density in the tank, σ_t, corresponding to equation [7.3.7] is (Bustin and Dukek, p. 18)

$$\sigma_t = \frac{i_s \tau [1 - e^{t/-\tau}]}{t v \pi d^2 / 4} \qquad [7.3.8]$$

where v is the average velocity in the fill pipe, and d is the fill pipe diameter.

Equation [7.3.8] at $t = 0$ reduces to equation [7.3.2], the charge density in the fill pipe. When $t \gg \tau$, the term in brackets in equation [7.3.8] approaches one, and σ_t approaches the fill pipe charge density multiplied by the ratio τ/t. Thus, the charge density in the tank should never exceed the charge density in the fill pipe.

Three separate effects can cause the actual charge density in the tank truck to significantly exceed the value calculated from equation [7.3.8]. First, a significant charge can accumulate on the free surface of the liquid. Secondly, sedimentation of water droplets sinking through the lighter hydrocarbon can generate a significant charge level. Third, splash filling with a fill pipe that ends above the liquid free surface can also generate high charge levels. When one or more of these three effects are present, the charge density and field strength in the tank can increase during filling as shown in Figure 7.6. Charge decay after filling can be considerably slower than predicted on the basis of exponential decay theory if there are sedimentation effects associated with the presence of water or particulates in the bulk liquid.

Once the charge density in the tank has been determined, the electric field strength can be calculated. The calculation is geometry and boundary condition dependent and can become complicated for actual tank shapes (see Appendix A of Bustin and Dukek). In the simple case of a one-dimensional (vertical) variation of field strength in a partially filled tank and no surface charge accumulation, the field strength, E, in the vapor space is (Klinkenberg and van der Minne, Equation VIII-50)

$$E = \sigma_t d_1 \frac{D}{2\varepsilon} \qquad [7.3.9]$$

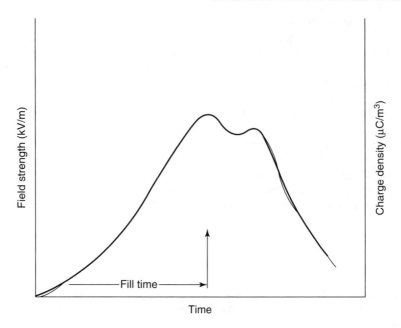

Figure 7.6. Electric field strength development and decay during tank filling

where
$$D = \frac{d_1}{d_1 + \varepsilon d_v}$$

d_1 is the liquid depth in tank (m), d_v is the vapor space in tank (m), and ε is the liquid dielectric constant relative to vacuum.

If the peak electric field strength in the tank vapor space exceeds about 3150 kV/m, a spark discharge can be expected. Maximum measured field strengths reported by Bustin and Dukek for tanker loading were about 50 kV/m. Experiments conducted by Klinkenberg and van der Minne pumping gasoline with 5% water at high flow rates into a large tank indicated peak field strengths of about 200 kV/m. Thus, it is difficult to generate the high field strengths needed for electrostatic discharge from the liquid free surface to the tank wall. On the other hand, sparks were observed when a grounded projection was inserted into the tank vapor space. The maximum safe charge density to preclude these discharges during tank truck filling is about 30 µC/m³ according to a study reported by Bustin and Dukek. Macksimov *et al.* note the greatest ignition hazard is due to probes with a diameter in the range 1–2 cm inserted just above the liquid surface. Their experiments indicate that the liquid surface electrostatic potential should be below 25 kV for negatively charged liquids, and below 54 kV for positively charged liquids, in order to prevent discharges to grounded probes in this diameter range.

Mancini has recently provided some interesting accounts of the influence of tank truck grounding on the potential generation and decay of electrostatic charge during tank truck loading. Grounding of the tank truck outer surface will not affect the electric field strength within the tank, but it will prevent the potential buildup of a high voltage on the truck external surface. Mancini described a tank truck fire that was mistakenly attributed to electrostatic ignition because the truck was not grounded. The actual ignition was probably due to vapors overflowing the open hatch and entering the operating engine of the truck. Bustin and Dukek quote some tank truck incident statistics that indicate 80% of electrostatic ignitions are associated with splash filling.

FLAMMABLE LIQUID IGNITABILITY AND EXTINGUISHABILITY 215

Other scenarios that promote electrostatic charge generation in liquids are tank washing, spray painting, and coating operations. Charge generation and decay mechanisms in these operations differ from those discussed here and warrant special analyses beyond the scope of this text. Concerned readers are referred to the analyses of Bustin and Dukek and Post *et al.* for tank washing, to Inculet and Post *et al.* for spray painting, and to Owens for coating operations. API 2003, NFPA 77, and the recent books by Pratt and by Britton provide additional practical guidance.

7.4 Pool and spill fire heat release rates

We turn our attention now from ignitability to fire severity as measured in terms of heat release rates. Since most available information on flammable liquid fire heat release rates pertain to a pool fire burning in a confined (diked) area or an open top tank, we consider this scenario before dealing with unconfined spills.

7.4.1 CONFINED POOL FIRES

The general equation for pool fire heat release rates with unlimited air access is

$$Q = m'' \Delta H_c x_{chem} \pi D^2 / 4 \qquad [7.4.1]$$

where Q is the chemical heat release rate (kW), m'' is the mass burning rate per unit surface area (g/m^2-sec), Δh_c is the net heat of combustion (kJ/g), x_{chem} is the combustion efficiency, and D is the pool fire diameter (m).

Data for m'' and x_{chem} are scenario dependent. Babrauskas (1986) and others describe the variation of m'' with D for pool fire diameters in the range 0.2 m (8 in) $\leq D \geq$ 5 m (16.4 ft) as

$$m'' = m''_\infty \{1 - e^{-k'D}) \qquad [7.4.2]$$

where m''_∞ is the asymptotic burning rate for large pools, and k' is an effective absorption coefficient, including the mean-beam-length correction factor.

Values of m''_∞, k', and ΔH_c are given in Table 7.4 for a number of common flammable liquids. Most of the liquids have values of k' in the range 1.1–3.6 m^{-1} (0.34–1.1 ft^{-1}). Thus, the mass burning rate is effectively equal to the asymptotic burning rate when D is approximately 2 m (6.6 ft) or larger. The hydrocarbon liquids have values of m''_∞ in the range 40–100 g/sec-m^2 (0.5–1.2 lb/ft^2-min) and heats of combustion in the range 40–45 kJ/g (17,000–19,400 Btu/lb). Thus, the theoretical heat release rate (based on perfectly complete combustion) per unit surface area is in the range 1.6–4.5 MW/m^2 (157–441 Btu/sec-ft^2), i.e. 3 MW/m^2 (294 Btu/sec-ft^2) ±50% for hydrocarbon pool fires larger than about 1 m. Since many hydrocarbon liquids have values of x_{chem} on the order of 0.9 for over ventilated fires, their actual chemical heat release rates are also 3 MW/m^2 (294 Btu/sec-ft^2) ±50%.

The alcohols listed in Table 7.4 have infinitely large effective absorption coefficients because their burning rates are not influenced by the relatively low levels of flame radiation from alcohol flames. Based on the data in Table 7.4, the chemical heat release rate from an ethanol or methanol pool fire is about 400 kW/m^2 (39 Btu/sec-ft^2), or roughly 10–25% of the heat release rate for the same size hydrocarbon pool fire.

There is some evidence that the tabulated values of m''_∞ overestimate the burning rates in very large pool fires on the order of tens of meters in diameter. According to Babrauskas (1986), this is probably due to inefficient mixing of vapor and air above these very large pools. He estimates that the actual mass burning rates per unit area are no less than 80% of the values based on his tabulated mass burning rate data.

Table 7.4. Pool fire burning rate data

Liquid	Mass burn rate, m''_∞ (g/m²-s)	ΔH_c (kJ/g)	ΔH_v (kJ/kg)	k' (m⁻¹)	Density (g/cc)	x_{chem}	x_c	x_r
Acetone	41	25.8	668	1.9	0.79	0.97	0.73	0.24
Benzene	85	40.1	484	2.7	0.87	0.69	0.29	0.40
n-Butane	78	45.7	362	2.7	0.57	0.95	0.68	0.27
Diethyl Ether	85	34.2	382	0.7	0.71	–	–	–
Ethanol	15	26.8	891	∞	0.79	0.97	0.73	0.24
Fuel Oil #6	35	39.7	–	1.7	0.94	–	–	–
Gasoline	55	43.7	330	2.1	0.74	0.92	0.61	0.31
Heptane	101	44.6	448	1.1	0.68	0.92	0.62	0.30
Hexane	74	44.7	433	1.9	0.65	0.92	0.61	0.31
Isopropanol	–	30.2	666	–	0.79	0.97	0.73	0.24
JP-4	51	43.5	–	3.6	0.76	–	–	–
Kerosene	39	43.2	670	0.82	–	–	–	–
Methanol	17	20.0	1195	∞	0.80	0.95	0.81	0.14
M.E. Ketone	–	31.5	444	–	0.81	0.97	0.67	0.30
Styrene	–	40.5	–	–	0.90	0.67	0.27	0.40
Toluene	–	40.5	–	–	0.87	0.67	0.27	0.40
Transformer Oil	39	46.4	–	0.7	0.76	0.84	0.56	0.28
Xylene	90	40.8	543	1.4	0.87	0.67	0.27	0.40

Data Sources
Mass burning rates, densities, heats of combustion, effective absorption coefficients and heats of vaporization are from Babrauskas (1988) for most liquids. Combustion efficiencies and some heats of combustion are from Tewarson (1988).

If both sides of equation [7.4.2] are divided by the liquid density, the result is an equation for the liquid regression rate in a pool fire. Liquid densities are also listed in Table 7.4. The calculated liquid regression rates are on the order of 0.1 mm/sec (0.004 in/sec) for most hydrocarbon liquids. The corresponding value for methanol and ethanol is about 0.02 mm/s (7.8×10^{-4} in/sec). Both calculated values are consistent with data reviewed by Mudan and Croce (1995).

Burning rates for flammable liquids not listed in Table 7.4 can be estimated with the following data correlation (Mudan and Croce, 1995):

$$m''_\infty = \frac{\Delta H_c}{[\Delta H_v + c(T_b - T_0)]} \qquad [7.4.3]$$

where ΔH_v is the liquid heat of vaporization (kJ/g), T_b is its boiling point, and T_0 is its initial temperature prior to ignition.

Figure 7.7 shows how well equation [7.4.3] correlates with the data reviewed by Mudan and Croce. Note that the correlation seems to be valid for liquefied gases and cryogenic liquids as well as the higher boiling point liquids. Other data obtained by Buch et al. (1997) also support the form of equation [7.4.3] for silicone fluids with values of $\Delta H_c/\Delta H_v^*$ in the range 7.5–97. However, their data were for m'' at $D = 0.30$ m, rather than for m''.

Burning rates obtained from Table 7.4 or equation [7.4.3] may lead to errors when the liquid is burning in a tank or dike with a large freeboard or lip height. Data reviewed by Babrauskas show that the freeboard can either decrease or increase the burning rate, with decreased burning rates being more common at freeboard heights greater than about 20% of the tank diameter. A theoretical model to account for this effect is described in Section 8.1.

Winds can significantly increase the effective pool fire diameter and corresponding mass burning rate. The wind tends to both tilt and increase the flame diameter in the downwind direction.

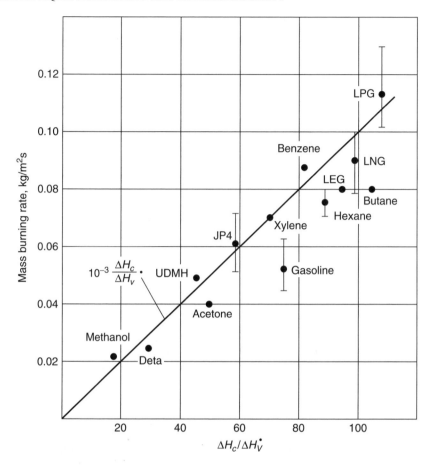

Figure 7.7. Mass burning rate versus ratio of heat of combustion to heat of vaporization (from Mudan and Croce, 1995. Reproduced by permission of Society of Fire Protection Engineers)

Mudan and Croce suggest the following correlation to estimate the increase in flame diameter:

$$D_w/D = 1.25(u_w^2/gD)^{0.069}(\rho_v/\rho_a)^{0.48} \qquad [7.4.4]$$

where D_w is the effective flame diameter in the presence of wind, u_w is the wind velocity (m/s) measured at an elevation of 10 m, g is the gravitational acceleration (9.8 m/sec²), and ρ_v and ρ_a are the densities of vapor and air, respectively.

Equation [7.4.4] should be used with caution at very large wind velocities or very small pool diameters where there is a strong possibility of flame blow off.

7.4.2 UNCONFINED SPILL FIRES

The preceding equations for pool fires exhibit a strong dependence on the pool diameter. In the case of an unconfined spill from a large reservoir, the pool spreads while it is burning. This spreading pool fire scenario has been analyzed by Cline and Koenig (1983) for the case of a specified spill depth, δ, and by Mudan and Croce (based on a 1981 paper by Raj) for the case of spill depth being determined from a balance between gravitational spreading and inertial forces.

In both cases, the liquid spreads until the pool burning rate is equal to the spill rate. The maximum pool diameter, D_{max}, determined by equating the spill rate to the burning rate is

$$D_{max} = 2(V_s/\pi y)^{1/2} \qquad [7.4.5]$$

where V_s is the volumetric spill rate (m³/s), and y is the liquid pool fire regression rate (m/s).

Equation [7.4.5] is based on the assumption of constant spill rate and constant burning rate per unit pool area. Cline and Koenig have compared pool diameters calculated from their analysis to those measured in some laboratory experiments in which 1.8 liter (0.5 gal) containers of JP-4 were spilled over a period of 170 sec and burned with an average liquid regression rate of about 0.02 mm/sec (7.9×10^{-4} in/sec). The measured value of D_{max} was about 25% larger than the value (80 cm or 2.6 ft) obtained from equation [7.4.5]. They attribute the difference to the cooling effect of the initially cold floor which seemed to reduce the burning rate early in the spill and allow it to increase toward the end of the spill. Pool diameters before and after the maximum occurred were in better agreement with the Cline and Koenig analysis for transient spreading and burning.

Substitution of equation [7.4.5] into equation [7.4.1] yields the following equation for the maximum heat release rate in a spill fire:

$$Q = \rho_L V_s \Delta H_{c\_chem.} \qquad [7.4.6]$$

In other words, the heat release rate in a continuous spill fire is simply the product of the mass spill rate, heat of combustion, and combustion efficiency.

The time at which the maximum heat release rate and maximum diameter are reached depends upon the pool spreading analysis. In Raj's 1981 analysis, the time is

$$t_{max} = 0.564 D_{max}/(gy D_{max})^{1/3} \qquad [7.4.7]$$

where t_{max} is the time at which $D = D_{max}$.

In the Cline and Koenig analysis for the case of a constant spill rate, the maximum diameter is reached asymptotically according to

$$(D/D_{max})^2 = 1 - e^{-ty/\delta} \qquad [7.4.8]$$

As an example, consider a 1 liter/sec (1 quart/sec) spill of a hydrocarbon liquid with a liquid regression rate of 0.1 mm/sec (0.0039 in/sec). The maximum diameter given by equation [7.4.5] is 3.6 m (11.8 ft). The time at which this occurs is about 13 sec according to equation [7.4.7]. This time is also consistent with equation [7.4.8] for $D/D_{max} = 0.96$ and $\delta = 0.5$ mm (0.019 in), or $D/D_{max} = 0.86$ and $\delta = 1$ mm (0.039 in).

In the case of an instantaneous spill of volume V_L (m³) ignited at $t = 0$, the maximum spill diameter is (Mudan, 1984)

$$D_{max} = 2[V_L^3 g/y^2]^{1/8} \qquad [7.4.9]$$

Values of D_{max} for different spill volumes are shown in Table 7.5 for a hydrocarbon liquid with $y = 0.1$ mm/sec (0.0039 in/sec).

The last column in Table 7.5 lists the diameter for an unconfined spill in which the liquid layer depth is 0.5 mm (0.019 in). In the case of a one gallon spill, Table 7.5 indicates that the maximum spill diameter would be approximately 3.2 m (10.5 ft), irrespective of when the spill is ignited. In the other extreme of a 10 m³ (2640 gal) spill, the maximum spill diameter is estimated to be 63 m (206 ft) for instantaneous ignition and 160 m (525 ft) for delayed ignition at the completion of the spill. These calculations neglect any vaporization prior to ignition and assume there is no

Table 7.5. Maximum pool diameters for instantaneous spills

Volume		Maximum diameter (m)	
(gal)	(m^3)	for $y = 0.1$ mm/s	for $y = 0$, $\delta = 0.5$ mm
1.0	0.0038	3.3	3.1
26.4	0.1	11.2	16
264.	1.0	26.6	50
2640.	10.0	63.1	160

puddling due to non-level floors. The reader is also reminded that these maximum diameters only exist for a short period during a fire, so that heat release rates based on D_{max} in an unconfined spill fire would be quite conservative.

The preceding discussion does not account for enclosure effects. These consist of an increase in burning rate due to radiation from heated walls and the hot ceiling layer, and eventual burning rate reductions due to oxygen vitiation if air inflow rates are limited. Data reviewed by Drysdale (1998, p. 340) suggest that the enclosure can temporarily enhance mass burning rates by several times the free air values determined by equation [7.4.1]. Whether or not, the burning will be ventilation limited depends on the air/fuel mass flow rate ratio compared to the stoichiometric air/fuel ratio. The air flow rate through a window of area A_w (m^2) and height H (m) is approximately $0.52 A_w H^{1/2}$ kg/s. This value should be compared to the product of the fuel burning rate times the stoichiometric air/fuel rate (approximately 15 for many hydrocarbon liquids). If the air flow rate into the enclosure is less than 15–20 times the liquid burning rate, the fire will eventually be ventilation limited. Numerous enclosure fire models are available to account for both radiation enhanced burning and oxygen vitiated burning.

7.5 Spray fires

The typical spray fire scenario involves a leaking or ruptured lubricating oil or hydraulic fluid line spraying flammable liquid onto a heated surface or electrical equipment. These fires have been a persistent problem at power plants, metal working plants, in sawmills, underground mines, and plastic injection molding operations. Several examples are cited in Factory Mutual Data Sheet 7-98 and for power plant turbine generator fires in EPRI NP-2843 (Westinghouse, 1983) and NP-4144. Paint spray operations also represent a serious spray fire hazard particularly when electrostatic systems are used to electrically charge the paint or coating. The typical ignition source in an electrostatic paint spraying system is an arc from an ungrounded solvent container or object being painted (FM Data Sheet 7-27).

Most of the large loss spray fires involved petroleum base liquids. Several organizations have developed spray ignitability tests intended to classify and/or approve less flammable hydraulic fluids and lubricating oils. Most tests involve spraying the fluid into a flame or a hot surface and observing the resulting flame length, duration, and propagation away from the ignition source. Table 7.6 lists the following test parameters in some of these tests: liquid temperature, nozzle pressure and (water) flow rate, and flame ignition source position downstream of the nozzle.

Although these tests do discriminate between petroleum based hydraulic oils and the less flammable fluids such as the phosphate esters and the water-glycol solutions, they have been criticized (Loftus et al., 1981; Holmstedt and Persson, 1986) for their lack of repeatability/reproducibility and their lack of quantitative results. The lack of repeatability and reproducibility is due in part to the results being very sensitive to fluid additives and contaminants

Table 7.6. Spray flammability test parameters

Organization (Ref)	Temperature (°C)	Pressure (MPa)[a]	Flow rate (1/min)	Ign position (cm)
ASTM D 3119 – 1983	65	?[b]	?[b]	10
British H.S.E. (Roberts)	40–65	1.7	0.3	15
Factory Mutual – 1975	60	6.9	18	15; 46
MSHA (NBSIR 81-2373)	65	1.0	0.22	45
Swedish N.T.I. (Holmstedt)	37	5–25	1–27	60; 200

[a] 1 MPa = 145 psi
[b] ASTM D 3119 uses a paint spray gun with a 6.4 mm slot width

which influence the spray drop size distribution. Some of the additives used by fluid manufacturers degrade due to wear in hydraulic system pumps, valves, filters, etc.

The lack of quantitative results in these tests can be corrected by utilizing instrumentation to measure temperature, velocity, heat flux, and heat release rate. The National Testing Institute in Sweden (Holmstedt and Persson, 1986) and the British Health and Safety Executive (Roberts and Brookes, 1981) have conducted such well-instrumented tests for a variety of fluids including mineral oil and 4–5 less flammable fluids. Their data can be used as a basis for estimating spray fire lengths and heat release rates as follows.

The spray fire heat release rate is the product of mass flow rate and actual heat of combustion, where the latter is theoretical heat of combustion multiplied by the combustion efficiency. Figure 7.8 shows the combustion efficiency measured for fluids with different theoretical heats of combustion. The Holmstedt and Persson data were relatively insensitive to nozzle size, pressure, and spray angle over the range of pressures and flow rates indicated in Table 7.6. The Roberts and Brookes data indicated that combustion efficiencies for the less flammable liquids were relatively

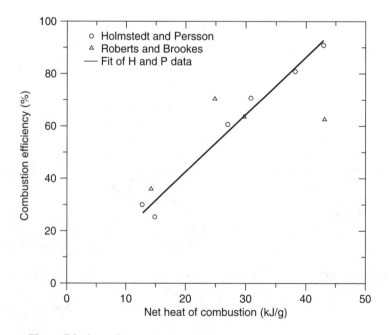

Figure 7.8. Spray fire combustion efficiency versus heat of combustion

constant over the temperature range tested (35–65 °C or 95–150 °F), but the combustion efficiency of mineral oil increased by about 25% over this range. The Holmstedt and Persson data fit shown in Figure 7.8 is given by

$$\chi = 2.2\Delta H_c + 0.56 \qquad [7.5.1]$$

where χ is combustion efficiency (%), and ΔH_c is net heat of combustion (kJ/g).

Equation [7.5.1] would appear to be an appropriate basis for estimating spray combustion efficiencies providing the fluid velocity is significantly high for the spray to have a drop size comparable to those used in the tests. The nozzle velocity at which this occurred in the Holmstedt and Persson tests was about 100 m/sec (224 mph), whereas Khan and Tewarson (1991) have measured asymptotically high combustion efficiencies with a hollow cone spray nozzle when velocities exceeded about 30 m/s (67 mph). The corresponding Froude numbers (u^2/gd) are 2.4×10^5 in the Khan and Tewarson tests and 1.5×10^6 in the Holmstedt and Persson tests. Some of the combustion efficiencies measured by Khan and Tewarson for fluids with low theoretical heats of combustion are higher than the correlation represented by equation [7.5.1] would suggest. Thus, combustion efficiencies for weakly flammable fluids are quite sensitive to the details of the spray formation; more so than the fluids with high heats of combustion.

Flame radiation from hydraulic oil spray fires were also measured by Holmstedt and Persson and Roberts and Brookes. Their data are shown in Figure 7.9 in the form of radiative fraction of heat release rate versus the net heat of combustion of the oil. The Roberts and Brookes radiative fractions are higher (45% versus 32%) than the Holmstedt and Persson data for the

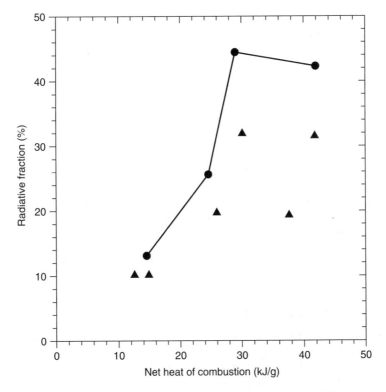

Figure 7.9. Spray fire radiative fraction versus heat of combustion. ●: Roberts and Brookes data; ▲: Holmstedt and Persson data

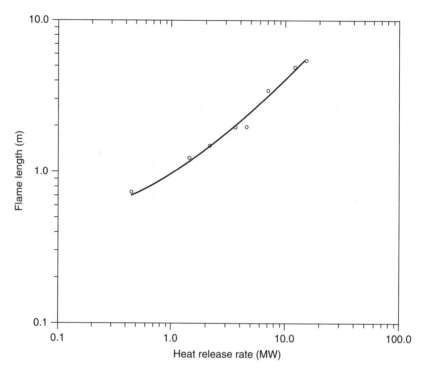

Figure 7.10. Spray fire flame length versus heat release rate

more combustible oils (mineral oil and phosphate ester) and would lead to more conservative estimates of spray fire flame radiation.

Spray fire lengths measured by Holmstedt and Persson are shown in Figure 7.10 as a function of heat release rate. The least-squares curve fit indicated in Figure 7.10 is given by

$$L_f = 0.578 Q^{0.824} + 0.42 \qquad [7.5.2]$$

where L_f is the flame length in meters, and Q is the flame heat release rate in MW.

Although L_f should depend upon drop size distribution as well as heat release rate, the current understanding of spray combustion (Faeth, 1987) does not include any convenient method for accounting for drop size effects, particularly when drop size is not known *a priori*. Data obtained by Khan and Tewarson (1991) for a hollow cone spray nozzle indicated a flame length significantly larger than would be obtained from equation [7.5.2]. More recent data by Yule and Moodie (1992) shows that spray fire flame lengths for marginally flammable fluids are sensitive to both the spray formation process and the strength of the ignition source as represented by the exposure flame size.

7.6 Water spray extinguishment

Water sprays can provide effective control and suppression of high flash point and water miscible flammable liquid fires. In some applications, water sprays can also be beneficial for fires of low flash point water immiscible liquids. Extinguishment mechanisms and data on required water application rates and extinguishment times are described here.

Figure 7.11 depicts the water spray extinguishment phenomena involved in a pool fire. If the spray is applied from above the fire, only a fraction of the droplets will penetrate the fire plume

FLAMMABLE LIQUID IGNITABILITY AND EXTINGUISHABILITY 223

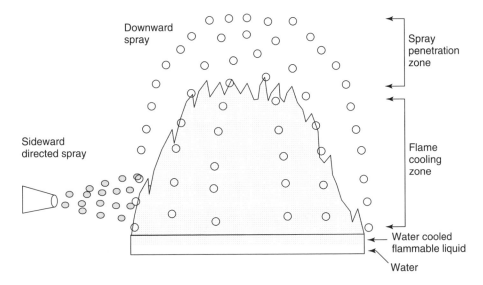

Figure 7.11. Water spray cooling of high fire point, water immiscible flammable liquid pool fire

to enter the flame zone. Many droplets will be vaporized and/or entrained into the plume. The droplets that do reach the flame will provide some measure of flame cooling and thereby reduce the radiant heat feedback from the flame to the liquid surface. Small droplets that enter and penetrate the liquid surface either by sideways application or by plume/flame penetration will cool the liquid and potentially reduce the surface temperature below the fire point. If the liquid is water miscible, the water can also dilute the liquid to produce a nonflammable solution.

Depending upon the water spray characteristics and flammable liquid properties, a variety of effects can result from water spray application to a pool fire. Visual observations of these effects, as reported by Kokkala (1992) are shown in Figures 7.12a–f.

Figure 7.12a shows shortened vertical flames because water spray cooling reduces the flammable liquid surface temperature and vaporization rate. This would correspond to a water spray with small drops and/or low velocity such that thermal effects are more important than spray aerodynamic effects. On the other hand, Figure 7.12b shows flattened flames caused by the downward flow of air entrained into the spray overcoming the fire plume buoyancy. This picture would correspond to a high velocity downward directed spray either applied close to the fire or containing large, heavy water drops.

Figure 7.12c shows what appears to be wind blown flames caused by an oblique spray being applied to the fire of a low flash point liquid such as heptane. Extinguishment in this case results when the flame is blown completely off the liquid pool. Figure 7.12d shows an intensified flame associated with the sputtering of water droplets reaching the surface of a liquid with a fire point much greater than 100 °C (212 °F). If the water drops are small and do not sink into the liquid, this sputtering occurs for a short period of time until the flammable liquid surface temperature is reduced and the flame extinguished. If the drops are sufficiently large to sink into the liquid and if the bottom portion of the liquid is heated (from either prolonged burning or from a spill onto a hot surface), the sputtering and intensified burning can continue for a relatively long period. Plate 8 is a photograph of a temporarily intensified flame due to water drop splashing and sputtering on the surface of a high flash point fatty acid.

Figure 7.12e shows what Kokkala calls ridge flames observed near extinction when most of the liquid surface is flame free. These ridge flames are observed with high velocity, small droplet,

Figure 7.12. Water spray application to high flash point flammable liquid (from Kokkala)

water sprays. Figure 7.12f shows rim flames also produced just prior to extinguishment because the container wall remains hot after the interior liquid free surface has been cooled by the spray.

Extinguishment data reviewed by Nash (1974) and Rasbash (1962) are presented in terms of three categories: high fire point water immiscible liquids, water miscible liquids, and low fire point water immiscible liquids. Their distinction between high and low fire point liquids is a fire point of 45 °C (113 °F).

7.6.1 HIGH FLASH POINT LIQUIDS

In the case of high fire point liquids, extinguishment usually results from water droplet cooling of the burning liquid. Heat transfer from the liquid to the cooler water is more efficient with small water drops and large differences between liquid and water temperatures. Nash and Rasbash's data on the minimum water application rates for extinguishment of kerosene pool fires, transformer oil pool fires, and "gas oil" tray fires can be correlated as

$$R''_{crit} = Kd/(T_{FP} - T_w) \quad [7.6.1]$$

where R''_{crit} is the critical water application rate (gpm/ft^2), d is the mass median drop size (mm) ($d \geq 0.4$ mm), T_{FP} is the liquid fire point (C), T_w is the water temperature (C), K is an empirical constant, and $T_{FP} - T_w \geq 40\,°C$.

The constant K depends upon the area burning and duration of burning prior to water application. A range of 6–12 gpm-°C/ft²-mm is consistent with the Nash and Rasbash data. The higher end of the range is more appropriate for kerosene which, according to Nash and Rasbash, has a critical application rate of about 0.14 gpm/ft² (5.7 L/min-m²) with 0.5 mm (0.019 in) drops and about 0.30 gpm/ft² (12.2 L/min-m²), with 1.0 mm (0.039 in) drops.

Values of R''_{crit} calculated from equation [7.6.1] are shown in Figure 7.13 for two drop diameters (0.5 mm and 1.2 mm) and for $K = 6 - 12$ gpm-°C/ft²-mm. More recent data obtained by Kokkala (1992) using a $\frac{1}{2}$ in orifice pendent spray sprinkler located 5 m (16.4 ft) above 1.6 m² (17.2 ft²) pool fires are also shown in Figure 7.13. The Kokkala data would be consistent with the Nash and Rasbash correlation if the volume median drop size produced by the $\frac{1}{2}$ in sprinkler was about 1.2 mm at a flow rate of 30–37 gpm (113.4–140 L/min), and about 2 mm at a flow rate of 16 gpm (60 L/min). In fact, drop size distribution data obtained by Factory Mutual Research and reported by Fleming (1995) for this sprinkler, indicate that the volume median drop diameter is 0.70–0.90 mm (0.027-0.035 in) at 30–37 gpm (113.4–140 L/min), and about 1.2 mm (3/64 in) at 16 gpm (60 L/min). In other words, the Kokkala data indicate that higher values of R''_{crit} are required than given by equation [7.6.1]. One possible explanation for this discrepancy might be that equation [7.6.1] is applicable for nozzles situated close to the burning pool of liquid, but that higher values of R''_{crit}, i.e. higher values of K in equation [7.6.1] are needed when the spray nozzles are situated about 5 m (16.4 ft) or further from the liquid pool. The higher water application rates are needed to achieve spray penetration through the fire plume. Thus, a conservative interpretation of equation [7.6.1] is that the values of R''_{crit} represent Required Delivered Densities (RDD), rather than water spray design densities. The interpretation is conservative in that the RDD values eliminate both flame and plume penetration, whereas the values of R''_{crit} include the water spray densities needed to penetrate the flame.

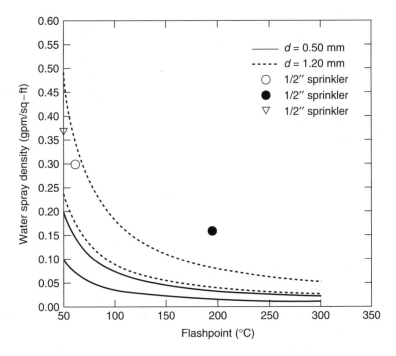

Figure 7.13. Critical water spray density for extinguishment versus flashpoint

When the water application rate is greater than the critical value, Nash suggests the following correlation for the time to extinguish:

$$t_{ex} = 5.2 \times 10^5 \times d^{0.85} \times R''^{-2/3}(T_{FP} - T_w)^{-5/3} \qquad [7.6.2]$$

where t_{ex} is the extinguishment time (s), and R'' is in units of L/sec-m^2.

When R'' is in units of gpm/ft^2, the coefficient 5.2 in equation [7.6.2] should become 6.7. Equation [7.6.2] indicates that t_{ex} is quite sensitive to the liquid fire point. For example, the extinguishment time for n-decane ($T_{FP} = 142\,°F$ or $61\,°C$) would be about 13 times as large as that for motor oil ($T_{FP} = 435\,°F$ or $224\,°C$). This correlation is at least qualitatively consistent with the more recent data of Kokkala (1992). The drop size, d, varies with water application rate for a given nozzle or sprinkler head and. In particular, d is proportional to the ratio $D_o^2/R''^{2/3}$, where D_o is the sprinkler heads orifice diameter (Fleming, 1995). Substitution in equation [7.6.2] reveals that t_{ex} actually varies with $R''^{-1.23}$ for a given orifice diameter.

There are two special reasons for extinguishing fires in high fire point liquids as quickly as possible when using water sprays. First, long water application times imply significant water accumulation under the burning liquid and the subsequent possibility of liquid overflow spreading fire over a large area. Second, the longer the liquid burns, the more likely a hot zone with temperatures over $100\,°C$ ($212\,°F$) can propagate down into the underlying water. If the hot zone reaches the water (assuming the water is heavier than the liquid and accumulating under it), sudden boiling can cause froth over or boil over with subsequent fire spread by burning liquid. Avoiding overflow and boil over is tantamount to designing for relatively early extinguishment, i.e. no longer than the tank/dike capacity at the design water application rate, and no more than 15–30 minutes to avoid hot zone formation (Nash, 1974). A prudent supplementary measure to avoid overflow and boil over would be to provide drainage of accumulated water to an environmentally isolated impoundment area.

In view of the limited 'throw distance' of small drop sprays, nozzles must be located relatively close to the pool fire hazard. In the case of $\frac{1}{2}$ in sprinklers, Factory Mutual recommends (see Bryan, 1990, Table 12.3) maximum nozzle standoff distances of 2.1 m (7 ft) for outdoor wide angle sprays and 3.7 m (12 ft) for narrow angle (60–80°) spray nozzles. The Factory Mutual recommended water application densities for extinguishment of kerosene fires are 0.50–0.70 gpm/ft^2 (20.3–28.5 L/min-m^2) and for high flash point transformer oils 0.36–0.47 gpm/ft^2 (14.6–19.1 L/min-m^2). These values would seem to be about twice the critical (minimum) water spray densities indicated by Figure 7.13. The extra water is presumably needed for both rapid extinguishment and to compensate for wind induced evaporation and spray deflection.

The water spray densities recommended in NFPA 15 for extinguishment/control of flammable liquid fires under pipe racks start with 0.25 gpm/ft^2 (10.2 L/min-m^2) for single level pipe racks and increase with additional levels. There is no provision in NFPA 15 to vary the water application rate with the flammable liquid flash point or spray drop size.

7.6.2 WATER MISCIBLE LIQUIDS

In the case of water miscible liquid pool fires, Nash maintains that the primary water spray extinguishment mechanism is water dilution. The amount of water dilution required for extinguishment depends on the liquid as indicated in Table 7.7. The Nash data for dilution ratios, and associated water volume fractions, are based on the volume, V_L, of liquid in the tank/dike at the initiation of water spray application.

The last column of Table 7.7 lists data reported by Khan (1997) on the pre-mixed volume fraction of water required to produce marginal flame stability (i.e. incipient extinguishment),

Table 7.7. Extinguishment dilution ratios for water miscibles

Liquid	Dilution ratio (Nash)	Water volume fraction	
		Nash	Khan
Acetone	30:1	0.967	0.71
Ethyl Alcohol	7:1	0.875	0.64
Ethylene Glycol	–	–	~0.10
Isopropyl Alcohol	9:1	0.90	0.68
Whiskey	1.5:1	0.60	–

which appears at a measured heat release rate of approximately 100 kW/m². Khan's water volume fractions are significantly lower than those of Nash because the Nash data accounts for the water evaporation that occurred in traveling through the flame. The Nash data also corresponds to actual observed extinguishment rather than incipient extinguishment.

The time to extinguish based on water dilution is

$$t_{ex} = \phi V_L / \{A(R'' + y\phi)\} \qquad [7.6.3]$$

where A is the tank/dike area, ϕ is the required dilution ratio, and y is the liquid regression rate due to burning.

Nash recommends use of a relatively small drop size (less than 0.4 mm or 1/64 in) spray for dilution in order to promote mixing and long residence times in the liquid layer.

7.6.3 LOW FLASHPOINT LIQUIDS

In the case of low fire point water immiscible liquids, water spray will only provide extinguishment if it lowers the flame temperature sufficiently to prevent radiant feedback and vaporization at the liquid surface. This mechanism is more effective with small drops because they tend to vaporize more rapidly than large drops upon passing through the flame. The drops that penetrate the flame and reach the liquid do not provide any benefit for this category of liquids.

Extinguishment data reviewed by Nash (1974) for gasoline indicates that the extinguishment time decreased from 100 sec to 10 sec when the spray mass median drop size decreased from 0.48 mm to 0.28 mm (0.019 in to 0.011 in) with a water application rate of 0.267 L/s-m² (0.39 gpm/ft²). Previous gasoline fire data obtained by Underwriters Laboratories and reproduced by Fleming (1995) indicates that the critical water spray/mist application rate for a 0.28 mm water drop is about 0.10 gpm/ft² (4 L/min-m²), whereas for a 0.40 mm (1/64 in) drop it is about 0.35 gpm/ft² (14.2 L/min-m²). It is generally not possible to suppress low flash point liquid (gasoline) fires with water drops larger than about 0.50 mm (0.019 in).

In view of the small drop size requirements for low flash point liquid fires and the unavailability (as of this writing) of suitable gaseous agents, there has been considerable work in recent years to develop commercially viable water fog suppression systems. Fire test data presented at the March 1993 Water Mist Fire Suppression Workshop at the National Institute of Standards and Technology indicate that water fogs (with drop diameters under 0.20 mm or 7.9×10^{-3} in) effectively extinguished 2–11 m² (6.6–36.1 ft²) pool fires of diesel fuel and higher flash point liquids in simulated ship engine rooms (Turner, 1993) equipped with mock-up equipment potentially obstructing the water mist. Nozzle placement relative to the burning is a key parameter determining the success or failure of these water fog suppression trials.

Compartment confinement effects can be significant with water fog systems. If the drops vaporize rapidly enough that the water vapor displaces oxygen in the compartment, the fire can be suppressed when the oxygen concentration falls below 11–13 vol% for most flammable liquid diffusion flames (Beyler 1995, p. 2–157). Even if the fire is not completely oxygen starved, it is more readily extinguished by water fog systems in compartments than in the open.

Another water fog design parameter relevant to compartment fires is the concentration of water droplets suspended in air/vapor. The relationship between fog application density, R'' expressed in liter/m²-min, and the water (liquid) concentration in the compartment is (Mawhinney, 1992)

$$c_w = \frac{R''}{v6 \times 10^4}$$

where c_w is the water droplet volume concentration (cm³/cm³), and v is the fog droplet average velocity (m/s).

Tests with fog nozzle systems (Mawhinney, 1992) have typically produced water fog concentrations in the range $5 \times 10^{-7} < c_w < 5 \times 10^{-5}$ cm³/cm³. If all those drops vaporize in the compartment, the volumetric concentration of water vapor would be about 1000 times as large as c_w, i.e. less than about 5 v%. This is not sufficiently high to produce oxygen inerting by itself, but the additional water vapor generated by the fire in the compartment could indeed lead to partial or even complete oxygen starvation. Indeed, preliminary tests reported by Carhart et al. (1992) demonstrated that a large (23 m² or 250 ft²) heptane pool fire in an 83 m³ (3000 ft³) enclosure could be rapidly suppressed (within 15–40 sec) with a water fog application rate of 0.075 gpm/ft² (3 L/min-m²) comprised of 80–100 micrometer (3–4 mil) drops.

If a water fog nozzle or fine mist application is not available, water cannot be relied upon to suppress low flash point liquid fires even with large water application rates. The author has conducted experiments with water application rates of 0.64 gpm/ft² (26 L/min-m²) from a water drip applicator a few centimeters above the burning surface of toluene (45 °F or 7 °C flash point) and mineral spirits (104 °F or 40 °C flash point). In both cases there was no extinguishment, but the heat release rates decreased by 50–75%. Thus, water application can provide significant benefits without actually suppressing the fire. The use of water application to cool tanks and drums exposed to fire is discussed in Chapter 8.

7.6.4 SPRAY FIRES

Spray fires can also be suppressed using water sprays providing the water is delivered to the right place in the right form. There have been some tests on the use of water sprays to suppress oil and gas well blowout fires (Pfenning and Evans, 1984) with water spray nozzles situated at or below the release site as shown in Figure 7.14. The configurations shown in Figure 7.14 were tested with methane jet flames with heat release rates of approximately 10 MW (9480 Btu/sec), i.e. methane flow rates of about 200 g/s (0.44 lb/sec). The ratio of water flow rate to methane flow rate required to achieve suppression ranged from 4 to 10, depending upon spray nozzle arrangement, as indicated in Figure 7.14. Pfenning and Evans scaled up the most effective configuration (two vertical nozzles at the release site) to 200 MW (190,000 Btu/sec) fires, and found that extinguishment with a mass flow ratio of 4 was relatively easy. When four nozzles were used instead of two, the minimum mass flow rate for extinguishment was reduced to 2.

If the water spray nozzles are situated far above the source of the flammable liquid spray it is very difficult to control or suppress the spray fire. This difficulty has been demonstrated with conventional ceiling sprinklers trying to provide protection for lubricating oil spray fires. The protection was ineffective for the spray fires even though the lubricating oil pool fires were rapidly suppressed.

If effective water spray nozzle locations for flammable liquid fires are needed, the Pfenning and Evans data for methane jet flames should also be roughly applicable to hydrocarbon liquid

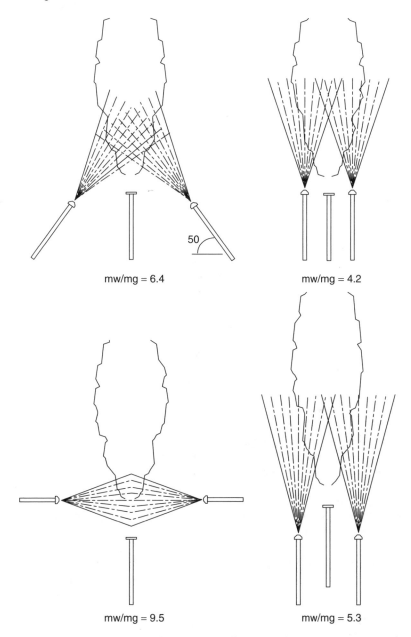

Figure 7.14. Water spray extinguishment of methane jet flames

spray fires with droplet diameters less than about 40 micrometers (1.6 mil). Another important data source is the recent report of spray fire tests conducted with water fog nozzles in simulated ship engine rooms. These data (Turner, 1993) demonstrate the rapid suppression (3–7 sec) of spray fires with water fog nozzles at floor level in the vicinity of the spray release site. Although an effective spray nozzle configuration may not be practical for all spray fire scenarios, it should be possible to install nozzles in situations where the most likely spray release site is known *a priori*. This would include spray painting and coating operations, and bearings and hydrogen seal oil units on turbine generators.

7.7 Foam extinguishment

Fire-fighting foam is generally acknowledged to be the preferred extinguishing agent for large flammable liquid spill fires and tank fires. The generic definition of a fire-fighting foam (NFPA 11) "is a stable aggregation of small bubbles of lower density than oil or water... It flows freely over a burning liquid surface and forms a tough, air excluding continuous blanket to seal volatile combustible vapors from access to air." Indeed, it is the blanketing and surface sealing effect which makes foam a uniquely suitable extinguishing agent for low fire point flammable liquids.

The most widely used foam for flammable liquid fires is Aqueous Film Forming Foam (AFFF). AFFF concentrates consist of fluorinated surfactants combined with foam stabilizing additives. The concentrate is usually diluted with water to form a 3% or a 6% solution. AFFF is usually incompatible with water miscible flammable liquids and polar solvents because it cannot maintain a blanketing effect over such liquids. Special 'alcohol type concentrates' or 'alcohol resistive concentrates' are used for water miscible liquids because they form an insoluble barrier at the liquid surface so as to resist breakdown of the bubble aggregate. These special alcohol compatible foams can also be used with most hydrocarbon liquids.

Depending upon the foam generator or nozzle, a wide range of expansion ratios (volume of air filled foam to foam solution) can be achieved. Low expansion foams, which usually have expansion ratios in the range 5–10:1, are most widely used for flammable liquid fires. Medium expansion foams (expansion ratios of 50–150) and high expansion foams (expansion ratios of 200–1000) are used in special situations described in Section 7.7.2, but are not usually as effective as the low expansion ratio foams for most flammable liquid spill fires.

7.7.1 LOW EXPANSION FOAM

Low expansion AFFF and alcohol compatible foam application rates required for extinguishment are based on tests such as those conducted by Corrie (1975), by Wesson and Associates (1983), and by Merritt (1983). Corrie's data for AFFF applied in toluene tray fire tests indicate that extinguishment did not occur with 6% AFFF application rates under 0.05 gpm/ft^2 (2 L/min-m^2), and there was a negligible reduction in time to extinguish at application rates greater than about 0.16 gpm/ft^2 (6.5 L/min-m^2). Thus, 0.16 gpm/ft^2 would seem to be an optimum application rate for these conditions, and is the design density specified in NFPA 16 for most hydrocarbon liquid pool fires.

Wesson and Associates (1983) conducted two different type of tests using aviation fuel and three different foams (AFFF, alcohol type concentrate foam, and a fluoroprotein foam) applied manually. The first type of test was a pool fire with fuel covering a 75 ft^2 (7 m^2) pit, in which a pipe rack was situated to partially obstruct foam application. Extinguishment occurred within three minutes with AFFF and alcohol concentrate applied at 0.05 gpm/ft^2 (2 L/min-m^2), and in 3–4.3 minutes with fluoro-protein foam applied at the same density. The second type of test involved a flowing fuel spill fire on an inclined board or chute. None of the foams were successful in controlling or extinguishing the flowing fuel fires.

Merritt's tests were conducted with a 3% fluoroprotein foam-water solution applied to heptane fires under/on a simulated chemical process structure. Two types of fires were tested. The first type of fire was a floor spill in which the floor of the process structure was first covered with 10–100 gal (38–380 L) of heptane and upon ignition a heptane flow rate of 10–20 gpm (38–76 L/min) was initiated as a continuing fuel source. The foam-water application at 0.16 gpm/ft^2 (6.5 L/min-m^2) did successfully control this type of floor spill fire. The second type of fire was an elevated spill of heptane from the 4 m (13 ft) level or the 6.4 m (21 ft) level of the 12.2 m (40 ft) high process structure. Foam-water application at all four levels of the test structure could not control the elevated spill fires. Thus Merritt's results with fixed foam application are consistent, in terms of success and limitations, to the Wesson and Associates tests with manual application.

One problem with application of foam in the form of a stream is that the foam can mix with and become saturated with fuel if the application velocity is sufficiently large. Briggs and Webb (1988) have conducted tests with a variety of foams applied at stream velocities up to 7 m/sec (16 mph), and have observed burn through and destruction of several of the foam blankets at these application velocities. Contamination of the foam by flammable liquid can also occur after the foam has successfully extinguished the fire and has begun to drain into the liquid below. There are several reported incidents of 'burnback' in which an extinguished fire has reignited and burned through a foam blanket because of this contamination effect. Some of the tests described by Briggs and Webb and those conducted by approval organizations include testing for burnback.

In the case of foam application via spray nozzles, there has been concern about the ability of the foam spray to penetrate a strongly buoyant fire plume and to migrate around major obstructions between the nozzle and the fire source. Hill (1986) has conducted tests in a simulated aircraft carrier elevator trunk to investigate these concerns. JP-5 fuel was spilled onto the bottom of the 15.2 m (50 ft) high elevator trunk and ignited before 6% AFFF foam-water sprays were applied at a rate of about 2 gpm/ft^2 (81 L/min-m^2) from the top of the structure. The foam had no difficulty in rapidly penetrating the plume to extinguish the fire under these conditions. In addition, the presence of a solid platform in the elevator trunk, such that only a 46 cm (18 in) gap existed for foam passage between the platform and the structure wall, caused only a minor delay in extinguishment time (35 s versus 20 s without the platform). Thus, foam sprays can penetrate fire plumes and pass around obstacles when applied at sufficiently high densities.

The preceding test data were obtained with open nozzles triggered manually or via some separate detector. Foam-water solutions can also be applied via closed head sprinkler systems with individual fusible elements. FMRC tests on the use of AFFF in closed head sprinklers were conducted by Young and Fitzgerald (1975), and later reported by Burford (1976) of the 3M Company. The test fuel was heptane delivered at a flow rate of 5–30 gpm (203–1220 L/min-m^2) through an open upturned elbow on the test floor. The ignition delay time from the onset of heptane flow was 15–25 sec corresponding to initial spill volumes of 1–12 gal (3.8–45.4 L). Sprinklers with 3/8 in or $\frac{1}{2}$ in orifices were located on the 9.1 m (30 ft) ceiling of the test bay.

Test conditions and key results for all 13 tests conducted by Young and Fitzgerald are reproduced in Table 7.8. The first sprinkler actuation times ranged from 22 sec to 41 sec with no obvious difference between the 71 °C (160 °F) sprinklers and the 138 °C (280 °F) sprinklers. The number of sprinkler heads opened ranged from 5–34, with the higher number of head openings occurring with 71 °C (160 °F) heads. Figure 7.15 is a plot of the number of heads opened versus heptane spill rate for the 71 °C (160 °F) heads and the 138 °C (280 °F) heads. The foam-water discharge density was 0.18 gpm/ft^2 (7.3 L/min-m^2) for both types of heads. These results demonstrate that heptane floor spill fires can be successfully controlled with AFFF solution delivered through closed head sprinklers, and that 138 °C (280 °F) heads will reduce the water demand (number of heads opened multiplied by the flow rate per head) compared to 71 °C (160 °F) heads. Additional tests on the use of closed head foam-water sprinklers to control flammable liquid fires are described in Chapter 8.

Perhaps the most challenging fire to suppress is a spray fire. Carhart et al. (1992) described tests involving manual application of AFFF (and other suppression agents) to two different JP-5 fuel spray fires. Pertinent AFFF results for two pairs of tests are shown in Table 7.9.

Table 7.8. Closed head AFFF sprinklered fire tests (data from Young and Fitzgerald, 1975)

Test number	1	2	3	4	5	6	7	8	9	10	11	12	13
Sprinkler Orifice Diameter[a] (in)	3/8	3/8	3/8	3/8	3/8	3/8	3/8	3/8	3/8	3/8	3/8	3/8	$\frac{1}{2}$
Sprinkler Temperature Rating (F)	160	160	160	160	160	160	280	280	280	280	280	280	286
Sprinkler Spacing	10' × 10'	10' × 10'	10' × 10'	10' × 10'	10' × 10'	10' × 13'	10' × 10'	10' × 10'	10' × 10'	10' × 10'	10' × 10'	10' × 10'	10' × 10'
Sprinkler Pressure (psig)	32.5	32.5	32.5	32.5	32.5	21.5	32.5	32.5	32.5	32.5	32.5	32.5	28.5
Agent Discharge Density (gpm/ft^2)	0.18	0.18	0.18	0.18	0.18	0.11	0.15	0.18	0.18	0.18	0.18	0.18	0.30
Heptane Spill Rate (gpm)	5	10	15	20	25	20	15	20	25	30	30	30	30
Ignition Delay Time (sec)	15	15	15	15	15	15	15	15	15	15	25	25	25
Time of First: Sprinkler Actuation (min:sec)[b]	0:41	0:27	0:27	0:22	0:23	0:27	0:35	0:33	0:30	0:28	0:24	0:22	0:24
Time of First Sprinkler Actuation (min:sec)[b]	0:52	1:06	0:53	1:08	0:45	1:01	0:50	0:53	0:38	1:14	1:03	1:13	1:19
Total Number of sprinklers Actuated	5	10	16	32	31	34	5	7	6	15	19	17	10
Control Time (min:sec)[b]	1:40	1:35	1:40	2:20	2:00	3:50	3:00[c]	1:50	1:50	2:20	2:00	c	2:25
Time for Max Temp. To Reduce to 200@F (Min:sec)[b]	1:12	1:07	1:06	1:21	0:51	1:27	1:53	1:33	1:21	1:21	1:55	1:33	2:08
Total Heptane spill Duration (min:sec)	3:20	4:45	5:20	4:00	3:30	4:50	4:10	3:20	4:05	3:40	3:05	5:08	3:15
Duration of Foam Discharge (min:sec)[d]	3:09	4:48	6:23	5:38	5:27	6:43	5:10	3:27	5:40	5:27	4:0	6:18	4:01

[a] The densities for 3/8 in sprinklers were calculated using k value based on hydraulic measurements that were made on heads used in there tests
[b] Measured from the time of ignition of spill
[c] See text
[d] Measured from the time of first sprinkler actuation

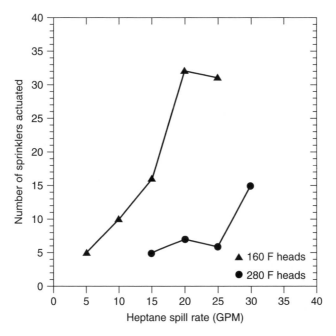

Figure 7.15. Number of sprinkler heads opened in AFFF closed head heptane fire tests of Young and Fitzgerald (1975)

Table 7.9. JP-5 Spray fire suppression tests with AFFF (Data from Carhart et al., 1990)

JP-5 flow rate (gpm)	Spray origin	Spray obstruction	AFFF flow rate (gpm)	AFFF/JP-5 ratio	Reduction in radiant heat flux (%)	Suppressed?
41	19 mm Orifice	None	60	1.5	64.3	No
41	19 mm Orifice	None	95	2.3	93.0	Yes
5.5	19 mm Slit	Steel Grate and Roof	60	10.9	81.8	No
5.5	19 mm Slit	Steel Grate and Roof	95	17.3	94.0	Yes

The first pair of tests listed in Table 7.9 shows the results of applying AFFF manually to a 41 gpm (155 Lpm) JP-5 unobstructed spray fire. Although the radiant heat flux from the flame was reduced by 64% when a flow rate of 60 gpm (227 Lpm) was applied, the fire was not extinguished until the AFFF flow rate was increased to 95 gpm (359 Lpm). Thus, the minimum AFFF/JP-5 application ratio for extinguishment is approximately 2 for an unobstructed spray with a single application nozzle. This is more effective than the water spray extinguishment of methane spray fires, which required a minimum water/fuel mass ratio of about 4 using two nozzles in their optimum locations and orientations.

When the spray fire was directed at an obstruction (in the form of a steel grate and an inclined roof above a debris pile), the data in Table 7.9 indicate that the minimum AFFF application rate for extinguishment is somewhere in the range 11–17 times the fuel spray flow rate. This minimum ratio is substantially higher than that for an unobstructed spray fire, and demonstrates the increased challenge caused partially by the flame holding effect of the obstruction, and partially by the difficulty in the AFFF reaching the obstructed portions of the spray fire.

7.7.2 MEDIUM AND HIGH EXPANSION FOAM

One noteworthy special situation lending itself to the use of medium expansion foam is the extinguishment of liquefied gas fires. Test data reported by Brown and Romine (1979) show the medium expansion foams to be significantly more effective than low expansion foams for this special application. The heat transfer (which varies inversely with foam expansion ratio) associated with foam solution on the surface of a burning liquid is beneficial for high fire point liquids because it cools the liquid surface; however, it is detrimental for liquefied gas which has a lower boiling point than the foam and can therefore be heated by the foam.

High expansion foam can be an attractive option in confined space flammable liquid fires, such as in an enclosed machinery space or a vault containing turbo-generators or internal combustion engines. An important consideration in these enclosures is to generate the foam with external, uncontaminated, air. If smoke laden, contaminated, air is used, the foam will be degraded because the contamination increases the surface tension of the foam solution and foam bubbles tend to collapse on the soot particles.

US Coast Guard tests (Richards, 1975) with high expansion foam (generated with outside air) applied to diesel fuel spray fires and bilge (pool) fires in a 100,000 ft^3 (2833 m^3) ship engine room, showed that foams with expansion ratios in the range 400:1 to 500:1 would extinguish these fires at foam application rates of 2 to 3 m^3 per min-m^2 (7–9 cfm per ft^2). This value is slightly higher than the required minimum application rate (3–6 cfm per ft^2) required to extinguish a pool fire in an unobstructed environment. According to Richards (1975), the increased foam application rates are required because of foam losses when the foam is generated externally and applied through vent openings or ducting into the fire enclosure. Other high expansion foam design considerations discussed by Richards for this type of application are the generator pressure required to inject the foam into the enclosure, and the maximum allowable foam free fall height (about 12 ft or 3.7 m).

7.8 Dry chemical and twin agent extinguishment

Dry chemical suppression agents used for flammable liquid fires include sodium bicarbonate, potassium bicarbonate and mono-ammonium phosphate. The concentration of dry chemical required for suppression depends on particle size, method of dispersing the agent, and the flammable liquid hazard. Heptane pool fire laboratory tests conducted by Ewing *et al.* (1992) produced data for minimum required extinguishment concentrations when the dry chemical is applied at maximum efficiency (negligible losses during delivery) to the fire and at an average particle size sufficiently small for all the dry chemical to be decomposed within the flame. Their data for these minimum concentrations (both in terms of agent per unit liquid surface area and agent per unit enclosure volume) and limiting particle sizes for three common dry chemical agents are shown in Table 7.10.

Ewing *et al.* note that the actual concentration required in most large-scale applications will be about two to four times as large as the minimum required concentrations shown in Table 7.10

Table 7.10. Minimum concentrations and particle sizes for dry chemical suppression of heptane pool fires (Data from Ewing *et al.*, 1992)

Dry chemical	Limit particle size (μm)	Required volumetric concentration (g/m^3)	Required surface concentration (g/m^2)
KHCO$_3$	20	35	49
NaHCO$_3$	16	46	64
NH$_4$H$_2$PO$_4$	29	54	75

because of losses and inefficiencies during agent delivery to the fire. Practical designs are usually based on specifications in NFPA 17 and/or the listed/approved equipment specifications of pre-engineered systems designed by manufacturers.

Agent delivery nozzles are divided into three types: total flooding, overhead local application, and tankside local application. Total flooding nozzles are designed to provide uniform concentrations throughout some maximum volume, while both types of local application nozzles are designed to provide uniform concentration over some maximum surface area. The number and location of nozzles is determined by dividing the enclosure or surface area into imaginary sub-volumes or sub-surfaces such that each unit can be covered within the limitations of the nozzle selected.

Applications of dry chemical systems at industrial facilities include spray paint booths (Figure 7.16), gasoline dispensing stations, and dip tanks. Another common application, but not necessarily at industrial facilities, is commercial cooking equipment. Actuation is automatic by means of thermal detectors complemented with a manual pull station provision. The spray paint booth system is particularly interesting because it entails total flooding for the booth (including any plenum area behind a filter in the booth), local application for the inlet and outlet openings for the product, local application for a short section of the product outlet conveyor in which the paint is still wet, and total flooding for the ventilation exhaust duct.

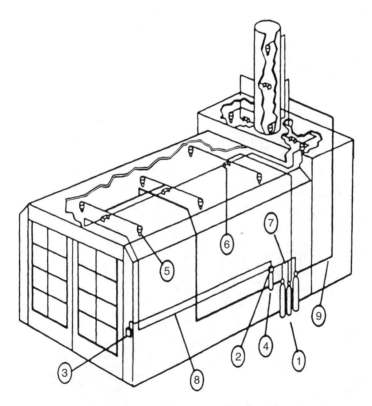

Figure 7.16. Dry chemical system layout for typical paint spray booth. 1. Model PCI Series extinguishing agent storage cylinders. 2. Model ECH or MCH Control Head. 3. Mechanical Remote Pull Station. 4. Model PAC-10 Multiple Cylinder Pneumatic Actuator. 5. Extinguishing agent discharge nozzle. 6. Thermal Detector. 7. Pneumatic copper tubing. 8. Electrical conduit protecting thermal detector input line. 9. Extinguishing agent discharge piping (from Kidde–Fenwal)

One common reservation about dry chemical systems is the need for cleaning the agent and reaction products. Another concern is the possibility of re-ignition after the dry chemical supply has been expended.

In some cases, the dry chemical is combined with AFFF in a twin manual delivery system that allows either agent or both to be applied simultaneously. The two agents complement each other in that the dry chemical can provide chemical inhibition of flames, while the AFFF can seal and cool the flammable liquid surface to prevent re-ignition. Large scale spray fire tests described by Carhart *et al.* (1992) showed that potassium carbonate application alone was not successful in extinguishing the spray fires, but simultaneous application of AFFF and dry chemical was very effective. Twin agent systems have been used predominantly in military applications, but now commercial twin agent systems are being promoted for industrial applications involving elevated spill hazards for which AFFF alone would not be expected to provide suppression.

7.9 Carbon dioxide suppression

Several industrial flammable liquid hazards are amenable to carbon dioxide suppression systems. Typical applications include flammable liquid mixing and storage rooms, dip tanks, commercial fryers containing cooking oils, liquid drainage trenches and impoundment areas, and printing presses utilizing inks with flammable liquid solvents (Figure 7.17). Some of these applications are sufficiently open to require local application systems, while others are more amenable to total flooding systems. The printing presses often use local application systems for the press itself and total flooding systems for the exhaust ducts.

Carbon dioxide total flooding systems are designed to achieve a specified minimum concentration throughout the protected enclosure. The minimum CO_2 concentrations specified in NFPA 12

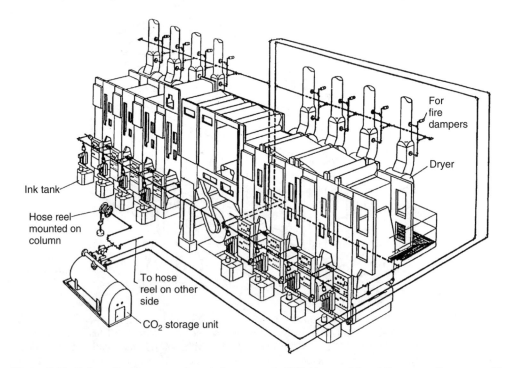

Figure 7.17. Carbon dioxide system for printing press. © 2002 Factory Mutual Insurance Company, with permission

Table 7.11. Minimum CO_2 design concentrations per NFPA 12

Flammable liquid	Min design concentration (v%)
Ethyl Alcohol	43
Gasoline	34
Hexane	35
Methyl Alcohol	40
Methyl Ethyl Ketone	40
Quench, Lube Oils	34

for flammable liquids are based on 120% of the concentrations required for inerting a worse-case vapor-air mixture as determined by laboratory flammability tests for gases and vapors. Values for representative flammable liquids are shown in Table 7.11.

Since these concentrations are well above the maximum tolerable values physiologically, a crucial aspect of carbon dioxide total flooding system design in potentially occupied enclosures is the provision for personnel evacuation prior to CO_2 discharge. This provision includes pre-discharge alarms, time delays prior to discharge, and rescue equipment and training. However, NFPA 12 prohibits the use of discharge abort switches because the standard writers believe the abort switches could cause confusion (and associated danger) about whether the system will or will not discharge depending on whether someone is or is not aborting it.

Local application CO_2 systems designed per the surface area method in NFPA 12 are based on manufacturer specific nozzle discharge rates as determined by testing conducted by listing/approval organizations. As with the dry chemical systems, both overhead and tankside local application nozzles are available for covering specified maximum surface areas.

NFPA 12 also has a provision for designing local application systems to provide coverage over an unenclosed volume surrounding the hazard. This unenclosed volume is specified to extend for a distance of at least 0.6 m (2 ft) away from and above the hazard being protected. One application for this type of design is the protection of internal combustion engines in a room or vault too large to be protected effectively by a total flooding system. Fire tests reported by the Coast Guard (Rynasko et al., 1986) have demonstrated that projection nozzles and horn nozzles can be situated to deliver the required concentrations of CO_2 while allowing sufficient clearance around the engine to allow personnel access during maintenance and normal operations in a ship engine room.

7.10 Halon replacement suppression agents

Total flooding suppression systems employing Halon 1301 replacement agents (so called clean agents) offer obvious advantages for flammable liquid hazards in normally or potentially occupied areas. The agent concentration requirements specified in NFPA 2001 for flammable liquids are based on 120% of the extinguishment concentration measured in the laboratory cup burner apparatus. Cup burner extinguishing data reported in NFPA 12 for heptane are summarized in Table 7.12.

It is clear from Table 7.12 that the cup burner concentrations of these three replacement agents are approximately two to four times the concentration for Halon 1301. However, NFPA 2001 allows a smaller margin between the cup burner concentration and the design concentration for the replacement agents (20%) than NFPA 12A allows for Halon 1301 (50% based on the 3.3 v% cup burner concentration). The higher concentrations and different physical properties of these agents require different delivery systems than those used for Halon 1301. These systems have

Table 7.12. Cup burner heptane flame extinguishment concentrations (from NFPA 2001)

Agent	Average reported concentration (v%)	Maximum reported concentration (v%)
Halon 1301 (CF_3Br)	3.3	3.9
HFC 227ea (CF_3CHFCF_3)	6.2	6.6
HFC 23 (CHF_3)	12.3	12.7
FC3-1-10 (C_4F_{10})	5.4	5.9

been developed and design software for engineered systems are available from system vendors. Additional clean agents besides those shown in Table 7.12 are also being tested and additional approvals/listings can be anticipated.

In applications where reflash fires can be anticipated subsequent to initial extinguishment, NFPA 2001 species that the agent design concentration should be 110% of the values determined from inerting tests. These inerting concentrations for propane are substantially higher (sometimes twice as high) as the heptane cup burner concentrations. Therefore, larger quantities of agent are required for those applications.

One other important consideration for some flammable liquid hazard applications is the generation of hydrogen fluoride by extinguishment with the fluorinated agents. Tests conducted by Sheinson *et al.* (1993) have shown the HF concentrations to be on the order of a thousand ppm even with rapid suppression. Delayed and extended application of the agent often results in even higher concentrations. These HF concentrations are of concern from both physiological and an equipment corrosion viewpoints.

Another family of clean agents is available without any fluorine in the molecule. These are blends of argon, nitrogen, and carbon dioxide. Three different blends are listed in NFPA 2001. They require substantially higher concentrations for suppression than the fluorinated agents because they rely entirely on cooling the flame, whereas the fluorinated agents have some chemical inhibition effects. The higher concentrations lead to substantially larger agent storage and piping requirements. Nevertheless, these halogen-free clean gaseous agents are gaining widespread acceptance in many European countries and other regions that are determined to avoid halogenated suppression agents.

References

'A Review of the Flammability Hazard of Jet A Fuel Vapor in Civil Transport Aircraft Fuel Tanks,' DOT/FAA/AR-98/26, June 1998.

Affens, W.A. and McClaren, G.W., Flammability Properties of Hydrocarbon Solutions in Air, *J. Chem. Eng. Data*, **17**, 1972.

API 2003, 'Protection Against Ignition Arising Out of Static, Lightning, and Stray Currents,' American Petroleum Institute, 1991.

ASTM E 502 - 84, 'Selection and Use of ASTM Standards for the Determination of flash Point of Chemicals by Closed Cup Test Methods,' American Society of Testing and Materials, 1984.

ASTM E 659 - 78, 'Autoignition Temperature of Liquid Chemicals,' American Society of Testing and Materials, 1984.

Babrauskas, V., Pool Fires: Burning Rates and Heat Fluxes, *Fire Protection Handbook*, Section 21, Chapter 6, NFPA, 1986.

Benedetti, R.P., ed, *Flammable and Combustible Liquids Code Handbook*, Third Edition, National Fire Protection Association, 1987.

Beyler, C., Flammability Limits of Premixed and Diffusion Flames, *SFPE Handbook of Fire Protection Engineering*, Section 2, Chapter 9, NFPA, 1995.

Black & Veatch, Engineers-Architects, 'Turbine Generator Fire Protection by Sprinkler System,' Electric Power Research Institute Report EPRI NP-4144, 1985.

Bright, A.W., Electrostatic Hazards in Liquids and Powders, *J. of Electrostatics*, **4(2)** January 1978.
Briggs, A.A. and Webb, J.S., Gasoline Fires and Foams, *Fire Technology*, **24**, 48–58, 1988.
Britton, L., *Avoiding Static Ignitions in Chemical Operations*, AIChE Publication G-67, 1999.
Brown, L.E. and Romine, L.M., Liquefied Gas Fires: Which Foam? *Hydrocarbon Processing*, **58(9)**, September 1979.
Bryan, J.L., *Automatic Sprinkler and Standpipe Systems*, Second Edition, National Fire Protection Association, 1990.
Buch, R., Hamins, A., Konishi, K., Mattingly, D. and Kashiwagi, T., Radiative Emission Fraction of Pool Fires Burning Silicone Fluids, *Combustion and Flame*, **108**, 118–126, 1997.
Burford, R.R., The Use of AFFF in Sprinkler Systems, *Fire Technology*, **12**, 5–17, 1976.
Bustin, W.M. and Dukek, W.G., *Electrostatic Hazards in the Petroleum Industry*, Research Studies Press Ltd (John Wiley & Sons Ltd), 1983.
Carhart, H.W., Leonard, J.T., Budnick, E.K., Ouellette, R.J., Shanley, J.H. and Saams, J.R., 'Manual Fire Suppression Methods on Typical Machinery Space Spray Fires', Naval Research Laboratory Memorandum Report 6673, 1990.
Carhart, H.W., Sheinson, R.S., Tatem, P. and Lugar, J., Fire Suppression Research in the U.S. Navy, *Proceedings First International Conference on Fire Suppression Research*, Stockholm, Sweden, 1992.
Cline, D.D. and Koenig, L.N., The Transient Growth of an Unconfined Pool Fire, *Fire Technology*, 149–162, 1983.
Corrie, J.D., 'The Fire Protection of Flammable Liquid Risks by Foams,' Inst of Chem Engrs Symp Series, 1975.
Denninger, V.L., 'A Method to Compare the Potential of a Flowing Liquid Hydrocarbon to Generate Static Electricity,' M.S. Thesis, Air Force Institute of Technology, NTIS # AD-753-386, 1972.
Drysdale, D., *An Introduction to Fire Dynamics*, Second Edition, John Wiley & Sons Ltd, 1998.
Ewing, C.T., Faith, F.R., Romans, J.B., Hughes, J.T. and Carhart, H.W. Flame Extinguishment Properties of Dry Chemical Extinction Weights for Small Diffusion Pan Fires and Additional Evidence for Flame Extinguishment by Thermal Mechanisms, *J. Fire Protection Engineering*, **4(2)**, 1992.
Faeth, G.M., Mixing, Transport, and Combustion in Sprays, *Progress in Energy and Combustion Science*, **13**, 293–345, 1987.
Factory Mutual Approval Standard No. 6930, 'Less Hazardous Hydraulic Fluid,' Factory Mutual Research Corporation, 1975.
Factory Mutual Loss Prevention Data Sheet 7-14, 'Fire & Explosion Protection for Flammable Liquid, Flammable Gas, & Liquefied Flammable Gas Processing Equipment & Supporting Structures,' FMRC 1982.
Factory Mutual Loss Prevention Data Sheet 7-27, 'Spray Application of Flammable and Combustible Materials,' FMRC 1977.
Factory Mutual Loss Prevention Data Sheet 7-32, 'Flammable Liquid Pumping and Piping Systems,' Factory Mutual Research Corporation, 1973.
Factory Mutual Loss Prevention Data Sheet 7-35, 'Flammable Liquids,' Factory Mutual Research Corporation, 1977.
Factory Mutual Loss Prevention Data Sheet 7-98, 'Hydraulic Fluids,' Factory Mutual Research Corporation, February 1981.
Flemming, R. P., Automatic Sprinkler Calculations, *SFPE Handbook of Fire Protection Engineering*, Section 4, Chapter 3, NFPA, 1995.
Gibson, N. and Lloyd, F.C., Electrification of Toluene in Pipeline Flow, *J. Physics D: Applied Physics*, **3**, 563–573, 1970.
Glassman, I. and Dryer, F. L, Flame Spreading Across Liquid Fuels, *Fire Safety Journal*, **3**, 123–138, 1980/81.
Hill, J.P., 'AFFF Protection of a Simulated Aircraft Carrier Elevator Trunk Against JP-5 Spill Fires,' FMRC JI 0M2N0.RR, 1986.
Holmstedt, G. and Persson, H., Spray Fire Tests with Hydraulic Fluids, *Proceedings of the First Intl Symp on Fire Safety Science*, pp. 869–879, 1986.
Inculet, I., Hazards in Electrostatic Painting, *Conference on Electrostatic Hazards in the Storage and Handling of Powders and Liquids*, Oyez Scientific, 1979.
Hshieh, F.-Y., Note: Correlation of Closed-Cup Flash Points with Normal Boiling Points for Silicone and General Organic Compounds, *Fire and Materials*, **21**, 277–282, 1997.
Hshieh, F.-Y. and Julien, C., Experimental Study on the Radiative Ignition of Silicones, *Fire and Materials*, **22**, 179–185, 1998.
Jeulink, J., Mitigation of the evaporation of Liquids by Fire-fighting Foams, *4th International Symposium on Loss Prevention in the Process Industries*, I. Chem. Eng. Symp. Series No. 80, 1983.
Kirby, D.C. and DeRoo, J.L., Water Spray Protection for a ChemicalProcessing Unit, *18th AIChE Loss Prevention Symposium*, 1984.
Khan, M. and Tewarson, A., Characterization of Hydraulic Fluid Spray Combustion, *Fire Technology*, **27**, 321–333, 1991.
Khan, M., Flame Extinction of Water Miscible Flammable Liquid/Water Solutions, *Proceedings, Fire Suppression and Detection Research and Application Symposium*, National Fire Protection Research Foundation, 1997.

Klinkenberg, A. and ven der Minne, J.L., *Electrostatics in the Petroleum Industry*, Elsevier, 1958.

Kokkala, M., Fixed Water Sprays Against Open Liquid Pool Fires, *Proceedings First International Conference on Fire Suppression Research*, Stockholm, May 1992.

Koszman, I. and Gavis, J., Development of Charge in Low Conductivity Liquids Flowing Past Surfaces, *Chem. Eng. Sci.*, **17**, 1013–1022, 1962.

Leonard, J.T., Static Electricity in Hydrocarbon Liquids and Fuels, *J. of Electrostatics*, **10**, May 1981.

Loftus, J. J., Maldonado-Rosado, A. and Allen, P.J., 'An Evaluation of the MSHA Temperature-Pressure Spray Ignition Test for Hydraulic Fluids,' NBSIR 81-2372, National Bureau of Standards, 1981.

Louvar, J., Maurer, B. and Boicourt, G., Tame Static Electricity, *Chem. Eng. Progress*, **90**, November 1994.

Macksimov, B., Obukh, A., Tikhonov, A. and Kapralov, V., Assessment of Spark-Free Conditions for Filling Tanks with Liquid Hydrocarbons, *J. of Electrostatics*, **23**, 137–142, 1989.

Mancini, R.A., The Use (and Misuse) of Bonding for Control of Static Ignition Hazards, *21st Annual AIChE Loss Prevention Symposium*, August 1987.

Mawhinney, J., Engineering Criteria for Water Mist Fire Suppression Systems, *Proceedings First International Conference on Fire Suppression Research*, Stockholm, Sweden, 1992.

Merritt, R.C., 'Foam-Water and Automatic Sprinkler Protection for Flammable Liquid Process Structures,' Paper No 8b, *17th Annual AIChE Loss Prevention Symposium*, 1983.

Modak, A.T., 'Ignitability of High-Fire-Point Liquid Spills', EPRI NP-1731, Electric Power Research Institute, March 1981.

Mudan, K.S., Thermal Radiation Hazards from Hydrocarbon Pool Fires, *Progress in Energy and Combustion Science*, **10**, 59–80, 1984.

Mudan, K.S. and Croce, P.A., Fire Hazard Calculations for Large Open Hydrocarbon Fires, *SFPE Handbook of Fire Protection Engineering*, Section 3, Chapter 11, NFPA, 1995.

NFPA 11, 'Low-Expansion Foam,' National Fire Protection Association, 1998.

NFPA 12, 'Carbon Dioxide Extinguishing Systems,' National Fire Protection Association, 1993.

NFPA 17, 'Dry Chemical Extinguishing Systems,' National Fire Protection Association, 1998.

NFPA 30, 'Flammable and Combustible Liquids Code,' National Fire Protection Association, 1996.

NFPA 77, 'Recommended Practice for Static Electricity,' National Fire Protection Association, 1993.

NFPA 497, 'Recommended Practice for the Classification of Flammable Liquids, Gases, or Vapors and of Hazardous (Classified) Locations for Electrical Installations in Chemical Process Areas,' National Fire Protection Association, 1997.

NFPA 2001, 'Clean Agent Fire Extinguishing Systems,' National Fire Protection Association, 1996.

Nash, P., The Fire Protection of Flammable Liquid Storages with Water Sprays, *Chemical Process Hazards*, Inst of Chem Engrs Symposium Series, No. 39, pp. 1–17, 1974.

Owens, J.E., Spark Ignition Hazards Caused by Charge Induction, *21st Annual Loss Prevention Symposium*, AIChE, August 1987.

Pfenning, D. and Evans, D., 'Suppression of Gas Well Blowout Fires Using Water Sprays; Large and Small Scale Tests,' American Petroleum Institute Annual Meeting, September 1984.

Post, L. Glos, M., Luettgens, G. and Maurer, B. The Avoidance of Ignition Hazards Due to Electrostatic Changes Occurring During the Spraying of Liquids Under High Pressure, *J. Electrostatics*, **23**, 99–109, 1989.

Putorti, A., '*Application of the Critical Radiative Ignition Flux Methodology to High Viscosity Petroleum Fractions*,' Master's Thesis, Worcester Polytechnic Institute, 1994.

Pratt, T., *Electrostatic Ignitions of Fires and Explosions*, AIChE publication G-58, 1997.

Raj, P., Models for Cryogenic Liquid Spill Behavior on Land and Water, *J. Hazardous Materials*, 1981.

Rasbash, D., The Extinction of Fires by Water Sprays, *Fire Research Abstracts and Reviews*, **4**, 28–53, 1962.

Richards, R.C., 'High Expansion Foam Extinguishment of Machinery Space Fires,' U.S. Coast Guard Report No. CG-D-83-76, 1975.

Roberts, A.F., Extinction Phenomena in Liquids, *15th Intl. Symposium on Combustion*, Combustion Institute, pp. 305–313, 1975.

Roberts, A.F. and Brookes, F.R., Hydraulic Fluids: An Approach to High Pressure Spray Flammability Testing Based on Measurement of Heat Output, *Fire and Materials*, **5**, 87–92, 1981.

Rynasko, J.D., Wolverton, C.D. Jr. and McLain, W.H., 'Localized Fire Extinguishing System (LFES): Low Pressure Carbon Dioxide Agent Tests (Horn and Projection Nozzles) with Large and Small Machinery Mockups,' U.S. Coast Guard Technical Note 052, May 1986.

Scarborough, D.R., Spray Finishing and Powder Coating, *NFPA Fire Protection Handbook*, Section 10, Chapter 7, 1986.

Sheinson, R.S., Eaton, H.G., Black, B.H., Brown, R., Burchell, H., Salmon, G., St. Aubin, J. and Smith, W.D., Total Flooding Fire Suppressant Testing in a 56 m^3 (2000 ft^3) Compartment, *Proceedings of the Halon Alternatives Workshop*, Albuquerque, NM, 1993.

Stubley, D. and Mulligan, D.J., The Effect of Aqueous Surfactant Films on the Ignitability of Aviation Kerosene, *Fire Technology*, **24**, 110–115, 1988.

Tewarson, A., Generation of Heat and Chemical Compounds in Fires, *SFPE Handbook of Fire Protection Engineering*, Section 1, Chapter 13, NFPA, 1988.

Thorne, P.F., Flash Points of Mixtures of Flammable and Non-Flammable Liquids, *Combustion and Flame*, 134, 1977.

Turner, A.F., Water Mist in Marine Applications, *Water Mist Fire Suppression Workshop*, National Institute of Standards and Technology, March 1–2 1993.

Westinghouse Electric Corporation, 'Fire Retardant Lubricant for Turbine Generators,' Electric Power Research Corporation Report EPRI NP-2843, 1983.

Wesson and Associates, Inc., 'Results of Tests Using Low Expansion Foams to Control and Extinguish Simulated Industrial Fires,' 1983.

Wray, H.A., Flammability of Chlorinated Hydrocarbons and Hydrocarbon Admixtures, *J. Coatings Technology*, **56**, 37–43, 1984.

Young, J.R. and Fitzgerald, P.M., 'The Feasibility of Using "Light Water" Brand AFFF in a Closed Head Sprinkler System for Protection Against Flammable Liquid Spill Fires,' FMRC Serial No. 22352, 1975.

Yule, A.J. and Moodie, K., A Method for Testing the Flammability of Sprays of Hydraulic Fluid, *Fire Safety Journal*, **18**, 273–302, 1992.

Zalosh, R., Explosion Hazards in Tanks of High Flash Point Liquids, *International Symposium on Fire Science and Technology*, Korean Institute of Fire Science and Engineering, 1997.

8 FLAMMABLE LIQUID STORAGE

The predominant issue in flammable liquid storage fire protection is container/tank integrity. In particular, the relevant fire protection goals are to prevent or delay container/tank failure and to mitigate the consequences of such failure. Engineering considerations involved in achieving these goals are discussed in this chapter, beginning with tank storage and working down in size (but not necessarily in difficulty) to the small liquid containers used for warehouse storage.

8.1 Storage tanks

8.1.1 GENERIC TANK DESIGNS

In keeping with National Fire Protection Association (NFPA) and American Petroleum Institute (API) standards, storage tanks are categorized by pressure rating and by roof design, with a special category for tanks designed to hold cryogenic liquids.

Atmospheric pressure tanks are designed to operate at pressures from atmospheric up to 0.5 psig (3.5 kPa) as measured in the vapor space at the top of the tank. Atmospheric tanks are usually vertical cylinders with either fixed roofs or floating roofs. The fixed roofs are either flat, as shown in Figure 8.1, or conical, as in the first sketch in Figure 8.2. Since they cannot withstand any substantial pressure increase or decrease (beyond about 0.1 psig or 690 Pa vacuum), atmospheric fixed roof tanks have so-called conservation vents to allow vapor outflow during filling and air inflow while emptying. As indicated in Figure 8.1, the conservation vent is equipped with a flame arrester to prevent flashback into the tank vapor space. A provisions for emergency venting is also needed to cope with more rapid pressure increases as might occur during fire exposure.

Atmospheric fixed roof tanks are used to store low volatility, high flash point liquids such as ethylene glycol, kerosene, Jet A fuel, and Number 6 fuel oil. The vapor pressures of these fuels are usually sufficiently low at ambient temperatures to maintain vapor concentrations below the lower flammable limit. However, fire exposure can cause sufficient vaporization to generate flammable mixtures in the tank vapor space. Flame entry into the tank during fire exposure could result in an explosion. Therefore, fixed roof atmospheric tanks designed in accord with API Standard 650 have weak roof-to-shell seams. An explosion in the vapor space should cause the roof to detach leaving the tank shell and its contents in place.

Atmospheric floating roof tanks have either pontoon or double deck floating roofs or some other approved flotation device. There is a flexible seal around the rim of the floating roof to prevent liquid seepage onto the top of the roof (API Standard 650, and API Publication 2021).

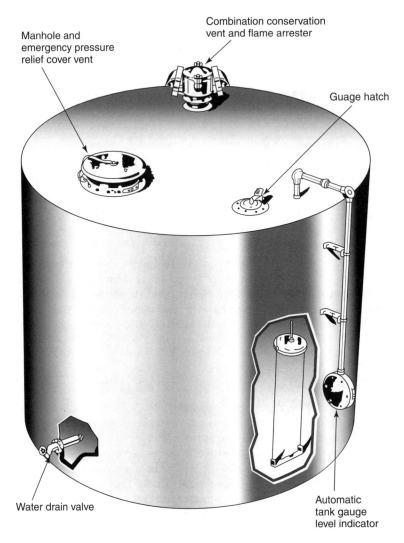

Figure 8.1. Fixed roof flammable liquid storage tank (courtesy J. Convery)

Sealing devices include rubber or foam tubes, spring-loaded fabric, and pantograph mechanisms. Vapor can and does pass through the rim seal. Although most floating roof tanks are open to the atmosphere, some are covered with a fixed roof at the top of the tank and ventilation provision between the fixed and floating roofs.

Crude oil and certain other volatile liquids such as gasoline and naptha are generally stored in floating roof tanks. Rim seal fires represent the most common fire scenario in open floating roof tanks. Foam application systems are often installed around the rim to cope with these fires. Covered floating roof tanks are less prone to rim fires but are more vulnerable to explosions because of the presence of flammable mixtures between the two roofs during and following filling.

Lifter-roof tanks allow the roof to slide up and down such that there is a variable vapor space above the liquid surface. A liquid channel at the top of the tank wall provides a roof-to-shell seal. The vapor space is usually fuel rich except when the liquid is pumped out sufficiently rapidly to

FLAMMABLE LIQUID STORAGE 245

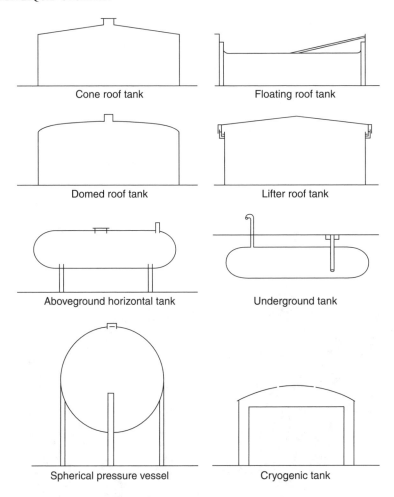

Figure 8.2. Storage tanks and vessels

necessitate air in-breathing. Lifter roof tanks have been used to store volatile liquids but they are few in number compared to floating roof tanks.

Low pressure tanks by NFPA 30 definition have a design pressure in the range 0.5–15 psig (3.5–103 kPa) measured at the top of the tank. They may be vertical fixed roof tanks, or horizontal tanks. The vertical fixed roof tanks usually have a conical or domed roof, which may or may not have a weak roof-to-shell joint design. If there is no deliberately weak roof-to-shell joint, the low pressure vertical tank may fail at the bottom-to-shell seam and rocket away from the fire area releasing its liquid contents in the process.

Vertical fixed roof low pressure tanks are commonly used to store ethanol, gasoline, heptane, methanol, methyl ethyl ketone, toluene, and many other liquids with boiling points in the range 38–110 °C (100–230 °F). These liquids have flash points below 22 °C (73 °F) and have vapor concentrations in the flammable range at most ambient temperatures. Figure 8.3 shows two tank farms comprised of vertical fixed roof tanks.

Horizontal low pressure storage tanks are often used for storage of gasoline, Number 2 and Number 4 fuel oil, and naptha. Aboveground horizontal tanks pose special problems in that:

Figure 8.3. Flammable Liquid Tank Farms

(1) their legs are vulnerable to collapse during direct fire exposure; (2) there is no high point in which to locate the emergency vent; and (3) upon failure they rocket horizontally over great distances. Below ground horizontal metal tanks are subject to corrosion and leakage unless special cathodic protection is installed and maintained (see NFPA 30, Section 2-3.1 and API Publication 1615). Vent design for buried horizontal tanks (see Figure 8.2) also entails special considerations to ensure that the vents are not obstructed and that vapors do not accumulate in occupied or confined areas.

Pressure vessel storage tanks are by decree in NFPA 30 and the ASME Boiler and Pressure Vessel Code designed to accommodate pressures above 15 psig (104 kPa). They may be spherical, toroidal, or horizontal cylinders with domed ends. They are commonly used for storing low boiling point liquids and liquefied gases under pressure. Emergency venting or failure of these vessels

FLAMMABLE LIQUID STORAGE 247

in the absence of fire exposure can cause enough vapor to be released to create a vapor cloud explosion hazard. Vessel failure in the presence of an exposure fire can result in a Boiling Liquid Expanding Vapor Explosion (BLEVE).

Cryogenic liquid storage usually entails use of special double wall tanks with either a vacuum or special low temperature insulation between the walls (last illustration in Figure 8.2). These vessels are commonly used to store Liquefied Natural Gas, liquid hydrogen, sometimes liquefied ethylene. As with the pressure vessel, massive release due to drain line or vessel failure can lead to a vapor cloud explosion hazard. Special precautions include the use of low thermal conductivity material on the dike floor to reduce liquid vaporization, and avoidance of spraying water into the liquid. Water contact with the cryogenic liquid will cause enhanced vaporization, which could in some cases be sufficiently rapid to result in a Rapid Phase Transformation explosion.

8.1.2 STORAGE TANK LOSS HISTORY AND FIRE SCENARIOS

A review of tank fire and explosion incidents can elucidate some of the dominant tank fire protection issues. The largest tanks and the largest losses occur in the petroleum and chemical industries. Table 8.1 is a listing of the 17 largest tank fire/explosion losses in the hydrocarbon and chemical industries, as measured by property damage. The incidents in Table 8.1 were taken from Garrison's updated compilation (1988) of the 100 largest losses at hydrocarbon or chemical industry facilities from 1957 through 1987.

The tank descriptions in Table 8.1 refer to the first tank involved in each incident. Every incident involved more than one tank, and every incident involved either a boilover/frothover or explosion/BLEVE as well as a fire. The fire preceded the boilover/explosion/BLEVE in eight incidents and followed in the other nine incidents.

Several incidents in which the fire preceded the tank failure began with a tank overflow during filling. This was the case in the 1985 incident in Naples, Italy even though the tanks at the marine terminal were equipped with high liquid level gauges. The overflow in that incident caused fire to quickly engulf 20 tanks over an area of $15,000\,m^2$ (3.7 acres). The 1983 Newark, New Jersey incident began with an overflow spill into a dike surrounding the tank. Gasoline vapors from the dike traveled about 305 m (1000 ft) downwind to an incinerator where they were ignited. The result was a vapor cloud explosion as well as a dike fire that spread to two adjacent internal floating roof tanks.

Three explosion-fire incidents listed in Table 8.1 were initiated by lightning. In the 1977 incident in Illinois and the 1979 incident in Texas, lightning initiated tank explosions which caused tank roof sections to rocket and strike other tanks. The result in both incidents was multiple tank fires.

The three fire-BLEVE incidents listed in Table 8.1 began with LPG released from a tank connection. These releases resulted from: improper sampling of tank liquid (1966 incident in France); unattended drainage of tank water (1972 incident in Brazil); and the rupture of a connecting line (1984 incident in Mexico). In all three incidents, vessel failure resulted from exposure to the pool fire from the released LPG. Blast waves and rocketing metal from vessel rupture damaged surrounding vessels and caused fire spread throughout the storage facility.

Boilovers were the primary cause of fire spread in four incidents. In the 1983 incident in Wales, the boilover caused fire spread over an area of $16,000\,m^2$ (4 acres), while in the 1972 incident in Italy and the 1982 incident in Venezuela the resulting fire spread over a distance of approximately 550 m (1800 ft).

The American Petroleum Institute (API) collects fire incident data provided voluntarily by member companies. The API collection of tank fire incident data for the period 1978–1987 is summarized in Table 8.2. The tabulation shows the distribution of the 514 involved tanks in terms of type of tank, type of liquid, and type of facility.

Table 8.1. Large-loss petrochemical tank fires (data from M&M Protection Consultants, '100 Large Losses,' 1988)

Date	Location	Loss (88)$	Tank description – No. Damaged	Liquid	Ignition Source	Fire/Explosion
05/22/58	California	$39M	50,000 bbl – 73% of Refinery	Hot Oil	?	Frothover-Fire
06/16/64	Japan	$80M	Floating Roof – 97	Crude Oil	Friction Spark	Fire & Explosion
01/04/66	France	$63M	12,580 bbl Sphere – 5	Butane	Vehicle	Fire-BLEVE
01/20/68	Netherlands	$89M	Refinery Slop Tank – 80	Hot Oil	?	Boilover-Fire
09/17/70	Texas	$19M	15,000 bbl – 16	Slop Oil	Lightning	Explosion-Fire
03/30/72	Brazil	$12M	Spheres, Cylinders – 5, 16	LPG	?	Fire-BLEVE
08/04/72	Italy	$11M	500,000 bbl Floating Roof – 5	Crude Oil	Dynamite	Expl-Fire-Boilover
08/17/75	Pennsylvania	$24M	60,000 bbl Internal Floating Roof – 4	Crude Oil	Steam Pipe	Fire-Explosion
09/24/77	Illinois	$14M	Cone Roof, Floating Roof – 2	Diesel Fuel, Gas	Lightning	Explosion-Fire
09/01/79	Texas	$97M	Tanker & 80,000 bbl Tank	Distillate, Ethanol	Lightning	Explosion-Fire
08/20/81	Kuwait	$48M	160,000 bbl Floating Roof – 8	Naptha	?	Explosion-Fire
03/09/82	New Jersey	$27M	25,000 gal Process Tank – 4	Cumene	Tank Heaters	Explosion-Fire
12/19/82	Venezuela	$54M	240,000 bbl Fixed Roof – 2	No. 6 Fuel Oil	?	Expl-Fire-Boilover
01/07/83	New Jersey	$37M	42,000 bbl Internal Floating Roof – 4	Gasoline	Incinerator	Fire-Explosion
08/30/83	Wales	$16M	600,000 bbl Floating Roof – 3	Crude Oil	Flare Soot	Fire-Boilover
11/19/84	Mexico	$21M	Spheres, Cylinders – 4, 44	LPG	Flare	Fire-BLEVEs
12/21/85	Italy	$43M	24 Tanks	Gasoline, Fuel Oil	?	Fire-Explosion

Table 8.2. API tank fire experience (1978–1987)

	Structural type of tank									
	Floating roof									
By class of product	Pantograph seal	Tube seal	Covered	Total	Cone roof	Dome roof	Sphere & spheroid	Horizontal	Other	TOTAL
Crude Oil	13	4	7	24	195	47	0	19	22	307
Products										
Under 100 F flash	22	9	12	43	28	8	2	0	6	87
100–200 F flash	6	0	1	7	24	4	0	1	1	37
Over 200 F flash	1	1	1	3	28	3	0	0	5	39
LPG	0	0	0	1	6	0	0	0	0	7
Other	0	1	0	1	30	3	0	0	3	37
Total	42	16	21	80	311	65	2	20	37	514
By location										
Producing properties	2	2	2	6	226	54	0	18	27	331
Pipeline operations	6	0	1	7	8	3	0	0	1	19
Refineries	21	11	13	45	63	4	2	0	2	116
Bulk plants & terminals	12	2	4	18	5	0	0	1	1	25
Other locations	1	1	1	3	9	4	0	1	6	23
Total	42	16	21	79	311	65	2	20	37	514

Total number of fires involving tanks (regardless of number tanks Involved) 414

Of the 514 tanks in the API tabulation, 311 (60%) had cone roofs. Most of those contained crude oil at producing facilities. There were 79 floating roof tanks involved (15.4%), most of them at oil refineries. There were 65 dome roof tanks reported (13%), mostly at producing properties. Since the API data come from an unknown population of tanks at risk, there is no obvious way to convert the loss numbers and percentages into excepted loss frequencies for the various tank categories. (To the author's knowledge, the only published data on tank fire and explosion frequencies is the value cited by Lees (1980) of one incident per 833 tank-years.)

An earlier compilation of API tank incidents (see FM Data Sheet 7-88, 1976) produced the following distribution of causes:

Lightning	43%
Static Electricity	11%
Unidentified Internal	11%
Spontaneous Ignition	8%
Cutting & Welding	7%
Exposure Fires	6%
Overfilling	6%
Tank Collapse	4%
Other	4%

This distribution of incident causes is probably unique to the petroleum industry. Other industries are less prone to lightning and to static electricity because the tanks are smaller relative to surrounding structures and because fewer tanks are filled at high flow rates of Class I flammable liquids.

The author is not aware of a comparable breakdown of incident causes for other industries, but there is reason to suspect that cutting and welding would be the leading cause of tank fires/explosions involving high flash point liquids. For example, Factory Mutual loss records include 52 tank cutting and welding ignitions during the period 1973 through November 1988. Many of these involved fuel oil storage tanks. The tanks are often initially free of flammable vapor but heat from the cutting and welding can vaporize residual liquid or sludge and create a flammable mixture in the tank.

Sometimes the cutting/welding is conducted on leaky tank connections which can release large quantities of liquid upon heat exposure.

Spontaneous ignition of tank vapors is usually attributed to pyrophoric iron sulfide deposits on the tank walls. The iron sulfide forms from a slow reaction between the tank wall and hydrogen sulfide present in some petroleum liquids. The reaction proceeds more readily under moist, oxygen deficient atmospheres in the tank. Sudden exposure of the iron sulfide deposits to dry air can raise the surface temperature above the ignition threshold of many vapor-air mixtures. Dimpfl (1985) has recently observed that an analogous reaction can occur with organic deposits in asphalt storage tanks. His measurements of vapor space composition in several asphalt tanks demonstrate that flammable compositions do in fact exist in many tanks even though the asphalt flash point is reported to be well above the tank temperature. The explanation for this apparent paradox is that the flammable vapors are not included in the liquid sample submitted for laboratory flash point tests. Dimpfl suggests that an oxygen deficient atmosphere should be maintained in hot asphalt tanks in order to prevent the flammable vapors from being ignited by pyrophoric iron sulfide or organic deposits. Davie et al. (1994) provided a review of 73 fire and explosion incidents involving asphalt storage tanks in the UK. Many of these incidents are associated with partial oxidation of the asphalt or asphalt deposits on the tank roof/wall.

Electrostatic ignitions have been attributed to electrostatic charging during tank filling and during tank washing. Charge generation and accumulation during tank filling is discussed in

Section 7.3. Explosions in large crude oil tankers due to electrostatic charge generation during tank washing are discussed in reports by the International Chamber of Shipping and by Kitagawa (1975). The charge generation mechanism during tank washing is believed to be water impingement against the tank walls. If water drops impinge against the wall and rebound in the form of small droplets, a charged mist is formed in the tank. Electrostatic fields associated with the charged mists are not as strong as those generated by the impact of large slugs of water against the tank walls. These slugs of water are produced by some high capacity water jet washers which have since been prohibited from use in flammable atmospheres on tankers. Recirculation of wash water, the entrainment of solid particles in the water, and the use of certain chemical additives, can exacerbate electrostatic charging during tank washing. Guidelines for washing and for spraying flammable liquids into grounded metallic storage tanks so as not to generate potentially incendive electrostatic fields have been developed by Post et al. (1989) on the basis of experiments with a variety of liquids and spray flow rates.

8.1.3 TANK BURNING RATES AND SPACING CRITERIA

Since large loss tank incidents usually involve fire spread to adjacent tanks, it is important to establish minimum separation distances between tanks so as to reduce the chance of fire spread. These minimum separation distances should be based on an appropriate design basis fire scenario. The most widely used scenario is a fire burning in a tank without a roof. The premise is that an explosion has blown off the roof in a fixed roof tank with a weak roof-to-shell seam, or that the roof in a floating roof tank has sunk, possibly because of accumulated water applied during the fire. This premise would not apply to other types of tanks such as horizontal cylinders or pressure vessels without a weak roof-to-shell seam. Furthermore, this design basis scenario cannot preclude fire spread in fires involving massive overfilling or boilovers such that burning liquid flows under and around the adjacent tanks.

The liquid burning rate in an open roof fire scenario depends on the liquid level beneath the top of the tank. As the liquid level recedes, the burning rate is reduced from the values described in Chapter 7 for a pool fire. This reduced burning rate is due in part to the reduction in flame heat flux to the fuel surface because of the tank ullage space, and in part to the reduced air flow rate into the combustion zone in the tank ullage. According to Wang (1991), the first effect dominates and can be accounted for by modeling the absorption and emission of radiant heat fluxes in the ullage. Wang's three layer radiation model produced the following equation for the liquid regression rate:

$$\frac{\dot{m}''}{\rho} = \frac{dH}{dt} = \frac{\alpha\sigma(T_f^4 - T_s^4)}{\rho(0.75 KH + \varepsilon_f^{-1} + \varepsilon_s^{-1} - 1)(\Delta H_v + c(T_s - T_0))} \quad [8.1.1]$$

where H = tank ullage height, K = fuel vapor radiation absorption coefficient, ε_f and ε_s are the emissivities of the flame and fuel surface, respectively, T_f and T_s are the temperatures of the flame and fuel surface, respectively, and ΔH_v and c are the heat of vaporization and specific heat of the liquid. The coefficient α is a type of geometric view factor as follows:

$$\alpha = \left(\frac{\sqrt{D^2 + H^2} - H}{D}\right)$$

where D is the tank diameter.

Equation [8.1.1] indicates that even relatively shallow ullages can cause a pronounced reduction in fuel burning rate. For example, a ullage height of 1 m in a 5 m diameter tank, with a vapor absorption coefficient of 5 m^{-1} and flame and fuel surface emissivities of 0.95, produces a 60%

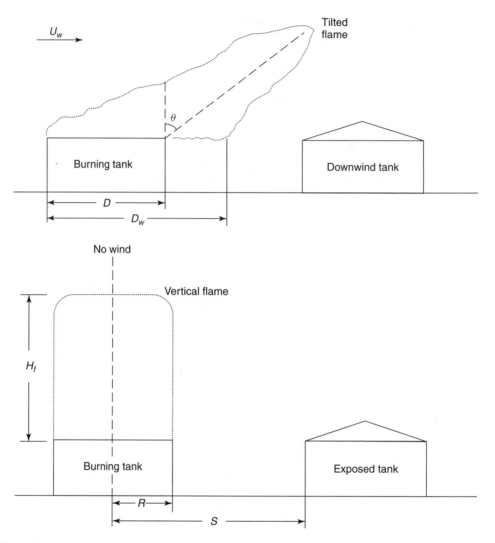

Figure 8.4. Tank spacing considerations. Top–wind blown flame threatening downwind tank; Bottom–flame radiation from vertical flame (no wind)

reduction in the fuel burning rate. However, equation [8.1.1] neglects any radiation from possibly hot tank walls onto the fuel surface inside the tank. The following discussion of flame tilt and drag ignores any tank ullage effects.

Figure 8.4 shows the open tank fire scenario in the absence of any significant wind and when there is a strong wind tilting the flame toward the exposed tank. In the case of the wind tilted flame, one criterion for tank separation would be that the flame does not impinge on the downwind tank. The downwind extended diameter of the fire, D_w, can be estimated from equation [7.4.4] if we specify the tank diameter, D, the wind speed u_w, at a 10 m (32.8 ft) elevation, and the vapor/air density ratio, ρ_v/ρ_a. For example, if $D = 30$ m (100 ft), $u_w = 10$ m/s (22 mph), $\rho_v/\rho_a = 2.0$:

$$D_w/D = 1.25\,(100/294.6)^{0.069}(2.0)^{0.48} = 1.60$$

Therefore, the exposed tank should be at least 18 m (60 ft) (30 × 0.6) away from the burning tank. If the adjacent tank diameter is different, this calculation should be based on the larger diameter so as to provide the most conservative separation.

Other approaches and correlations are available to account for downwind flame drag and tilt. One simple observation reported by Lees (1980, p. 755) is that a 2 m/s (4 mph) (surface) wind speed produces a 50% downwind extension of flame diameter and a 45° flame tilt. It would seem on the basis of both these simple observations and the preceding calculation that a shell-to-shell spacing of approximately one-half the larger tank diameter should suffice to avoid direct flame impingement.

Most engineering estimates of minimum required tank spacing consider flame radiant heat flux levels on the exposed tank. A maximum tolerable heat flux is specified and the corresponding tank separation is calculated from the flame surface emissive power and the radiation view factor. However, there are significant uncertainties involved in specifying or calculating all three parameters as explained below.

Most recent calculations (Robertson, 1976; Lees, 1980, p. 755; Das, 1983) use a maximum tolerable tank heat flux of 36–38 kW/m² (11,400–12,000 Btu/hr-ft²). Swiss Reinsurance (1987) has specified 36 kW/m² (11,400 Btu/hr-ft²) for water cooled tanks and 12 kW/m² f (3800 Btu/hr-ft²) or uncooled tanks. The engineering basis for these values is not apparent and warrants review. If the criterion is to avoid thermal or structural damage to the tank, the critical heat flux might be selected such that the tank wall or roof is not heated to the weakening point, which is roughly 540 °C (1000 °F) for steel. In the absence of any water cooling, the tank wall above the liquid level and the tank roof will be cooled by natural convection to the air (external surface) and to the vapor space (internal surface) and by reradiation from the outer wall and roof. After the wall has been heated via flame radiation to a steady state temperature, T_w,

$$q_r'' = h_{out}(T_w - T_a) + h_{in}(T_w - T_v) + \varepsilon_w \sigma T_w^4 \qquad [8.1.2]$$

where q_r'' is the flame radiant heat flux on the tank wall or roof, h_{in} and h_{out} are the natural convection heat transfer coefficients at the inside and outside surfaces of the tank wall, and T_a and T_v are the air and tank vapor temperatures, respectively, ε_w is the wall emissivity, and σ is the Stefan–Boltzmann constant (5.67×10^{-11} kW/m²-°K).

A logical approach to determine a critical heat flux for steel weakening is to evaluate equation (8.2) at the steel critical temperature ($T_w = 540\,°C$ or $212\,°F$) and T_v equal to the liquid boiling point, e.g. $100\,°C$ ($212\,°F$) if water were in the exposed atmospheric pressure tank. The value of h_{in} and h_{out} depends upon whether we specify the tank wall or roof. In the case of a 10 m (32.8 ft) 'freeboard' on the tank wall, h is estimated to be about 20 W/m²-C (9.8×10^{-4} Btu/sec-ft²-F) based on the heat transfer correlations for turbulent free convection along a vertical wall. The value of wall emissivity depends on the condition of the steel outer surface. According to the emissivity data in Holman (1990), mild steel has an emissivity in the range 0.20–0.32 at wall temperatures of interest here, whereas sheet steel with a 'rough oxide layer' has an emissivity of 0.80.

Using $\varepsilon_w = 0.20$ and the other cited parameter values, the critical radiant heat flux, q_r'', is from equation [8.1.2]

$$q_r'' = 19 + 5\,\text{kW/m}^2 = 24\,\text{kW/m}^2$$

Using $\varepsilon_w = 0.80$,

$$q_r'' = 19 + 20\,\text{kW/m}^2 = 39\,\text{kW/m}^2$$

The latter value is about equal to the critical value used by Robertson, Das, etc. It should be noted that, since the value of h for a horizontal surface, such as the tank roof, is significantly lower than the 20 W/m²-C (9.8 × 10⁻⁴ Btu/sec-ft²-F) selected for the tank wall, a correspondingly lower value of q_r'' would have resulted. The calculated value of $q_r'' = 24$ kW/m² (7600 Btu/hr-ft²) for a clean steel wall is also lower than the critical radiant heat fluxes in the literature for steel storage tanks.

Values of flame emissive power for flammable liquid pool fires have been reviewed by Mudan and Croce (1988). They explain that the data for emissive power depends on whether or not the flame from the large pool fire is obscured by smoke on the outer periphery of the fire. This has been the case for large fires of gasoline, kerosene, and JP-4, and the relationship between E and D for these fires is shown in Figure 8.5. This relationship is not applicable to hydrocarbons with carbon-to-hydrogen ratios less than about 0.4, nor is it applicable to oxygenated organics. Liquids not prone to smoke obscuration of the flame surface, would have emissive powers of 100–120 kW/m² (31,700–38,000 Btu/hr-ft²) even at diameters on the order of 10 m (32.8 ft) and higher. This is also the range of emissivities observed for gasoline, kerosene, and JP-4 when the flame is not obscured by the surrounding smoke.

The calculation of radiant view factor is described in Appendix A for vertical flames such as would occur under calm conditions. The more complicated geometry of a tilted flame can be evaluated from the view factor equations and graphs given, for example, by Mudan and Croce (1988) and Das (1983). Most tank separation calculations have been based on a vertical cylindrical flame assumption, with either an assumed flame height (Robertson, 1976, assumes it equal to twice the tank diameter) or by using flame height correlations such as those in Appendix A.

As an example, consider the spacing of 30 m (98 ft) diameter tanks of kerosene. Since Figure 8.5 is applicable to kerosene, the flame emissive power for a 30 m (98 ft) diameter open top tank fire would be about 23 kW/m² (7300 Btu/hr-ft²). This is roughly equal to the critical heat flux calculated previously for a tank wall emissivity of 0.20. Thus a kerosene tank fire should not jeopardize adjacent tanks if the wind does not tilt the flame and if the smoke reduces the flame emissive power as indicated by Figure 8.5. Based on a calculated flame height of 32 m (105 ft)

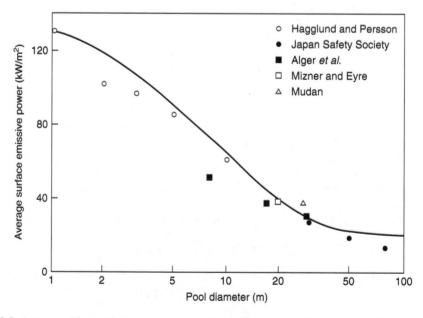

Figure 8.5. Average surface emissive power for gasoline, kerosene, and JP-4 pool fires (from Mudan and Croce. Reproduced by permission of Society of Fire Protection Engineers)

FLAMMABLE LIQUID STORAGE

for this situation and using the view factors in Figure A.5, the view factor at a target distance of one diameter (30 m or 98 ft) from the flame axis (i.e. a shell-to-shell spacing of 15 m or 50 ft) is estimated to be about 0.22, corresponding to an estimated heat flux of $5\,kW/m^2$ (1600 Btu/hr-ft^2). The radiant heat flux measured by Yamaguchi and Wakasa (1985) at that location was $6\,kW/m^2$ (1900 Btu/hr-ft^2), which is remarkably close to the calculated value. These values are well below the estimated thresholds for tank wall damage.

Das (1983) has performed radiant heat flux calculations for the more complicated and more conservative case of a wind tilted flame threatening the downwind adjacent tank as illustrated in Figure 8.4. He calculated view factors and corresponding radiant heat fluxes for different tank diameters and flame tilt angles. According to the most relevant correlation given by Mudan and Croce (1988), the flame tilt angle, θ, can be found from

$$\cos\theta = \begin{cases} 1 & \text{for } u^* \leq 1 \\ u^{*-0.5} & \text{for } u^* > 1 \end{cases} \qquad [8.1.3]$$

where $u^* = u_w/[gm''D/\rho_v]^{1/3}$, and the other symbols are as defined in Chapter 7.

Consider the implications of equation [8.1.3] for a liquid such as heptane burning in a large storage tank. Using $m'' = 0.10\,kg/m^2$-s (241 lb/ft^2-hr), and $\rho_v = 4.5\,kg/m^3$ (0.28 lb/ft^3) for heptane, equation (8.3) indicates that a wind speed of 5 m/s (11 mph) (evaluated 1.6 m or 5.2 ft above the burning liquid as specified in the correlation), and a tank diameter of 91.4 m (300 ft) would result in $u^* = 1.8$ and $\theta = 43°$. Smaller diameters would result in slightly larger tilt angles, with $\theta = 45°$ being a representative value for a large range of diameters at this wind speed. The threshold wind velocity to begin tilting the flame from the vertical is 2.7 m/s (6 mph), according to equation [8.1.3] and the specified values for heptane and large tank diameters.

Based on the preceding discussion, it is of interest to consider the radiant heat fluxes emanating from tank fires with flame tilt angles of 45°. Tank spacing implications of 45° tilted flames with an emissive power of 118 kW/m^2 (37,400 Btu/hr-ft^2)(corresponding to a black body flame temperature of 1200 °K or 2160 R) are shown in Table 8.3 based on the tilted flame radiation view factors given by Das (1983).

The two values of q_r'' in Table 8.3 correspond to tank wall emissivities of 0.80 and 0.20 as described previously. The wall-to-wall separations in Table 8.3 are nondimensionalized by tank diameter. Comparing the calculated values to the NFPA 30 specifications, the calculated separations are larger than the NFPA requirements except for the first case ($D = 91.4$ m or 300 ft, $q_r'' = 37\,kW/m^2$ or 11,700 Btu/hr-ft^2). However, the NFPA 30 requirements are based on risk judgements as well as some approximate considerations of flame radiation.

Table 8.3. Tank separation criteria

D (m)	qr" (kW^2)	Wall-to-wall Separation Flame Radiation[a]	(x/D) NFPA 30 Specs
91.4	37	0.625	0.50–0.67[b]
45.7	37	0.75	0.33
18.8	37	0.65	0.33
91.4	24	1.30	0.50–0.67
45.7	24	1.30	0.33
18.8	24	1.40	0.33

[a]Radiant Heat Fluxes based on Emissive Power = 118 kW/m^2, Flame Tilt Angle = 45°
[b]Lower value in NFPA 30 is for remote impoundment of spills; larger value is for diked tank

8.1.4 TANK EMERGENCY VENTING

Emergency venting provisions on flammable liquid storage tanks are supposed to relieve the internal pressure increase caused by exposure fires. The pressure increase is due primarily to the boiling of liquid and secondarily to the heating of vapor above the liquid surface. An effective emergency vent should maintain pressures below the tank design strength by venting the tank vapors at a rate equal to their generation rate due to boiling, i.e.

$$V_v = V_{boil} \qquad [8.1.4]$$

where V_v is the vapor flow rate through vent, and, V_{boil} is the vapor generation rate due to boiling.

The vapor generation rate during fire exposure is

$$V_{boil} = Q_L/(L\rho_v) \qquad [8.1.5]$$

where Q_L is the rate of heat absorption by liquid (kW), L is the liquid latent heat of vaporization (kJ/kg), and ρ_v is the vapor density (kg/m^3).

Although tank emergency venting is conceptually straightforward, its implementation requires realistic estimates of flame heat input and vent flow rates including the possibility of two phase vapor-liquid flow. Flame heat input rates are based on large-scale pool fire tests with a simulated tank or similar large object immersed in the flames. The following discussion includes a review of recent vent flow and heat input data as well as the test data that provided the basis for the current emergency vent guidelines in NFPA 30 and similar codes and standards.

The National Academy of Sciences (NAS) 1973 study of pressure- relieving systems provided a comprehensive compilation of test data and vent guidelines that is still relevant. (According to Benedetti (1987), emergency vent guidelines in the current editions of NFPA 30 and API 2000 remain essentially unchanged since the 1963 and 1966 revisions, respectively.) The NAS study concluded that the rate of heat absorption, Q_L, should be calculated from the following equation:

$$Q_L = q''FEA_w \qquad [8.1.6]$$

where q'' is the fire heat flux per unit wetted surface area exposed to fire (kW/m^2), F is the reduction in absorbed heat flux due to tank insulation, water spray protection, etc., E is the fraction of total wetted surface area exposed to fire, and, A_w is the tank surface area wetted by liquid (m^2).

Table 8.4 is a reproduction of the NAS compilation of tank and vessel fire exposure test data that form the basis of the NAS recommended value for q''. The vessel total surface area, wetted surface area, heat flux to wetted surface area, percent of wetted surface area exposed to fire, and heat flux per unit exposed surface area (q'') are listed for nine fire tests involving petroleum base liquids. The values of q'' in Table 8.4 range from 25,900 to 36,500 Btu/hr-ft^2 (82–115 kW/m^2) with an average of 32,400 Btu/ft^2-hr (102 kW/m^2). The NAS committee recommended "As a first and good approximation $q'' = 34,500$ Btu/hr-ft^2 can be selected as a reference heat flux." The committee intended the reference value to be refined as more test data become available and as the application differs in terms of fuel, vessel, environment, etc. from the existing test database.

Table 8.5 lists results of flammable liquid fire exposure heat flux tests conducted since the NAS compilation. The tests fall into the following three categories: (1) large objects (surface area >4 m^2 or 43 ft^2) exposed to petroleum liquid fires; (2) small objects exposed to petroleum liquid fires; and (3) large tank exposed to alcohol fire. The values of q'' for the first category range from 27,000 to 36,500 Btu/hr-ft^2 (85–115 kW/m^2) which is very close to the range for q'' in Table 8.3. Thus, the recent data further support the NAS recommended reference

FLAMMABLE LIQUID STORAGE

Table 8.4. Summary of fire exposure tests as of 1973 (data from National Academy of Sciences, 1973)

	Underwriters Laboratories[a]	Duggan et al. Test No 1[b]	Duggan et al. Test No 4[b]	Rubber Reserve Test No 17[c]	Rubber Reserve Test No 17[c]	Rubber Reserve Test No 17[c]	API Project Test No 1[d]	API Project Test No. 2[d]	Standard Oil Of California[e]
Type of exposure	Water flowing over plate	Water flowing over tank	Water heated in tank	Heating Water in tank	Generating steam in tank	Water flowing inside 3/4-in std pipe	Heating water in tank	Heating water in tank	Heating water in tank
Fuel	Gasoline	Propane gas	Propane gas	Gasoline	Gasoline	Gasoline	Kerosene	Kerosene	Naphtha
Heat flux to wetted surface ~ Btu/(hr)(ft^2)(Q/A)	32,500	25,900	12,630	23,200	21,000	30,400	15,700	16,800	32,000
Observed fire exposure, % wetted surface	100	100	48	60–70	60–70	80–90	46	46	100
Temp receiving surface, (F)	76	136	120	300	—	—	~300	~320	70–212
Heat flux to expose wetted surface, Btu/(hr)(ft^2)	32,500	25,900	26,300	35,700	32,300	35,800	34,200	36,500	32,000
Total surface, ft^2	24	242	330	568	568	9.0	16.2	16.2	206
Wetted surface, ft^2	24	242	176	400	400	9.0	6.13	6.13	105
Nominal vessel capacity, bbl	—	72	72	119	119	—	0.88	0.88	33

[a] Underwriters' Laboratories (1935)
[b] Duggan et al. (1944)
[c] Rubber Reserve Co (1944)
[d] American Petroleum Institute Project Test No. 1 and 2, report to API Subcommittee on Pressure-Relieving Systems at University of Michigan, June 1947, unpublished Flame was subjected to wind
[e] Standard Oil of California tests. April and June 1925 Results unpublished

Table 8.5. Other large pool fire immersed object tests

Reference	Pool size	Fuel	Test object	q'' (kW/m²)	(Btu/hr-ft²)
Gregory et al. (1987)	162 m²	JP-4	Horizontal Cylinder, 1.4 m Diam 6.4 m Long	115	36,500
			20 cm Sphere	150	47,600
Wachtell and Langhaar (1966)	81 m²	JP-4	1 × 1.2 × 1.5 m steel cask	85	27,000
Russell and Canfield (1973)	11.5 m²	JP-5	Horizontal Cylinder, 22 cm Diam, 91 cm Long	158	50,100
Anderson et al. (1974), Townsend et al. (1974)	216 m²	JP-4	Tank Car, 3.0 m Diam, 18.4 m Long	99	31,400
Lougheed (1989)	10.5 m²	Test 1 Ethanol	Vertical Tank, 1.8 m diameter, 4.3 m high	31	9,900
		Test 6 Hexane	Vertical Tank, 1.8 m diameter, 4.3 m high	169	54,300

value of approximately 34,500 Btu/hr-ft². The new data also suggest that q'' should be substantially higher (approximately 50,000 Btu/hr-ft²) for small objects immersed in petroleum liquid fires, and should be substantially lower (about 10,000 Btu/hr-ft²) for tanks immersed in alcohol fires.

NFPA 30 and most other consensus standards use different heat input correlations instead of equation (8.6). (The NAS report contained a compilation of 52 different formulas for sizing emergency relief vents.) Figure 8.6 shows the NFPA 30 recommended values for Q_L as a function of A_w, as well as the value corresponding to equation (8.6) for $F = 1$, $E = 1$, and $q'' = 34,500$ Btu/hr-ft².

The lower portion of the NFPA 30 recommended curve for Q_L corresponds to a value of q'' of 20,000 Btu/hr-ft² (63 kW/m²). The NFPA 30 values of Q_L are significantly lower than the NAS values particularly for large tanks/vessels. Benedetti (1987) explains that the reduced values of q'' in the large area portion of the NFPA 30 curve reflect the NFPA 30 Technical Committee's judgement that large tanks are not expected to be completely immersed in flame, and large atmospheric pressure tanks are likely to self vent via failure of the heated portion of the tank above the liquid surface.

Figure 8.6 includes key test data that form the basis of the NFPA 30 and API 2000 recommended values of Q_L for tank surface areas in the range 9.3–93 m² (100–1000 ft²). The Rubber Reserve data, Duggan et al. data and Standard Oil data are shown in Table 8.4. They are plotted in Figure 8.6 with values of E assumed equal to 1, even though Table 8.4 shows that E was actually equal to 0.48 and 0.60–0.70 for several tests. When the observed values of E for these tests are accounted for, equation [8.1.6] actually fits the data better than the NFPA 30 recommended correlation.

If equation [8.1.6] is to be used for emergency vent design, some guidance is needed on the appropriate values for E and F. Values of E can be estimated from a comparison of tank height and flame heights calculated with the correlations in Appendix A. If the calculated flame height exceeds the tank liquid surface elevation, E can be taken as 1. If the flame height is lower than the liquid surface elevation, E can be reduced proportionately. Although both the

FLAMMABLE LIQUID STORAGE

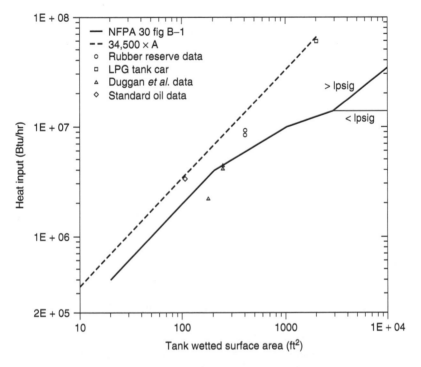

Figure 8.6. Absorbed heat versus tank wetted area

liquid surface elevation and the tank wetted surface area are dependent on tank filling practices, NFPA 30 provides the following guidance in relating tank surface area to actual wetted surface area:

Tank geometry	Wetted/surface area
Vertical Cylinder	lower 30 ft of exposed area
Horizontal Cylinder	75% of exposed area
Sphere	55% of exposed area

NFPA 30 also provides the following guidance on values of F:

Tank protection	F
Drainage to Remote Impoundment	0.50
Water Spray at 0.25 gpm/ft² exposed area	0.30
Insulation with conductance <4 Btu/hr-ft²-F	0.30
Drainage + Water Spray + Insulation	0.15

The basis for the drainage/impoundment credit is apparently judgmental although a minimum drainage slope of 1% and a minimum impoundment area distance of 50 ft from the tank are specified.

The basis of the water spray credit factor is the Rubber Reserve Company tests in which a water spray of 0.20 gpm/ft² (8.1 L/min-m²) of exposed vessel surface area resulted in an absorbed heat flux of 6000 Btu/hr-ft² (19 kW/m²), or about 30% of the absorbed heat flux measured for a free burn. The additional 0.05 gpm/ft² (2 L/min-m²) needed for a F value of 0.30 is due to the

NFPA Committee's evaluation of water losses associated with wind effects, incomplete surface coverage, etc. API 2000 maintains that these effects and the inherent reliability issues associated with a water spray system render the use of values of F less than 1 questionable. On the other hand, Fritz and Jack (1983) suggest that contemporary tank and vessel construction (specifically the use of welded construction as opposed to riveted walls) is more amenable to water film stability than the old rusty vessels used in the Rubber Reserve tests.

The value of F for water spray protection should account for the fuel burning as well as the water application rate and tank surface condition. Recent tests sponsored by the distillers industry (Lougheed and Crampton, 1989) provide data on the value of F for an ethanol tank protected with ceiling sprinklers. Values in the range 0.33 to 0.71 were obtained for $\frac{1}{2}$ in orifice sprinklers with a design density of 0.30 gpm/ft^2 (12.2 L/min-m^2), of floor area and a ceiling height of 20 ft (6.1 m). Installation of an additional sprinkler on the underside of the raised tank reduced the value of F to 0.19 for the ethanol exposure fire.

The NFPA recommended value of F for insulation is based on the Rubber Reserve tests with two inches of gunnite insulation with an effective conductance of 4 Btu/hr-ft^2-F (23 W/m^2-C). The 1973 LPG tank car tests (Townsend *et al.* 1974) provide another important datum for the effect of insulation. Application of a 1/8 in (3.2 mm) thick coating of Korotherm insulation reduced the net heat flux into the tank car by 50% compared to the bare shell value of q''.

The NAS 1973 study included some theoretical heat transfer calculations of the effects of tank insulation. Results from those calculations are reproduced here in Table 8.6. The parameters in Table 8.6 are the percent coverage of the tank with insulation, the assumed flame temperature, and the insulation thermal conductivity (also equivalent to the conductance for a one inch layer of insulation). The calculated value of F corresponding to a reference value of q'' of 34,500 Btu/hr-ft^2 is shown in the last column, labeled ratio. In the case of a conductance of 4 Btu/hr-ft^2 (23 W/m^2-C), the calculated value of F is 0.14 to 0.18 for a 100% insulated tank, and 0.28 to 0.31 for a 90% insulated tank of propane. Values of F for lower conductances (corresponding to thicker or lower conductivity insulation) can be obtained from Table 8.6.

Once the appropriate value of Q_L is available, we return to equation [8.1.4] to calculate the required vent flow rate which can be expressed as

$$V_v = C_D A_v v \qquad [8.1.7]$$

where C_D is the vent discharge coefficient, A_v is the vent area (m^2), and v is the vapor velocity through vent (m/s).

Table 8.6. Comparison of heat transfer rates between insulated and uninsulated containers (material of lading, propane, insulation thickness = 1 in. [$8'' = 34,500$ Btu/(hr)(ft2)]d) (data from NAS, 1973)

Flame Temp. F	Mean thermal conductivity, Btu/(hr-ft2)/(*F/in)	Outside temp. of insulation	Heat Flux q Btu/hr-ft2	Ratio (percent insulated)			
				100	95	90	80
1500	0.6	1468	797	0.023	0.108	0.173	0.286
1500	1.8	1408	2280	0.066	0.149	0.212	0.322
1500	4.0	1312	4688	0.136	0.216	0.276	0.380
1850	0.6	1800	1000	0.029	0.114	0.178	0.291
1850	1.8	1710	2830	0.082	0.164	0.227	0.335
1850	4.0	1560	5680	0.164	0.244	0.303	0.405
2000	0.6	1940	1080	0.031	0.116	0.180	0.293
2000	1.8	1834	3050	0.088	0.170	0.232	0.341
2000	4.0	1666	6096	0.176	0.254	0.314	0.411

Substitution of equations [8.1.5, 8.1.6] and [8.1.7] into [8.1.4] yields

$$C_D A_v v = q'' E F A_w / (L \rho_v) \qquad [8.1.8]$$

The appropriate equation for v depends upon the applicable flow regime for the vent situation. In the case of atmospheric pressure tanks, the incompressible flow version of Bernoulli's equation would be valid, such that

$$v = \sqrt{2(P_v - P_a)/\rho_v} \qquad [8.1.9]$$

where P_v is the tank pressure at full vent opening (usually specified as 110% of the vent set pressure), and P_a is atmospheric pressure.

Substituting equation [8.1.9] into [8.1.8] and solving for A_v,

$$A_v = \frac{q'' E F A_w}{(L C_D \sqrt{2(P_v - P_a) \rho_v})} \qquad [8.1.10]$$

Although equation [8.1.10], can be used to calculate the required emergency vent area if the vent discharge coefficient is known, vent capacities are often specified in terms of air flow rate as a function of pressure drop across the vent. It can be seen from equation [8.1.9] that the ratio of vapor flow rate to air flow rate at the same pressure drop is

$$C_D A_v'' \text{vapor} = C_D A_v v_{air} \sqrt{\rho_a/\rho_v} \qquad [8.1.11]$$

Substituting equation [8.1.11] into [8.1.8] and noting that density ratios at a specified pressure and temperature are equal to molecular weight ratios,

$$C_D A_v v_{air} = \frac{q'' E F A_w}{L \rho_a \sqrt{M/29}} \qquad [8.1.12]$$

where M is the vapor molecular weight and 29 is the molecular weight of air. If the air flow rate through the vent is specified at 60°F and 1 atmosphere, $\rho_a = 0.076\,\text{lb/ft}^3$ and $C_D A_v v_{air} = 70.5 q'' E F A_w / (L\sqrt{M})\,\text{ft}^3$ per hr when q'' is in Btu/hr-ft², A is in ft², and L is in Btu/lb.

For a given value of q'', equation [8.1.12] indicates that the required vent air flow varies inversely as $L\sqrt{M}$.

The vent air flow rates specified in Table 2.8 of NFPA 30 correspond to the $L\sqrt{M}$ value for hexane, and should be modified accordingly for other flammable liquids.

In the case of pressure vessels with storage pressures at least one atmosphere above atmospheric pressure, the vent flow rate corresponds to the following critical, or choked, flow rate given by (NAS p. 99)

$$V_v = (\text{cfm}) = (18.34 Q_L / L f)(Z T / M^{0.5}) \qquad [8.1.13]$$

where (cfm) is the required vent air flow capacity for critical flow, f is the function of vapor ratio of specific heats, Z is the compressibility of vapor at saturated conditions, and T is the vapor boiling point (°R) at vent set pressure.

The function f, which is given on page 52 of the NAS study, is not very sensitive to the actual value of γ. Setting $f = 350$ will be accurate to within 4% for γ in the range 1.2 to 1.5.

The actual vapor velocity in the vent when $P_v/P_a > 1.89$ (for $\gamma = 1.4$) is

$$v = [2\gamma R_0 T / M(\gamma + 1)]^{1/2} \qquad [8.1.14]$$

where R_0 is the universal gas constant (8314 m²/s²-°K).

If equation [8.1.14] is substituted in equation [8.1.8], the following solution for A_v can be obtained for high pressure storage vessels:

$$A_v = q'' EFA_w / [LC_D \rho_v (2\gamma R_0 T / M(\gamma + 1))^{1/2}] \qquad [8.1.15]$$

In the case of low pressure storage tanks for which the vent pressure is too high for equation [8.1.9] to be applicable ($P_v/P_a > 1.2$) and too low for equation [8.1.15] to apply ($P_v/P_a < 1.89$), the more complicated equations of subcritical compressible flow should be utilized (as given, for example, by Forrest in the all vapor flow example in his 1985 paper). Alternatively, the reader can use API 2000 or Table 2.8 of NFPA 30.

The preceding equations and the guidelines in NFPA 30 and API 2000 assume that no liquid is entrained into the vent flow. If liquid does get carried into the vent, vapor volumetric flow rates will be significantly lower than calculated above. Emergency vents sized for all vapor venting will be undersized for two phase vapor-liquid vent flow. Therefore, it is important to understand when two phase vent flow will occur and to establish a two phase vent design basis.

Grolmes and Epstein (1985) have analyzed the conditions required for all vapor venting of vertical tanks exposed to a uniform heat flux as illustrated in Figure 8.7. Boiling is assumed to occur in the boundary layer along the vessel walls and the vapor contained in the boundary layer causes the liquid level to increase in the tank. When the two-phase upward flow in the boundary layer reaches the liquid free surface, the liquid fraction recirculates in the tank while the vapor fraction exits through the vent providing that the recirculation velocity, U_c, is smaller than the bubble rise velocity, U_∞ as shown in Figure 8.8a. If the recirculation velocity is greater than the bubble rise velocity, Grolmes and Epstein hypothesized that vapor will be recirculated along with the liquid, as shown in Figure 8.8b.

Experiments conducted by Fauske et al. (1986) with electrical wall heaters generating a heat flux of about 35 kW/m² (11,100 Btu/hr-ft²) indicate that tank swell levels and overpressures are

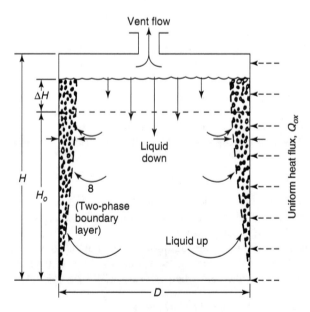

Figure 8.7. Illustration of boiling two-phase boundary layer in an upright cylindrical storage vessel subjected to a uniform external heat flux (from Grolmes and Epstein, 1985)

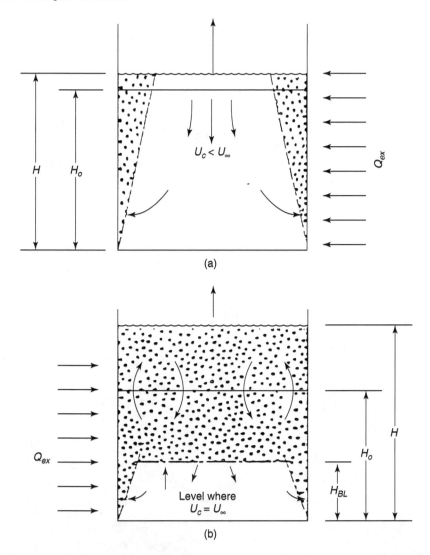

Figure 8.8. Upright cylindrical vessel with uniform external heat flux Q_{ex}. (a) Recirculating liquid flow U_C is less than the bubble rise velocity U_∞. The boundary layer region H_{BL} is equal to H; (b) recirculating flow at H_o exceeds U_∞. Vapor is carried downward to a level H_{BL}, where $U_C = U_\infty$. Above H_{BL} the flow of both vapor and liquid undergoes intense recirculation approximating uniform volume heating. All Vapor Venting (above) and two-phase V in vertical cylinder storage tank fire (from Grolmes and Epstein, 1985)

more consistent with the theory of vapor disengagement from the liquid (as in Figure 8.8a) than with vapor carry under at the recirculation velocity (Figure 8.8b). They conclude that all vapor venting is more likely than two-phase venting except for foamy liquids, very high liquid fill levels, and large vent velocities as explained below. Tanks supported on legs, as shown for example in Figure 8.9, such that the bottom of the tank is also exposed to flame, are more prone to large swell levels and two phase emergency venting.

Fauske et al. (1986) also suggest that a different type of two-phase vent flow will occur if the vapor velocity in or immediately above the liquid exceeds a critical value needed for liquid

Figure 8.9. Vertical tank supported on legs

entrainment. This critical entrainment velocity, U_{ent}, is given by

$$U_{ent} = 3.0[\sigma g \rho_L / \rho_v^2]^{1/4} \qquad [8.1.16]$$

where σ is the liquid surface tension, and g is the gravitational acceleration.

The value of U_{ent} for water at one atmosphere is approximately 21 m/s (47 mph). If the vent velocity given by equations [8.1.8] or [8.1.14] is greater than U_{ent}, and if the liquid fill level is sufficiently high, liquid can be entrained into the vent flow along with vapor. Epstein et al. (1989) have assumed the vapor flow field in the vicinity of a vent on a vertical tank would correspond to a hemispherical sink flow through the vent. Their solution indicates that all vapor venting will

FLAMMABLE LIQUID STORAGE

occur if the height, h, of the vapor space above the liquid is greater than a critical value given by

$$h_{crit} = d(v/U_{ent})^{1/2}/(2)^{1.5} \qquad [8.1.17]$$

where d is the vent diameter and $v > U_{ent}$.

El-Iskandarani (1994) has developed a computer code that uses the Fauske et al. criteria for two phase vent flow including liquid swell effects associated with the vapor volume fraction in the thermal boundary layer on the tank wall. His calculations indicate that two phase flow will occur for tanks completely engulfed in flames when the liquid fill levels exceed about 87% (of the tank height) in the case of heptane, and about 94% in the case of methanol.

Vent sizing to accommodate two phase flow depends on the flow regime as specified by (1) the volume fraction of vapor in the flow entering the vent, and (2) the relative velocity of the liquid fraction to the vapor fraction. If the vapor volume fraction is small (including the limiting case of an all liquid flow partially flashing into vapor as its pressure is reduced at the vent outlet), the required vent area to prevent tank pressures from increasing beyond the vent actuation pressure is (Crowl and Louvar, 1990, p. 298):

$$A_v = \frac{Q_L m_L v_{fg}}{G_T V L} \qquad [8.1.18]$$

where

$$v_{fg} = \frac{1}{\rho_v} - \frac{1}{\rho_L}$$

and m_L is the mass of liquid in the tank, V is the volume of the tank, and G_T is the two phase mass flux through the vent, as given by,

$$G_T = 0.9 \frac{L}{v_{fg}\sqrt{c_p T_b}}$$

T_b is the boiling point of the liquid at the vent set pressure.

If the vapor volume fraction is large, such as in the case where liquid is entrained into the vapor vent flow in the form of suspended droplets, different equations would be needed for the mass flux through the vent. El-Iskandarani has included these equations in his two phase vent flow computer code. The increase in required vent area, beyond that required for all vapor venting, is substantial. For example, some of El-Iskandarani's results indicate vent areas more than twice as large when there is substantial liquid entrainment and homogeneous two phase flow.

Example – An emergency vent is needed for a 9.1 m (30 ft) diameter, 6.1 m (20 ft) high xylene storage atmospheric pressure tank. The maximum fill level in the tank will be 90% of tank capacity, and the emergency vent is to have a full open set pressure of 0.36 psig (2480 Pa) and a discharge coefficient of 0.50. There is no tank insulation, but water spray protection at 0.25 gpm/ft^2 (10.2 L/min-m^2) of tank surface area will be provided along with drainage. Calculate the required vent area using the following property values for xylene:

$$\rho_L = 48\,\text{lb}_m/\text{ft}^3,\ \rho_v = 0.20\,\text{lb}_m/\text{ft}^3,\ L = 150\,\text{Btu/lb}_m,\ \sigma = 17\,\text{dyne/cm} = 0.0374\,\text{lb}_m/\text{ft}^3$$

The wetted tank surface area, A_w, is $0.9\pi(30)(20) = 1700\,\text{ft}^2$.

If we use the NAS recommendations,

$$q'' = 34{,}500\,\text{Btu/hr-ft}^2$$

$$Q_L = 34{,}500(1)(0.3)(1700) = 1.76 \times 10^7\,\text{Btu/hr} = 4888\,\text{Btu/s}$$

The vent velocity from equation (8.8) is

$$v = [2(0.36)(32.2)(144)/0.2]^{1/2} = 129 \text{ ft/s}$$

The required vent area, according to equation (8.9) is

$$A_v = 4888/[150(0.5)(129)(0.2)] = 2.5 \text{ ft}^2$$

which corresponds to a vent diameter of 1.8 ft.

We note that if the NFPA 30 recommended value of Q_L was used, as obtained from Figure 8.6, the vent area would be 20% of the value calculated above.

The entrainment velocity for the onset of two phase flow of xylene is, from equation [8.1.16],

$$U_{ent} = 3[0.0374(32.2)(48)/(0.2)^2]^{1/4} = 18.5 \text{ ft/s}$$

The freeboard (vapor space) height needed to avoid two phase vent flow in this case is from equation [8.1.17]:

$$h > 1.8[129/18.5]^{1/2}/2.83 = 1.7 \text{ ft}$$

which is less than the freeboard height at a 90% fill level (2 ft). Thus, all vapor venting should occur in this example.

Finally, we note that these calculations demonstrate that the use of a tank with a weak roof-to-shell seam would provide more than adequate vent capacity with or without two phase flow.

8.1.5 TANK FIRE SUPPRESSION

Foam systems are generally acknowledged to be the only suitable suppression systems for petroleum liquid storage tanks. Historically, most storage tank fires have been extinguished with portable foam equipment (Herzog, 1974). However, portable systems have several limitations including relatively long response times, limited foam discharge rate capability, and difficulties in maintaining a continuous foam supply over the required duration of discharge. (Many successful suppression incidents have entailed delays due to the need to obtain agent and equipment from remote locations, sometimes requiring international deliveries.) Fixed installations avoid most of these limitations, and are required in several standards (including NFPA 30) for large fixed roof tanks containing Class I flammable liquids.

Fixed application systems for storage tanks usually have permanently installed piping terminating in foam discharge outlets (sometimes called foam pourers or vapor seal boxes) on the tank shell or roof. Fixed roof tanks often have multiple foam riser pipes and discharge outlets offset from the tank shell in order to minimize damage caused by blown tank roofs.

Most foam application standards for storage tanks call for a 3% foam solution and an expansion ratio in the range 5:1 to 10:1. Recommended application rates vary; Kaura's (1980) comparison of application rates called for in fixed roof tank standards from five nations showed a range of 2.5 to 5 liter/min-m² (0.06–0.12 gpm per ft² of liquid surface area). NFPA 11, the American standard, calls for 0.10 gpm/ft² (4.1 liter/min-m²) for fixed roof tanks and 0.16–0.50 gpm/ft² (6.5–20.3 liter/min-m²) of annular seal area in floating roof tanks. Design basis calculations for foam and water supplies are described by Hickey (1988).

Another foam application technique that has been successful particularly with fixed roof tanks is subsurface injection of foam solution. Pignato (1978) has described tests conducted in several countries to demonstrate fire extinguishment using subsurface injection of AFFF at an injection rate of 0.10 gpm per ft² of liquid surface area. Tank diameters for the subsurface injection tests (some topside application tests are also reported) varied from 2–8.5 m (6.5–28 ft), and liquid depths were in the range 1.2–6.1 m (4–20 ft). Time required for control were in the range 1:00 to 2:45 min in all but one of the ten tests. Extinguishment times varied from 2:45 min to 12:40 min. There was no consistent variation of control/extinguishment time with liquid depth, but Pignato did note that these times did depend on the duration of the preburn and the method of foam injection. In this regard, it should also be noted that extinguishment times in actual fire incidents are often on the order of hours instead of minutes. This may be due to the extended free burn times heating the tank liquid or possibly to the less efficient foam-liquid mixing in actual incidents involving large tanks.

Although AFFF and fluoroprotein foam are usually recommended for petroleum liquid tanks, other suppression agents are needed for water miscible liquids. In fact, Herzog maintains that one of the two leading causes of foam suppression failure is the use of an incompatible agent (the other cause is too low an application rate) for the tank liquid. Alcohol compatible foam concentrate is available for water miscible liquids and foam manufacturers and listing agencies have conducted tests to establish recommended application rates and concentrations. Water spray systems and automatic sprinkler systems (for indoor tanks containing alcohols) can also be effective for water miscible liquids. Water application rates and expected extinguishment times are discussed in Chapter 7.

8.1.6 PORTABLE TANKS AND INTERMEDIATE BULK CONTAINERS

Flammable liquids are often transported in portable tanks that can be temporarily located in industrial processing or storage areas. One particular type of portable tank that is growing in use is the Intermediate Bulk Container (IBC). IBCs intended for flammable liquid transport/storage are rigid portable containers with capacities up to 693 gal (3,000 liter) and designed for material handling and stacking. Portable flammable liquid tanks are constructed of metal, and have been used with capacities up to 5,500 gal (20,818 liter), but should have a maximum allowable capacity of 660 gal (2,857 liter) when stored in indoor warehouses per Chapter 4 of NFPA 30. IBCs for flammable liquids are constructed of either metal, rigid plastic, or composite materials. Figure 8.10 is a photograph of a composite IBC consisting of a steel outer frame for handling and impact resistance, and a plastic inner container for flammable and/or corrosive liquids. Other types of composite IBCs have outer frames made from aluminum or rigid plastic, or steel mesh.

Portable tank and IBC design and construction are regulated by the United Nations "Recommendations for the Transport of Dangerous Goods, Chapter 16," and the U.S. Department of Transportation (49 CFR Chap. 1, Oct 1995). These regulations include a requirement for a pressure relief device that will prevent breaching of the container/IBC when internal pressures exceed the hydrostatic test pressure of the container, or 9 psig (65 kPa) for metal IBCs. The size of the pressure relief device is not specified in the DOT/UN regulations, but NFPA 30 emergency venting requirements (as described in Section 8.1.4 of this text) are applicable to tanks/IBCs with capacities over 660 gal. Certification organizations, such as Underwriters Laboratories, Inc. (UL), provide testing and inspection services to verify that tanks/IBCs are in compliance with the regulations. However, as of this writing (Feb 1999), UL does not have a specific standard for IBCs

Prolonged fire exposure to plastic and composite IBCs will eventually lead to breaching unless the IBCs are protected with an appropriate fire suppression system. Tests reported by Scheffey (1997) for water filled IBCs indicated that the most common breaching mode is thermoplastic softening and eventual collapse without apparent pressure damage. Massive releases of liquid were a rare occurrence, with most breaches producing small liquid release rates. Preliminary sprinkler protection scoping tests (conducted with a Required Delivered Density water applicator) indicated that two-stack high steel-encapsulated IBCs probably can be protected with

Figure 8.10. Composite type intermediate bulk container with D.o.t. exemption for transporting flammable/corrosive liquids

substantial sprinkler discharge densities (at least 0.6 gpm/ft^2), but stacked un-encased plastic IBCs probably cannot be protected without experiencing multiple breaches. Additional testing is expected.

8.2 Drum storage

8.2.1 DRUM DESIGNS AND STORAGE MODES

Steel drums seem to be omnipresent in flammable liquid warehouses and process facilities. The most common size drum for both storage and transport in the United States is the 55 gal (208 L) drum. Although there are several designs, most 55 gal (208 L) drums have a diameter of approximately 58 cm (23 in) and a height of about 89 cm (35 in) as shown in Figure 8.11. There are usually two plugs or bungs in the drum head for liquid filling and pumping. Bung diameters are typically 5 cm (2 in) and 1.9 cm (3/4 in) as shown in Figure 8.11. Head designs, wall thickness, and pressure ratings for 55 gal (208 L) drum vary with the intended application and relevant

FLAMMABLE LIQUID STORAGE

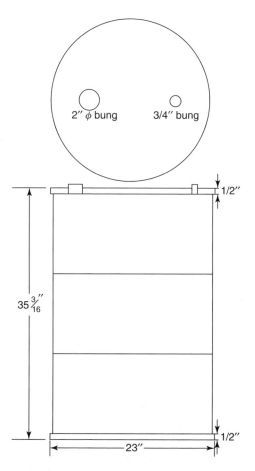

Figure 8.11. Typical (DOT-17E) 55 gal drum

Table 8.7. 55 gallon drum specifications (from Code of Federal Regulations Title 49, Section 178)

Material	D.O.T. Spec	Head	Hydrostatic Pressure[a]	Wall Gauge	Restrictions
Steel	17E	Permanent	15 psig	18	Single Trip
Steel	5B	Permanent	40 psig	16	Single Trip
Steel	17H	Removable	15 psig	18	Chime on Head
Steel	17C	Removable	20 psig	16	Bolted Ring on Head
Steel	17C	Permanent	40 psig	16	Single Trip
Polyethylene	34 Special Permit	Permanent	15 psig	–	No Class I Liquids

[a]Hydrostatic pressures must be withstood without any leakage

government regulation. Department of Transportation specifications for several steel and plastic drums are shown in Table 8.7.

Although the government does not require these drums to be hydrostatically tested to the point of structural failure, there have been several such tests reported in the literature. Results of some of these hydrostatic rupture tests are summarized here:

Drum type	Failure pressure (psig)	Reference
17E	61	G.E. (1965)
17E	74	Richards and Munkenbeck
17C	101	G.E. (1965)
5B	125	Richards and Munkenbeck

The predominant failure mode observed in the Richards and Munkenbeck hydrostatic tests was bottom seam failure. If this failure mode were to occur during a fire, the drum would be expected to rocket. Of course, failure modes and pressures during fire exposure can be significantly different from those in hydrostatic tests because of nonuniform heating thermal stress effects. Fire exposure failure mode and pressure data are discussed in Section 8.2.3.

NFPA 30 allows polyethylene 55 gal (208 L) drums only for Class II and III flammable liquids (flash points over 100 °F or 38 °C). The maximum size plastic drum for Class I liquids is 5 gallons (19 L) (1 gal or 3.78 L for Class 1A).

Drum storage practices vary greatly depending on the industry involved and the local building codes and insurance requirements. Outdoor yard storage is common at facilities that don't require frequent drum handling and have adequate space to satisfy minimum separation distances from the property line (as specified for example in Table 4.8 of NFPA 30). However, minimum separation distances in most codes and standards do not preclude damage and injuries from rocketing drums as observed in some of the incidents described in Section 8.2.2. Two other aspects of yard storage that may require special consideration are the threat of accelerated corrosion due to exposure to wet weather and the potential for pollution of groundwater by spilled liquid. The groundwater issue is also applicable to indoor storage, but may be more manageable indoors.

Indoor storage of drums takes many forms including palletized storage and rack storage either on-end or on-side. Fixed protection for indoor storage varies from no fixed suppression in small, remote, expendable buildings to elaborate special suppression systems for facilities with severe space limitations. Some of these special systems have been described by Capizzani 1984 and in Chemical Processing (1982). Protection effectiveness using more conventional automatic sprinkler and closed head foam systems is discussed in Section 8.2.4.

8.2.2 LOSS EXPERIENCE AND FIRE SCENARIOS

The following collection of anecdotal loss summaries is presented in the spirit of providing the reader with an added appreciation of drum fire scenarios and problems encountered in providing adequate protection. These losses are by no means all inclusive but should be representative of the large loss potential of drum storage fires.

Temporary storage of naphtha drums in a can manufacturing warehouse led to a 1951 fire described in Appendix E of NFPA 30. Small punctures in two drums allowed naphtha to leak and be ignited by some unidentified source. Drum rupture at the bottom seam created a fireball that opened 272 sprinklers in the warehouse. None of the other drums ruptured and the only other combustible material in the fire was the wood crates used to store can ends. Property damage was estimated to be about $200,000 in 1951 dollars.

A 1971 fire in a distribution warehouse for automotive supplies overpowered the sprinkler system because of the extensive storage of flammable liquids in 5 gal containers, 55 gal (208 L) drums, and aerosol cans. Storage was 4.6–5.2 m (15–17 ft) high on racks, and the water supply was only capable of delivering 0.20 gpm/ft^2 (8.1 L/min-m^2) over the hydraulically most remote

186 m² (2000 ft²) in the 6230 m² (67,000 ft²) warehouse. Rupturing aerosol cans and drums contributed to early roof collapse and failure of sprinkler piping, and inhibited manual fire fighting attempts (NFPA 30, Appendix E).

A 1975 fire in a general purpose warehouse started in the storage area for 55 gal (208 L) drums containing Class 1B and 1C liquids. About 100 drums were stored three high on pallets under the 4.6 m (15 ft) high roof. Several drums ruptured creating large fireballs. One drum rocketed through the roof and landed 230 m (750 ft) from the building. The roof partially collapsed disabling a significant portion of the sprinkler system. Property damage was estimated to be $3.3 million (NFPA 30, Appendix E).

Yard storage fires in 1976 and 1977 illustrate the importance of providing ample separation of drum piles from buildings and other structures. The 1976 fire occurred in a paint factory storage yard and involved rocketing drums that exploded during the fire (Capizzani, 1984). The 1977 fire occurred in 4.9 m (16 ft) high piles of drums stored in the yard of a pharmaceutical plant. About 1900 drums containing Class I flammable liquids were alongside the wall of a main building. One drum of tetrahydrofuran (6 °F or -14 °C flash point) was punctured as it was being loaded onto a lift truck. Leaking liquid and vapors were ignited by the truck. The resulting fire involved drum BLEVEs, thermal damage to the wall and roof of the main building, and wind blown flames warping one leg of a nearby gravity water tank used for the plant sprinkler system. Property damage was $670,000 (including 750 destroyed drums) and would have been significantly larger if the plant emergency organization and public fire department had not responded promptly and applied as much as 12,000 gpm (45,360 L/min) of water to exposed structures (FM Data Sheet 7-29).

A 1977 fire in a plastic products factory started when an electrostatic discharge ignited alcohol vapors near a 55 gal (208 L) drum. The fire blew the top off the drum spreading the fire over a large section of the factory. Damage was estimated at $385,000 according to a Marsh and McLennan compilation of large loss fires.

A 1984 fire in a gasket manufacturing plant started with an overheated 55 gal (208 L) drum of lubricating oil. The oil was supposed to be heated to 93 °C (200 °F) by an electric heater and thermostat that may have malfunctioned. The hot oil ignited and caused widespread fire damage totaling $4.7 million (Redding and O'Brien, 1985).

The 1986 fire in a chemical storage facility in Basel, Switzerland involved drum storage of flammable liquids with flash points of 30 °C (86 °F) and higher. Solid chemicals in sacks and plastic drums were also stored near the fire origin. The metal and plastic drums were stacked up to 8 m (26 ft) high in the unsprinklered building. The plant fire chief was prepared to sacrifice the building without any extensive suppression attempt until he saw drums exploding and crashing through the roofs of neighboring buildings. At that point the plant and public fire brigades directed about 8000 gpm (30,240 L/min) of water onto the burning warehouse. Water runoff contaminated with insecticides, herbicides, and a potent fungicide from the burning warehouse overflowed into the Rhine River where it caused an environmental disaster (Wackerlig, 1987).

One recent fire at a chemical manufacturing plant demonstrated another important aspect of flammable liquid drum fires. This fire involved a spill fire next to a nominally empty drum containing residual vapors at a concentration within the flammable range. The drum exploded because either flame entered the drum or the wall temperature exceeded the mixture autoignition temperature. Two other drums also exploded, but it is not clear whether these were combustion explosions or the more common BLEVE associated with filled metal drums.

8.2.3 DRUM FAILURE TIMES AND FAILURE MODES

How long can a filled metal drum be exposed to a spill fire before it can be expected to BLEVE or fail in some less dramatic manner? How does this time compare to the failure time and failure mode of a plastic drum? How do these times and failure modes depend on the volatility of the liquid in the drum? Is a pressure relief device in the bung opening effective in preventing a BLEVE? Test data to provide answers to these questions are discussed here.

The US Coast Guard (Richards and Muckenbeck, 1977) conducted an extensive series of tests in 1975–1976 to determine the relative hazards of steel drums and polyethylene drums during exposure to large JP-4 spill fires. Tests included various flammable liquids in Specification 17E and 5B 55 gal (208 L) steel drums and in various size Specification 34 polyethylene drums. The steel drums failed first by jetting flame at or near the drum head and later in some cases by exploding. (One test conducted with an array of twelve 5B drums containing ethyl ether resulted in ten drums failing via jet flames only and the other two failing via a BLEVE explosion.) The polyethylene drums failed by melting, softening, or burning and releasing their liquid content into the fire. This liquid release would start as a leak and then suddenly intensify as the polyethylene drum collapsed inward. Similar tests reported by Gordon (1989) for composite fibre drums with polyethylene interior liners showed they failed in a similar manner to the polyethylene drums.

Steel drum failure times and failure pressures measured by Richards and Muckenbeck in their single drum tests are shown in Table 8.8. The times and pressures at drum breaching (i.e. emission of jet flame) are shown along with the corresponding values for final drum rupture (often in the form of a BLEVE). Data obtained at Factory Mutual (Yao, 1968; Newman *et al.*, 1975) are also in Table 8.8. In the 17E steel drum tests with various flammable liquids in the drums, the average time to breaching/jetting was 1 min 49 sec, and the average time to drum rupture was 3 min 56 sec. The average drum pressure at breaching/jetting was 13 psig (90 kPa) (5 psig or 34.5 kPa less than the required room temperature hydrostatic pressure without leakage), while the average rupture pressure was 25 psig (172 kPa). The considerable scatter in the data for a series of tests with a particular drum type and liquid was attributed by Richards and Muckenbeck to random variations in drum strength and integrity.

Variations in liquid content did affect the time to steel drum breaching and rupturing. The average time to breaching in the 5B drum tests was 2:01 min when ethyl ether was in the drum and 4:23 min when JP-4 was used. The more volatile ethyl ether also caused earlier failures than JP-4 in the 17E drum tests, but so did lubricating oil which is the least volatile of the liquids tested. The lubricating oil tests results in jetting at a surprisingly low drum pressure (7 psig or 48 KPA) and never escalated to produce a BLEVE. The longest times to breaching in the 17E drums were measured in the tests with water filled drums.

Polyethylene drum failure times measured by Richards and Munkenbeck are listed in Table 8.9. In the tests with JP-4 in the drum, the failure times increased consistently with increasing drum size. The failure time for the 55 gal (208 L) drum (1:40 min) was a full minute longer than for the 5 gal (19 L) drum with JP-4. There was some variation in failure time with differing drum contents and, as with the steel drums, lubricating oil produced the shortest time to failure.

Comparing the data in Tables 8.8 and 8.9, it is clear that, on average, polyethylene 55 gal (208 L) drums fail earlier than steel drums. The average breach time for the 17E steel drums (109 sec) is 29 sec (36%) longer than the average failure time for 55 gal (208 L) polyethylene drums; *the average 17E drum rupture time (3:56 min) is three times as long as the polyethylene drum failure time.* This is the primary reason the Department of Transportation and NFPA 30 prohibit the use of 55 gal (208 L) polyethylene drums for all Class I flammable liquids.

Factory Mutual Research has conducted single drum fire tests using a special 17H steel drum equipped with a domed head, a manually operated pressure relief valve, and other modifications to allow it to be reused and avoid the randomness associated with drums tested to the point of breaching (Delichatsios, 1981). Instrumentation in the FMRC tests included a pressure transducer

Table 8.8. Drum failure data from fire exposure tests

Drum type	Liquid	Failure time (min:sec)			Failure pressure (psig)		Reference
		Minimum[a]	Breach[b]	Rupture[b]	Breach[b]	Rupture[b]	
5B	JP-4	3:14	4:23	4:49	56	61	Richards & Munkenbeck[c]
5B	Ethyl Ether	0:56	2:01	2:24	27	53	Richards & Munkenbeck[c]
17E	JP-4	3:12	4:17	5:47	8	32	Richards & Munkenbeck[c]
17E	Acetone	1:25	1:25	2:47	–	–	Richards & Munkenbeck[c]
17E	Lube Oil	1:14	1:58	None	7	None	Richards & Munkenbeck[c]
17E	Ethyl Ether	1:25	1:34	2:28	20	24	Richards & Munkenbeck[c]
17E	Heptane	1:51	2:45	5:21	18	18	Yao (1968)[e]
17E	Average		1:49	3:56	13	25	
17E	Water	3:513	4:38[c]	–	23[c]	–	Newman et al.[d]
17E	Water	–	16:00	–	25	–	Richards & Munkenbeck
17E	Water	–	–	11:33	–	43	Yao (1968)

[a] Breach failure time is the average time for drum breaching, while rupture time is the average time for massive/explosive failure. Minimum time is the earliest observed breach time
[b] Breach pressure and rupture pressure are the averages for replicate tests
[c] The Richards & Munkenbeck data are for 90 ft^2 JP-4 exposure fires
[d] The Newman et al. data are based on maximum drum pressures in Test 5
[e] Yao's data are for 1 gpm heptane burner exposure fires

Table 8.9. Polyethylene drum failure times during fire exposure (data are for 90 ft^2 JP-4 pool fires as reported in Table 2 of Richards and Munkenbeck, 1977)

Drum size (gal)	Average Failure Time (min:sec) for liquid Contents				
	Acetone	Ethyl Ether	JP-4	Lube Oil	Average
5	1:16	–	0:39	–	0:58
15	1:43	–	0:54	–	1:19
30	2:00	–	1:27	–	1:44
55	1:31	1:17	1:40	0:53	1:20

and thermocouples to measure the drum inside wall temperature and the liquid surface temperature during fire exposure. Tests have been conducted with a wide variety of liquids including water, several other homogeneous flammable and nonflammable liquids with a wide range of boiling points, and liquid mixtures with and without suspended solids (Tavares, 1982). Representative results are discussed here, along with the Coast Guard data, with the intention of providing insight into the phenomena governing liquid heating and steel drum pressure development for various flammable liquids.

The simplest possible hypothesis regarding steel drum pressurization and breaching is that the liquid is heated uniformly until it reaches its boiling point, at which time sudden vaporization causes a rapid pressure increase and drum breaching. Based on this hypothesis, the rate of increase in drum wall temperature and in liquid temperature is given by

$$\rho c_L V \frac{dT_L}{dt} = A_D h_L (T_s T_L) \qquad [8.2.1a]$$

$$m_s c_s \frac{dT_s}{dt} = q'' A_D A_D h_L (T_s T_L) \qquad [8.2.1b]$$

where ρ is the liquid density (m³), c_L is the liquid specific heat (kJ/kg-C), V is the liquid volume in the drum (m³), T_s is the steel drum temperature (°C), T_L is the liquid temperature (°C), t is the fire exposure time (s), h_L is the natural convection heat transfer coefficient at interface of liquid and drum interior wall (kW/m²-°C), q'' is the fire exposure heat flux (kW/m²), A_D is the drum wetted surface area exposed to fire (m²), m_s is the mass of empty steel drum (kg), and c_s is the steel specific heat (kJ/kg-°C).

The solution for T_L from equations [8.2.1a,b] for a constant heat flux is

$$T_L - T_0 = \frac{q'' A_D}{\rho c_L V} \{t - \tau[1 - e^{-t/\tau}]\} \qquad [8.2.2]$$

where T_0 is the initial temperature of the drum prior to fire exposure,

$$\tau = \frac{m_s c_s}{A_D h_L}$$

Equation [8.2.1] can be solved numerically for the value of t corresponding to the time (t_b) required for liquid to be heated from T_0 to its boiling point T_B. If the initial (ambient) temperature is below the liquid flash point, the drum vapor space will be in the flammable concentration range during at least part of this heating. If the drum wall is at a temperature above the vapor-air mixture ignition temperature at this time, an internal combustion explosion could occur well before the liquid reaches its boiling point. The time, t_f, for the liquid to be heated to its flash point can also be obtained from equation [8.2.2].

Calculations of t_b and t_f for the flammable liquids tested by Richards and Munkenbeck and by Delichatsios (1981) have been carried out for filled 55 gal drums with lateral surface areas exposed to a heat flux of 100 kW/m² (31,720 Btu/hr-ft²). Results are compared in Figure 8.12 to the observed time to drum breaching. There is approximate agreement between breach time and t_B for four liquids (acetone, ethyl ether, water, and JP-4), but the time-to-breaching is significantly overestimated for other liquids, particularly ethylene glycol and lubricating oil.

The uniform heating hypothesis implicit in equations [8.1.18] and [8.2.1] is not valid for liquids such as ethylene glycol and lubricating oil because these liquids do not readily conduct heat across the thermal boundary on the interior tank wall. Delichatsios (1981) has analyzed the ratio of the heat flux across the liquid thermal boundary layer (see Figure 8.6) to the heat flux imposed by the exposure fire. This ratio is

$$[k(T_b - T_0)/\delta]/q''$$

where k = liquid thermal conductivity, and, δ = thermal boundary layer thickness.

Delichatsios used the following equation for the natural convection thermal boundary layer adjacent to a vertical wall:

$$\delta = H(\upsilon/\alpha)^{1/2}[k\alpha\upsilon/g\beta q'' H^4]^{1/4} \qquad [8.2.3]$$

where H = drum height, υ = liquid kinematic viscosity, α = liquid thermal diffusivity, and, β = liquid coefficient of thermal expansion.

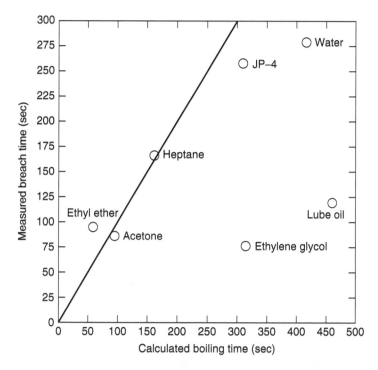

Figure 8.12. Comparison of measured and calculated 55 ga drum breach times

Substitution of equation [8.2.3] into the preceding ratio of heat fluxes yields

$$k(T_b - T_0)/\delta q'' = (\alpha/\upsilon)^{1/2} De^{1/4} \qquad [8.2.4]$$

where De is equal to

$$[k^3(T_b - T_0)^4 g\beta]/\upsilon\alpha q''^3 \qquad [8.2.5]$$

When both sides of equation [8.2.4] are smaller than one, the wall heat flux is primarily absorbed within the thin liquid boundary layer and rapid temperature increases and vaporization can be expected. Conversely, large values of each side of equation [8.2.4] imply relatively slow vaporization and pressure buildup corresponding to uniform heating of the liquid during the early period of fire exposure.

Delichatsios showed that the magnitude of De determines whether the drum lower wall temperature will equal or exceed the liquid boiling point. When De is large, the drum wall remains at the liquid boiling point, but when De is of the order of one or smaller, the wall temperature is much higher than the boiling point and vigorous nucleate boiling can occur throughout the liquid. The liquid surface temperature rise (normalized by $T_b - T_0$) for liquids that are heated primarily in the wall boundary layer is inversely proportional to the right hand side of equation [8.2.4]. Values of De and the right hand side of equation [8.2.4] for several liquids tested by Tavares and Delichatsios (1981) are given below for $q'' = 65 \text{ kW/m}^2$ (20,600 Btu/hr-ft²), which was the wall heat flux measured by Delichatsios for 1.5 m² (16 ft²) heptane pool fires in which the drums were immersed.

Liquid	De	$(\alpha/\upsilon)^{1/2} De^{1/4}$
Freon-TF	0.58	0.301
Heptane	33	1.4
Ethylene Glycol	360	0.38
Water	475	1.79

The fact that $(\alpha/\upsilon)^{1/2} De^{1/4}$ for ethylene glycol is smaller than it is for heptane and water implies that ethylene glycol is more prone to boundary layer heating/vaporization. This is consistent with ethylene glycol's large deviation from the uniform heating hypothesis as evidenced by the data in Figure 8.12.

Although Delichatsios' analysis does provide insight into the early exposure heating of the liquids, it does not suffice to predict the entire pressure history either qualitatively or quantitatively. Until a more complete understanding of these complicated phenomena becomes available, testing similar to the FMRC single drum tests will be needed to sort out the relative hazards of various flammable liquids and mixtures stored in metal drums.

One other interesting aspect of steel drum response to fire exposure is the testing of candidate pressure relief devices for drum heads. Tavares and Delichatsios (1981) conducted tests with various plastic and fusible alloy plugs inserted into the bung openings in the special FMRC 17H and ordinary 17E drum heads. They demonstrated that there is at least one plastic plug that can successfully vent heptane filled and JP-4 filled drums such that the drums did not BLEVE and the pressure remained under 15 psig (103 KPa) in the heptane test and under 25 psig (172 KPa) in the JP-4 test. The venting did produce jet flaming which could significantly enhance the severity of an exposure fire to any adjacent drums that may be present. Hence, drum venting would not suffice as protection unless there is also an effective fire suppression system.

8.2.4 FIRE SUPPRESSION SYSTEMS FOR DRUM STORAGE

The primary basis for drum storage protection guidelines is the 1949 FMRC test program conducted with water discharged through ceiling sprinklers (NFPA 30 Appendix E) and the 1974 FMRC test series conducted with AFFF discharged through ceiling and in-rack sprinklers (Newman et al., 1975). These test programs and some more recent tests with special systems are summarized here along with the current recommendations on suppression systems in NFPA 30 and Factory Mutual Data Sheet 7-29.

The 1949 test series employed old style sprinklers on a 4.6 m (15 ft) high ceiling. The sprinklers were open and were manually activated with water either at the start of the fire or after a delay estimated to simulate the first sprinkler actuation in response to the fire. Sprinklers were spaced either at 9.3 m² (100 ft²) per head with a water density of 0.22–0.28 gpm/ft² (8.9–11.4 L/min-m²), or at 4.6 m² (50 ft²) per head to produce a density of 0.44–0.56 gpm/ft² (17.9–22.8 L/min-m²).

Drum storage in the 1949 test series was palletized three high in some tests and on-side rack storage up to four tiers high in other tests. Some drums contained benzene or gasoline, and were equipped with temperature and pressure sensors and manual vents to avoid BLEVEs. Drums in the third and fourth tiers were either empty or contained water. Gasoline poured onto the floor or discharged at a higher level to simulate a leaking drum was used for the exposure fire. Small discharge rates (1–2 gpm, or 3.8–7.6 L/min) and small spills (5 gal or 19 L) produced more severe local exposures than the larger spills because the smaller spills did not produce sufficiently high ceiling temperatures to operate sprinklers early in the fire.

The 1949 tests demonstrated that a water density of 0.56 gpm/ft^2 (22.8 L/min-m^2) discharged from the 5.6 m (15 ft) ceiling was needed to prevent excessive drum pressures in multi-tiered storage. The rate of temperature rise in the drums on the lowest tier was three to five times as high in the multi-tier rack storage tests as in the one-tier high tests. The drums and pallets in the upper tiers obstructed the flow of water needed to cool the drums at the lower levels. As a result of these tests, Factory Mutual recommended that volatile liquids in drums should only be stored one high unless rack storage with in-rack sprinklers at every level was installed. Multi-level palletized storage of Class II and Class III liquids in drums is allowed with a ceiling density of 0.60 gpm/ft^2 (24.4 L/min-m^2) (lower densities are allowed for IIIB liquids according to Data Sheet 7-29).

The 1974 test series was conducted with water filled drums some of which had pressure transducers to determine if and when the drum pressure exceeded 15 psig (103 KPA), the designated failure threshold. Two tests were conducted with a palletized storage array and 9.1 m (30 ft) high ceiling protection only, and the other three tests were conducted with five tier high rack storage protected by in-rack sprinklers as well as 138 °C (280 °F) ceiling sprinklers. A 1% AFFF foam solution flowed through both the ceiling and in-rack sprinklers. The exposure fire consisted of a 38 L (10 gal) heptane spill prior to ignition and a 2–15 gpm (7.6–56.7 L/min) heptane discharge from the top tier of storage and initiated at ignition. The specific conditions and sprinkler actuation summaries for each test are listed in Table 8.10.

The AFFF sprinkler system was effective in controlling the floor spill fires in both rack storage tests but was only partially effective in the palletized tests because the pallets blocked the flow of foam on the periphery of the sprinkler discharge (see Figure 8.13). The elevated spill fire was not extinguished in either the palletized or the rack storage tests (see Figure 8.14). Several drums exceeded the 15 psig (103 KPA) pressure threshold in the palletized tests but only one drum in the lowest tier (there weren't any in-racks at the bottom tier) exceeded 15 psig (103 KPA) in each of the rack storage tests. Protection was least effective in the two-tier high palletized test (Test 5) both from the standpoint of drum cooling (all eight monitored drums exceeded 15 psig or 103 KPA and 37 other drums were bulged) and in terms of spill fire suppression. The explanation for the particularly poor sprinkler system performance in Test 5 was the large clearance (27 ft or 8.2 m) between the ceiling sprinklers and the spill discharge elevation, which delayed first sprinkler operation until 3:27 min after ignition (Table 8.10).

Since test conditions were different in the 1949 tests and the 1974 tests, there is no direct comparison of water based protection versus foam protection of drum storage. Both types of sprinkler systems are used but opinions differ on the extent to which the discharge density, demand area, and water supply duration can be reduced when a foam-water closed head system is utilized. It is clear that foam-water systems are more effective than water based systems in extinguishing floor spill fires, but they have not been able to suppress elevated spills (of course, neither can water systems), and it is not clear that they provide any better cooling of the fire exposed drums. As of this writing, additional tests are being planned to determine the relative effectiveness of water based versus foam-water systems for drum storage.

One generally recommended technique for limiting the size of a spill fire in a flammable liquid warehouse is the use of drainage trenches and sloped floors. Figure 8.15 shows one arrangement for using trenches, sloped floors, and in-rack sprinklers for on-side drum storage on racks. One advantage of storing the drums on-side in metal cradles rather than on pallets is that drums on pallets have been observed to topple when the pallets burn out from under them. High expansion foam and water deluge systems have also been shown to provide effective drum protection in special situations (Capizzani, 1984). The high expansion foam system was tested for a product with less than 20% flammable solvent. The system is actuated with a rate-of-rise heat detector and is backed up with a ceiling sprinkler system (0.35 gpm/ft^2 or 14.2 L/min-m^2) for secondary

Figure 8.13. Persistent spill fires in flammable liquid drum sprinklered fire test. © 2002 Factory Mutual Insurance Company, with permission

Figure 8.14. Spill fires before (a) and after (b) sprinkler activation in rack storage drum test (from Newman et al., 1975). © 2002 Factory Mutual Insurance Company, with permission

FLAMMABLE LIQUID STORAGE

Table 8.10. 1974 Drum test storage test summary (data from Newman et al., 1975)

	Test conditions				
Test number	1	2	3[c]	4[c]	5
Sprinkler Orifice Diameter (in.)	1/2	17/32	1/2	1/2	17/32
Sprinkler Temperature Rating (F)	280	286	280	280	286
Sprinkler Spacing	10' × 10'	8' × 10'	10' × 10'	10' × 10'	8' × 10'
Sprinkler Pressure (psig)	28.5	36	28.5	28.5	36
Agent Discharge Density (gpm/ft)	0.30	0.60	0.30	0.30	0.60
Heptane Spill Rate (gpm)	2	2	2	15	2
Initial Heptane Spill (gal)	10	10	10	10	10
Storage Array	Palletized	Palletized	Rack	Rack	Palletized
Levels of Storage	4	3	5	5	2
	Test results				
Time of First Ceiling Sprinkler Actuation (min:sec)[a]	0:34	1:20	0:40	0:11	3:27
Time of Final Ceiling Sprinkler Actuation (min:sec)[a]	1:12	1:23	–	1:46	3:52
Total Number of Ceiling Sprinklers Actuated	4	2	1	8	2
Total Number of In-Rack Sprinklers Actuated	–	–	5	5	–
Time for Max. Temp. to Reduce to 200 F (min:sec)[a]	10:22	6:42	30:54	5:45	4:05
Total Heptane Spill Duration (min:sec)	29:00	30:00	30:00	4:00	30:00
Duration of Foam Discharge (min:sec)[b]	35:26	39:55	35:02	9:49	43:03
Wind Speed (mph)	To 8	To 10	To 2	To 6	To 2

[a] Measured from the time of ignition of spill
[b] Measured from the time of first sprinkler actuation
[c] Designed Protection is for ceiling sprinklers only

protection. The water spray deluge system tests indicated that it would protect multiple level palletized storage only when supplemented with wide angle fan nozzles discharging down through a 15 cm (6 in) gap between the pallet loads. This special spacing and storage arrangement is difficult to maintain in normal warehouse operations. Capizzani concluded his summary of the various special protection systems by calling for additional large scale testing to further investigate these and other new protection methods. In the meantime, NFPA 30 and other standards (e.g. FM Data Sheet 7-29) provide guidance on the use of conventional systems for limited size/height drum storage arrays.

8.3 Flammable liquids in small containers

8.3.1 CONTAINER TYPES

We are concerned here with metal, plastic, and glass containers ranging in size from one pint (0.47 L) to 5 gal (19 L). They are used to store a variety of flammable liquid products and intermediates including paints, solvents, lubricating oils, automobile antifreeze, insecticides, wood

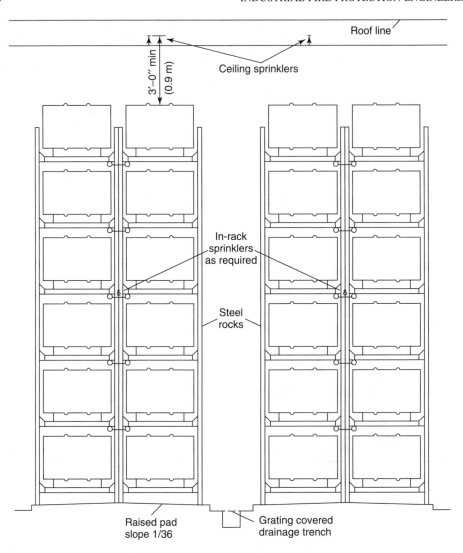

Figure 8.15. Storage of flammable liquids in drums on side in metal racks (Note: 6 in (150 mm) minimum separation is provided between top of drum an in-rack sprinkler deflector. In-rack sprinklers spaced 8 ft (2.4 m) on centers. No drum is more than 2 ft (0.6 m) horizontally from the nearest sprinkler. A typical drainage arrangement is also shown). © 2002 Factory Mutual Insurance Company, with permission

preservatives, hydraulic oils, cooking oils, alcohol, distilled spirits, personal care, and pharmaceutical products. As the product list implies, they are stored both in flammable liquid warehouses with larger containers, and in general purpose warehouses alongside ordinary solid combustibles. The 5 gal containers are usually stored in polyethylene stretch wrap while the smaller containers are often stored in cardboard cartons (at least in the US).

Pint and quart containers used for consumer products usually have a threaded cap without any special vent. The 1 and 5 gal containers are made with a myriad of different closures including friction lids, clamp on lids, retractable spouts, threaded caps, approved safety containers with vents, and smaller versions of the metal and plastic drums described in Section 8.2. Most of the containers are required to pass drop tests and sometimes hydrostatic pressure tests specified in government shipping regulations.

Metal containers are made from either steel or aluminum. High density polyethylene is the most prevalent plastic container material, but nylon, polyvinyl chloride, and polyethylene terepthalate are also used. Glass containers are restricted in size according to NFPA 30; from a one pint limit for a Class IA liquid on up to a 5 gal (19 L) limit for a Class III liquid.

8.3.2 LOSS EXPERIENCE

Several of the incidents described in Section 8.2.2 involved flammable liquids in small metal containers as well as in drums. However, it is difficult to sort out the contribution to the fire from the smaller containers since the drum failures are so much more dramatic. The following two incident descriptions provide a better understanding of the behavior of small containers in flammable liquid warehouse fires.

A 1978 fire in a paint and caulking compound manufacturing plant involved one gal and five gal cans of paint stored in 5.5 m (18 ft) high racks. In-rack sprinklers were installed in every tier, and ceiling sprinklers were provided with a discharge density of 0.30 gpm/ft^2 (12.2 L/min-m^2) over 560 m^2 (6000 ft^2). Fire broke out in the racks one night when an arsonist slashed open several cans and ignited the resulting spill. The fire was controlled by seven ceiling sprinklers, seven in-rack sprinklers, and the hose line water application of the public fire department responding to the sprinkler alarm. Property damage amounted to $106,000 including 15–20 pallet loads of product (FM Data Sheet 7-29).

The 1987 fire in an automotive paints distribution warehouse in Dayton, Ohio received a great deal of attention because of the fire severity and threat to the public water supply. Approximately 1.5 million gal (5.7 ML) of flammable liquids in the form of paints, solvents, lacquers, and thinners were stored in various metal and plastic containers ranging in size from pints to 5 gal. Aerosol cans of paint were also stored near the unpressurized containers. Storage was palletized up to 4.9 m (16 ft) high and in racks up to about 6.1 m (20 ft) high in the 8.5 m (28 ft) high warehouse.

The fire started when approximately ten cartons of one gal cans of Class IB liquid (a product containing methyl ethyl ketone, toluene, and xylene) tipped over while a fork lift operator was placing two pallet loads of product on the storage pile. The resulting 5–10 gal (19–38 L) spill from the friction lid cans was ignited by the lift truck and a rapidly spreading fire ensued. The fire quickly overpowered the sprinkler system which included 0.35 gpm/ft^2 (14.2 L/min-m^2) ceiling sprinkler protection and in-rack sprinklers in the second and third tiers of the rack storage. When the fire department arrived (about nine minutes after ignition), fireballs 30.5–45.7 m^2 (100–150 ft) in diameter were emerging through the roof, along with thick black smoke and flames extending several hundred feet above the roof.

The fire destroyed the entire 16,730 m^2 (180,000 ft^2) warehouse despite the presence of a four hour fire wall which was breached within 20 minutes. The fire department curtailed its attempt to control the fire with water and shut off the water flow to the damaged sprinkler system because of the fear that the water runoff would contaminate an underground aquifer used for the city potable water supply. Property damage was estimated at $50 million (Isner, 1988).

In addition to these losses in flammable liquid warehouses, there have been several losses involving small flammable liquid containers in general purpose warehouses. In most cases, the flammable liquid containers are not the first material ignited and it is difficult to ascertain their contribution to fire severity. A recent NFPA study (Hall, 1994) indicates that there are an estimated 2800 fires in approximately 285,000 general purpose warehouses in the US each year. The probability of each of those fires being a flammable liquid product spill fire is only 0.0039, with a large uncertainty (about a factor of 5) associated with these probability statistics. One example of such a spill fire is described here.

This 1978 fire in a department store merchandise warehouse started when 30–35 cartons of rubbing alcohol (70% isopropyl alcohol) in small plastic containers toppled while the storage pile was being rearranged. Alcohol vapors from some ruptured containers were ignited by either an electric pallet jack or an LP-gas lift truck. The waist high flames from the alcohol fire soon involved adjacent plastic products. As the fire spread and two ceiling sprinklers operated, the palletized storage collapsed causing a smoldering, burrowing fire that was extinguished by the fire department. Property damage amounted to $622,000, corresponding to about $1 million in 1988 dollars (Hall, 1994; FM Data Sheet, 7-29).

A 1984 fire in a general purpose warehouse in Australia is worth noting because of the different ignition scenario. An employee was using a flame type heat gun to shrink wrap a pallet load of plastic bottles of kerosene because the plastic stretch wrap machine was down for repairs. A leaky bottle was ignited and the resulting fire damaged about $600\,m^2$ ($6500\,ft^2$) of storage before it was extinguished by the Plant Emergency Organization.

8.3.3 CONTAINER FAILURE TIMES AND FAILURE MODES

Fire exposure tests have been conducted for single containers of 1 and 5 gal (3.8 and 19 L) capacity and for cartoned containers up to 1 gal in capacity. The single container tests have been smaller scale versions of the drum fire exposure tests described in Section 8.2.3. The cartoned multiple container tests have been conducted with the cardboard providing the exposure fire instead of a pool fire. Data from both types of tests are described here in order to provide a basis for estimating the times and modes in which flammable liquid is released from the containers so as to increase fire size and heat release rate.

The Bureau of Mines (Perzak *et al.*, 1985) developed a single container fire exposure test in which the container is placed in the center of a $1\,m^2$ ($11\,ft^2$) tray containing 1 gal (3.8 L) of kerosene so as to provide a 4 minute burn time. A variety of metal and plastic 5 gal (19 L) containers with 1 gal of either petroleum base hydraulic oil or kerosene was tested, as well as some 1 gal containers with 1 pint of hydraulic fluid. A container was considered to pass the test successfully if it did not spill its liquid in any of seven fire exposures.

Results of the 5 gal (19 L) container tests are summarized in Table 8.11. All the containers were breached/vented within two minutes, but most of the metal containers did not spill the hydraulic oil. The most common failure modes for the metal cans were blown caps, burned cap gaskets, or top seam failures as illustrated in the bottom sketch in Figure 8.16. On the other hand, all the polyethylene containers failed by melting, burning and releasing all the hydraulic oil.

When kerosene was placed in the containers in place of the less volatile hydraulic oil, the metal containers were less successful in retaining liquid. Three containers vented successfully, but two other designs failed by either overturning (because of a bulged bottom) or failing at the bottom seam and rocketing. Pressure and temperature data obtained by Perzak *et al.* suggest that the container failures with kerosene inside probably were due to vapor-air explosions in the container vapor space. The three designs that vented without rocketing or spilling kerosene were the DOT 5L unvented container, the tin plated steel square container, and the DOT 17E container with a 9.5 mm (3/8 in) plastic vent.

Perzak *et al.* tested three 1 gal (3.8 L) metal containers with hydraulic oil inside. The containers ruptured violently in less than 23 seconds. The seam ruptures that occurred in the 1 gal (3.8 L) container tests are shown in the top sketch in Figure 8.16.

Perzak *et al.* concluded that the most fire resistant container design was the metal container with a plastic friction-fit cap or plug seal such as the one used on the vented DOT 17E container. They recommended avoiding all plastic (polyethylene) containers because of their consistent softening, collapsing, and spilling of liquid contents soon after fire exposure.

Table 8.11. Fire exposure tests for 5 gal containers with hydraulic oil (data from Table 4 of Perzak et al., 1985)

Container shape	Vent time (min)	Failure	Comments
METAL – No Vent			
DOT 17E cylinder	1.7	0/7	Cap Gasket Burned & Top Seam Failed
DOT 5L rectangular	1.1	0/7	Cap Gasket Burned
Tin Plate square	0.6	2/15	Cap Blew Off; 2 cans tipped over
METAL – Vented			
DOT 37B cylinder	0.6	0/7	Vent Blew Off
DOT 17E cylinder	0.5	0/7	Vent Blew Off
METAL – Open Head			
DOT 37A60 cylinder	0.3	1/1	Liquid Burned in Can after Lid Blew
PLASTIC			
Open Head cylinder	0.6	1/1	Melted and Burned
Safety Can cylinder	0.2	1/1	Melted and Burned
5-Gallon Metal Containers with Kerosene			
DOT 17E unvented	1.0	3/10	Top Seam Failed; Container Overturned
DOT 5L unvented	1.8	0/1	Gasket Vented
Tin plate square no vent	0.5	0/1	Cap Blew Off
DOT 37B vented	0.5	3/3	Bottom Blew Off
DOT 17E vented	0.4	0/2	Plastic Vent Blew

Hill (1991) reported on a series of free burn metal container fire exposure tests conducted with heptane and isopropanol as the container fluids and the exposure pool fire. Single container and multiple container tests were conducted with one gal and five gal metal containers. The unvented, screw cap containers produced jet flames about 7.6 m (25 ft) high when they breached. Table 8.12 shows the breach times, failure modes, and flame heights for the 16 free burn tests. The data demonstrate that friction lid (1 gal or 3.8 L) and lug cover (5 gal or 19 L) containers breach in a way that produces rim or spill fires as compared to the more spectacular jet flame fires in screw cap containers. In both cases, the times to breaching and release of liquid contents are significantly longer than observed with flammable liquids in plastic containers.

Early concern about the fire hazard of flammable liquids in plastic containers in general purpose warehouses led Factory Mutual Research to conduct several test programs including some free burns to determine when and how the liquid contributes to fire severity. Three pairs of tests were conducted with 12 gal (45 L) of liquid burned first as a pool fire and later in 1 gal (3.8 L) polyethylene containers in cardboard cartons. Each test was conducted with the liquid in a 1.1 m^2 (12 ft^2) metal pan located under the FMRC Fire Products collector instrumented to measure heat release rates. The liquids tested were mineral spirits, isopropyl alcohol, and corn oil (Dean, 1986).

The carton and container arrangement used in Dean's double carton tests is shown in Figure 8.17. Six 1 gal (3.8 L) polyethylene containers were placed in each carton and a standard FMRC igniter was placed in the 7.6 cm (3 in) gap between the cartons. The cartons burned through

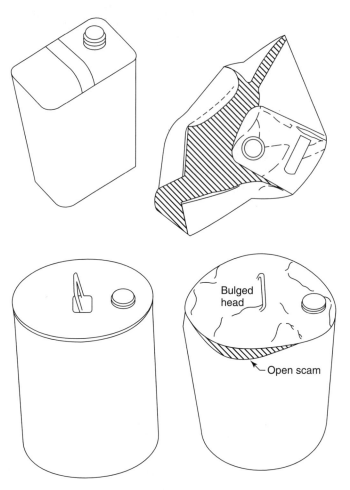

Figure 8.16. Metal containers in Burrow of Mines Fire Exposure tests. Above: one gallon cans before and after (235) tray fire exposure; Below: five gallon DOT17E containers before and after (2 min) fire exposure

and allowed the containers to soften and release liquid in about 1 min 20 sec with mineral spirits and corn oil, and in about 3 min 40 sec with alcohol.

The convective heat release rates measured in the double carton tests with mineral spirits and with isopropyl alcohol are compared in Figure 8.18 to the corresponding tests with the liquid ignited as a pool fire in the pan. It is clear from Figure 8.18 that the liquids eventually were released in sufficient quantity for the heat release rate to reach approximately the same steady burning value that was measured in the pool fire tests without the containers and cartons. There is about a two minute delay for complete liquid involvement in the mineral spirits test and about a six minute delay in the alcohol test. The corn oil was not heated sufficiently either in the pool configuration (even with alcohol floating on it) or in the double carton to reach self sustained burning.

Several other free burn calorimetry tests have been conducted with other types and sizes of plastic containers and cardboard cartons. For example, the Society of the Plastics Industry (SPI) sponsored some exploratory tests in 1986 using pint (0.47 liter) and gal (3.8 liter) containers of paint thinner and pint and quart (0.94 liter) containers of charcoal lighter fluid. The results

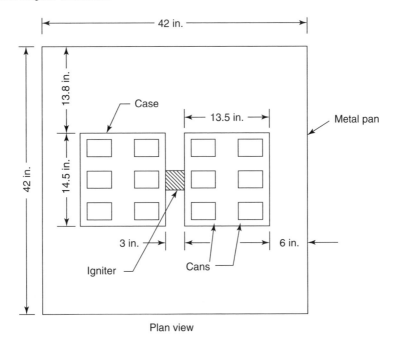

Figure 8.17. Setup in double case tests with flammable liquid in polyethylene containers (from Dean, 1986)

demonstrated that the smaller containers produced significantly lower peak heat release rates (at least 40% lower) than the gal containers. The advantages of thermal insulation such as vermiculite (expanded mica) in the carton to delay container/liquid involvement were also demonstrated in the SPI tests. Some benefits were also achieved with fire retardant cartons and with coextruded nylon/polyethylene containers. Subsequent tests in the National Flammable Liquid Container Research Program indicate how these measures influence sprinkler protection effectiveness as described in Section 8.3.4.

8.3.4 SPRINKLER PROTECTION FOR FLAMMABLE LIQUIDS IN SMALL CONTAINERS

8.3.4.1 Small metal containers

Until recently, most of the protection guidelines for flammable liquids in metal containers were based on loss experience and relatively few large-scale fire tests.

The first documented test (NFPA 30 Appendix E) was conducted in 1957 with 3.2 m (10.5 ft) high palletized storage of paint in 1 gal friction lid cans in cartons. The paints had flash points in the range 40–77 °C (105–170 °F) representing Class II and IIIA liquids. Protection was provided by 71 °C (160 °F) 1/2 in sprinklers discharging at 0.23 gpm/ft^2 (9.4 L/min-m^2) from the 4.6 m (15 ft) high ceiling.

The fire was controlled by six sprinklers even though about 500 cans had their lids blown off, and 20 cans had burst seams. Apparently, most of the paint remained in the open/damaged cans. This test result is probably the basis for NFPA 30 (Appendix D) specifying a sprinkler discharge density of 0.25 gpm/ft^2 (10.2 L/min-m^2) for palletized storage of Class II (one pallet high) and Class III flammable liquids in metal containers.

Another 1 gal metal container test was conducted in 1987 at Southwest Research Institute (sponsored by the American Iron and Steel Institute), this time with paint thinner (100 °F or

Table 8.12. Test conditions and results for Task 3A

Test No.	Container No.	Container Vol. (gal)	Style	Flammable liquid	Container failure mode	Breach time (min:sec)	Primary liquid burn mode	Flame height (ft)
1	1	1	F.L.[a]	Heptane	Lid Blew Off	2:50	In-can	2
2	1	5	L.C.[b]	Heptane	Cover Blew Off	3:35	Spill Spray	20
3	1	5	T.H.[c,d]	Heptane	Cover Released	5:15	Jet Flame	40
4	2	10	T.H.[d]	Heptane	Cover Released	Bottom: 4:30 Top: 6:00	Spill Jet	Bottom:9 Top: 40
5	2	10	L.C.	Heptane	Cover Leaked & Blew Off	Top: 2:30 Bottom:4:30	Rim Leak Spill	25
6	1	1	F.S.[e]	Heptane	Cover Leaked & Blew Off	3:25	Jet	25
7	2	2	F.L.	Heptane	Lid Blew off	Top: 2:00 Bottom: 2:15	Spill	6
8	1	1	F.S.	Isopropanol	Cap Leaked & Blew Off	5:10 Blow Off	Rim Fire Jet Flame	12
9	2	10	L.C.	Isopropanol	Cover Leaked & Blew Off	2:30 7:00	Rim Fire Spill Fire	8
10	12	12 (cartons)	F.S.	Heptane	Cap Leaked & Blew Off	5:20 7:30	Spill Spray	12-20
11	1	5	T.H.	Heptane	Spout Leaked & Blew Off	Vapor 2:30 Liquid 4:00	Rim Fire Jet	25
12	2	10	T.H.	Heptane	Spout Leaked & Blew Off	Vapor 2:30 Liquid 4:35	Spray Fire	25
13	1	5	L.C.	Heptane	Cover Blew Off	4:00	Spill/Pool Fire	15
14	2	10	T.H.	Isopropanol	Spout Leaked & Blew Off	4:10 4:38	Jet	15
15	12	12	F.S.	Isopropanol	Cap Leaked & Blew Off	4:00 4:50	Spill Jet	12-15
16	8	40 (with spout)	T.H.	Heptane	Spout Leaked & Blew Off	2:08 4:00	Jet Pool	25

[a]F.L. = 1 gal Friction Lid
[b]L.C. = 5 gal Lug Cover
[c]T.H. = 5 gal Tight Head
[d]Without plastic spouts
[e]F.S. = 1 gal F-Style

38 °C flash point) in tin plated steel cans similar to those shown in Figure 8.16. Thirty cartons, each with six cans, were placed on a pallet and ignited with two standard 7.6 cm (3 in) long FMRC igniters. None of the cans were breached and none of the sprinklers on the 9.1 m (30 ft) ceiling actuated. The test is noteworthy by comparison to the equivalent test with plastic containers as described in Section 8.3.4.2. It also raised a question (addressed below) about the effect of the ignition scenario on fire development and protection effectiveness.

According to Nugent's 1994 compilation of test results, between 1991 and 1993 there were 42 additional sprinklered fire tests involving flammable liquids in small containers. Two sprinklered fire tests involving spill fire ignition of heptane in metal containers were conducted at Underwriters Laboratories in 1991 (Nugent, October 1991, November 1991). The water discharge density of the 1/2 in orifice, 141 °C (286 °F) sprinklers on the 5.5 m (18 ft) high ceiling was 0.29 gpm/ft^2 (11.8 L/min-m^2) in both tests. In the first test, the 1 gal (3.8 L) and 5 gal (19 L) metal containers were stacked to a height of 1.2 m (4 ft), i.e. one pallet load high. This array, exposed to a 5 gal (19 L) heptane spill fire, was successfully protected without any liquid release from the containers. In the second test the storage array was 2 m (6 ft 8 in) high and exposed to a 10 gal (38 L) heptane

FLAMMABLE LIQUID STORAGE

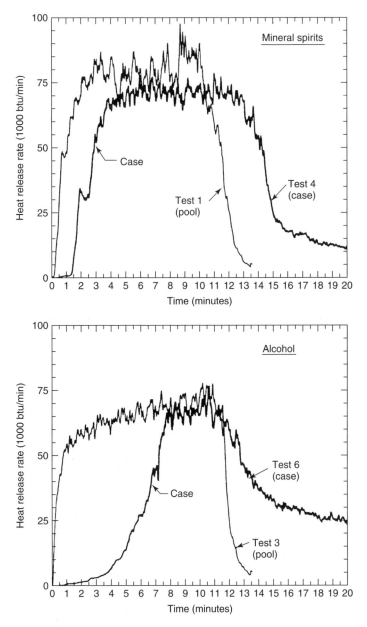

Figure 8.18. Convective heat release rates–pool vs. case (from Dean, 1986). © 2002 Factory Mutual Insurance Company, with permission

spill fire. The same sprinkler protection was not effective for this array. The water did not cool the cans sufficiently to prevent several from releasing liquid, and one 1 gal (3.8 L) can from rupturing. Thus array height is a critical factor in water cooling effectiveness. A similar storage array and exposure spill fire was successfully protected with a foam-water sprinkler system with the same sprinkler parameters (Underwriters Laboratories, 1992).

In view of the varying levels of success of sprinkler protection in these metal container tests, questions remain about the effects of other liquids, larger spill fire exposures and other types of containers. Some of these questions can be answered by comparing calculated sprinkler actuation times to container failure times presented in Section 8.3.3. For example, consider a 10 gal (38 L) spill of a hydrocarbon liquid with a chemical heat release rate per unit area of 3.3 Btu/hr-ft^2 and a corresponding liquid regression rate of 7.6 mm/min (0.3 in/min) (representative values for hydrocarbon liquids as discussed in Section 7.4). Sprinkler activation times for different spill areas can be calculated as described in Appendix D. Results for a ceiling height of 9.1 m (30 ft) and sprinklers with 141 °C (286 °F), 166 (m-s)$^{1/2}$ (300 (ft-s)$^{1/2}$) RTI links situated 2.2 m (7.1 ft) from the fire plume centerline are shown in Figure 8.19. A minimum spill area of about 2 m^2 (21.5 ft^2) is needed to actuate the sprinklers in this situation. Any spill larger than 3 m^2 (32.3 ft^2) will actuate the sprinklers before the liquid burns out.

The range of 5 gal (19 L) container failure times, as measured by Perzak *et al.* for metal containing kerosene (Table 8.11) and by Richards and Munkenbeck for plastic containers with JP-4, are also indicated in Figure 8.19. If these containers were engulfed in the spill fire, the comparison in Figure 8.19 indicates these cans would breach and release liquid before the sprinkler actuates if the spill area is smaller than 4–10 m^2 (43–108 ft^2). Within that range of spill areas, sprinkler actuation times and 5 gal (19 L) container failure times are comparable, so there is a chance that some containers will contribute to the fire before the sprinkler actuates. In the case of the Southwest Research Institute 5 gal (19 L) container sprinklered fire, the sprinkler actuation time was reported to be 26 seconds, which would be consistent with Figure 8.19 if the spill area were about 10 m^2 (108 ft^2). In the UL tests, the first ceiling sprinklers actuated in about 35 sec, corresponding to a 6 m^2 (65 ft^2) spill area in Figure 8.19.

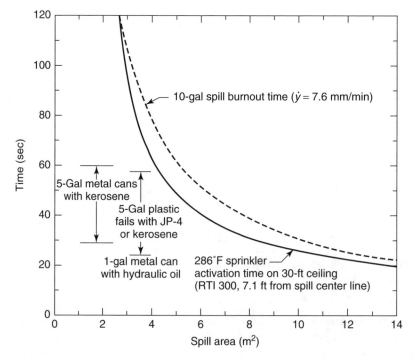

Figure 8.19. Comparison of sprinkler activation, 10 gal spill burnout, and container failure times ($\dot{Q} = 3.3$ MW/m^2)

The 1 gal (3.8 L) metal container average failure time measured by Perzak et al. is also indicated in Figure 8.19 for comparison with sprinkler opening times. The comparison indicates that the 1 gal (3.8 L) metal cans of paint thinner tested by Southwest Research probably would have been breached before sprinkler actuation if a hydrocarbon spill of 2–10 m^2 had been used for ignition instead of the two 7.6 × 7.6 cm (3 × 3 in) igniters. There is some uncertainty about this interpretation because of the insulating effects of the carton and surrounding cans. The UL water only sprinkler test with the two pallet high array resulted in a 1 gal container rupturing at 3–4 minutes because of the cooling effect of the water and the cartons.

The more recent sprinklered fire tests have led the NFPA Flammable Liquids Committee to propose significantly revised NFPA 30 sprinkler protection requirements for flammable liquids in small metal containers. The new requirements in the 1996 edition of NFPA 30 are presented in four complex tables with ceiling sprinkler area/density requirements depending on the NFPA flash point categorization (Classes I, II, and III and subclasses) and water miscibility, the container size and pressure relief provision, storage and ceiling heights, and sprinkler orifice size and link response. Detailed requirements for in-rack sprinklers are also provided for rack storage.

Small plastic containers

Current concern about the hazard of flammable liquids in small plastic containers stored in general purpose warehouses (Benedetti, 1987) originated from a series of five sprinklered fire tests conducted by FMRC (Dean 1986, and FM Data Sheet 7-29). Test conditions and key results are summarized in Table 8.13. Each test involved one or two pallet loads of product packaged in ordinary cardboard cartons. Four tests involved liquids in polyethylene gal containers while the other involved hair spray in 8 oz (0.24 L) containers. Sprinkler protection in four tests was $\frac{1}{2}$ in, 138 °C (280 °F), sprinklers discharging at a density of 0.30 gpm/ft^2 (12.2 L/min-m^2). The test with mineral spirits (petroleum base liquid with flash point in the range 94–108 °F, or 34–42 °C) used 0.64 in orifice sprinklers at a discharge pressure of 75 psig (518 KPA), corresponding to a water density of 0.83 gpm/ft^2 (33.8 L/min-m^2).

The tests with hair spray and with corn oil each opened only two sprinklers, but the tests with paint thinner, mineral spirits, and isopropyl alcohol in gal containers each opened 41–45 sprinkler heads. Photographs of the fire development in the isopropyl alcohol test are shown in Figure 8.20. The paint thinner, mineral spirits, and isopropyl alcohol tests were considered failures in the sense that only two pallet loads would have easily overtaxed the sprinkler system for palletized storage in a general purpose warehouse (NFPA 231, for example, currently specifies 0.30 gpm/ft^2 (12.2 L/min-m^2) over only 186 m^2 (2000 ft^2), i.e. 20 heads at 9.3 m^2 (100 ft^2) per head, for 138 °C (280 °F) heads and 6.1 m (20 ft) high palletized storage of a Class IV commodity). The mineral spirits product fire overwhelmed the protection.

A virtual repeat of the FMRC paint thinner test (Test No. 1 in Table 8.13) was conducted by Southwest Research in 1987 (when the 1-gal metal container test described in above was conducted). Only 24 sprinklers were installed in the Southwest Research test, and 16 of the 24 were activated within 3.5 minutes from ignition. Test observers reported that the sprinkler discharge appeared to be spreading the fire at this point. The test was terminated via hoseline water application at 14 min 23 sec when the liquid seemed to reflash after a dormant period of several minutes.

FMRC concluded on the basis of the five tests summarized in Table 8.13 that most flammable liquids in plastic containers should not be stored in general purpose warehouses without being isolated from the rest of the warehouse. The exceptions to account for the successful results of the hair spray and corn oil tests were Class IIIB liquids, and water miscible liquids in 16 oz (0.48 L) or smaller containers. The Class IIIB liquid exception has subsequently been questioned because corn oil has a closed cup flash point (510 °F or 266 °C) significantly higher than many

Table 8.13. Fire test summary (from Dean, 1986)

Test Commodity	Test No. 1 Paint Thinner	No. 2 Mineral Spirits	No. 3 Hair Spray	No. 4 Isopropyl	No. 5 Corn Oil
Arrangement	Single Pallet	1 × 2 × 1	1 × 2 × 1	1 × 2 × 1	1 × 2 × 1
Stack Height, Ft-in. (m)	3–9 (1.14 m)	4–7 (1.4)	4–4 (1–3)	4–7 (1.4)	4.7 (1–4)
Clearance to Sprinklers, Ft-in. (m)	25–9 (7.8)	24–10 (7.6)	24–10 (7.5)	24–10 (7.6)	24–10 (7.6)
Sprinkler Orifice Size, in. (mm)	$\frac{1}{2}$ (12.7)	0.64 (16.3)	1/2 (12–7)	1/2 (12–7)	1/2 (12–7)
Sprinkler Temperature Rating F (C)	280 (138)	160 (71)	280 (138)	280 (138)	280 (138)
Sprinkler Spacing ft^2 (m^2)	100 (9.3)	100 (9.3)	100 (9.3)	100 (9.3)	100 (9.3)
Constant Water Pressure psi (kPa)	29 (200)	75 (517)	29 (200)	29 (200)	29 (200)
First Sprinkler Operation min:sec.	4:31	1:53	16:26	2:57	After 30:00
Last Sprinkler Operation min:sec	8:43	3:47	25:18	6:56	After 30:00
Total Sprinklers Opened	42	45	2	41	2
Total Sprinkler Discharge gpm (l/sec)	1275 (80.4)	3750 (236)	60 (3.8)	1240 (78.2)	
Peak Gas Temp. F (C)	2275 (1246)	1150[a] (621)	360 (182)	1275 (690)	170 (77)
Peak Steel Temp F (C)	1220 (660)	210 (99)	280 (138)	475 (246)	145 (63)

[a] Wetted thermocouple

other Class IIIB liquids (minimum of 200 °F or 93 °C), and because the small size of the array did exposure fire to heat the corn oil.

The concern about Class IIIB liquids in plastic containers is that a sufficiently large exposure fire associated perhaps with other commodities in the warehouse can raise the temperature of the Class IIIB liquid above its fire point such that it will generate a much more severe fire than was observed in the two pallet load corn oil test. (The corn oil was only marginally burning throughout most of that test.) Therefore, a series of mixed storage tests was conducted with motor oil (closed cup flash point of 380 °F or 193 °C) in polyethylene quart containers in cardboard cartons. A variety of other commodities were located next to the motor oil. Tests conducted under the FMRC Fire Products Collector revealed that the heat release rates in this mixed storage configuration (with and without water application) were higher than those of the ordinary combustible and plastic commodities alone. A sprinklered fire test with two to five tiers of palletized storage and a sprinkler discharge density of 0.40 gpm/ft^2 (16.3 L/min-m^2) indicated that the motor oil was causing the fire to escalate significantly until there was a major collapse of the storage array. These results indicate that many Class IIIB liquids in plastic containers also warrant special sprinkler protection.

A large-scale sprinklered test with seal oil (flash point equals 206 °F or 96 °C) in polyethylene containers has confirmed that Class IIIB flammable liquids also pose a severe challenge to conventional sprinkler protection. The seal oil (similar to mineral oil) was stored in cartons on racks to a height of 5.8 m (19 ft). In-rack sprinklers were installed in each of the lower three tiers, and ceiling sprinklers were installed with a design density of 0.30 gpm/ft^2 (12.2 L/min-m^2). Eighty-seven sprinklers (79 on the ceiling and 8 in-racks) opened without controlling the fire at all once the seal oil became involved, which occurred about 100 seconds after ignition.

FLAMMABLE LIQUID STORAGE

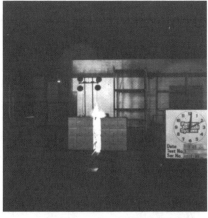
Test 8, alcohol at 1 min 13 S. 3958–37

Test 8, alcohol at 2 min 55 S. 3958–39

Test 8, alcohol at 1 min 45 S. 3958–38

Test 8, alcohol at 4 min 10 S. xxx

Figure 8.20. FMRC sprinklered fire test with isopropyl alcohol in polyethylene containers. © 2002 Factory Mutual Insurance Company, with permission

A major effort has been launched under the auspices of the National Fire Protection Research Foundation (NFPRF) to provide a better understanding of how the overall fire hazard of flammable liquids in plastic containers can be reduced through product and packaging modifications. An extensive series of Fire Products Collector tests has been conducted at Underwriters Laboratories to determine the heat release rates of different product/packaging combinations. Most tests were partial pallet load configurations (either 7 cartons or 15 cartons) with a delivered water density of 0.30 gpm/ft^2 (12.2 L/min-m^2) applied to the top of the cartons at a time calculated to correspond to the sprinkler actuation time of 74 °C (165 °F), 166 (m-s)$^{1/2}$ (300 (ft-s)$^{1/2}$) or RTI sprinklers on a 7.6 m (25 ft) ceiling.

Data from the NFPRF/UL tests (Przybyla and Ghandi, 1990) indicate that pint size (0.47 L) containers of Class II and Class IB liquids can be protected when they are packaged in sturdy fire retardant cartons. Figure 8.21 shows the heat release rates measured with kerosene in pint polyethylene containers in one test and in gal polyethylene containers in the other test. In both tests the containers were in fire retardant cartons that delayed fire development for about 15 minutes.

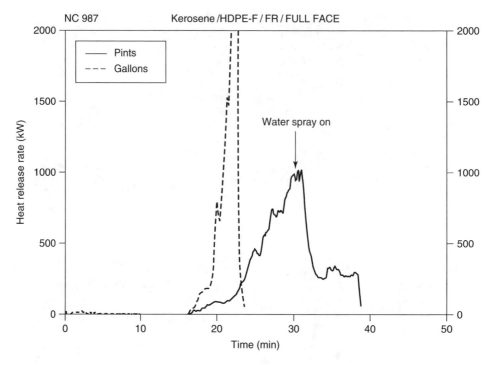

Figure 8.21. Heat release rates measured in UL Fire Products Collector tests for kerosene in polyethylene pint containers and gallon containers in fire retardant cartons containers

Once fire penetrated the cartons, the gal containers rapidly spilled kerosene that caused the heat release rate to increase too rapidly for the water spray to be activated before the test had to be terminated. By contrast, the pint containers released the kerosene sufficiently slowly for water spray to easily control the fire.

There have been two sprinklered fire tests to confirm the protection effectiveness indicated in the UL Fire Products Collector tests. The pint coextruded nylon/polyethylene container and fire retardant carton used in the first sprinklered confirmation test are shown in Figure 8.21. Kerosene was placed in the containers for the test which involved four pallet loads of product. Each pallet load contained a total of 68 gal (257 L) of kerosene in 45 cartons. Sprinkler protection on the 7.6 m (25 ft) ceiling consisted of 17/32 in orifice, 165 °F (74 °C), 287 (ft-s)$^{1/2}$ 158 (m-s)$^{1/2}$ RTI sprinklers discharging water at a density of 0.30 gpm/ft^2 (12.2 L/min-m^2). Only three of the 16 installed sprinklers activated and the maximum temperature of gas at the ceiling was 126 °C (259 °F). The second sprinklered confirmation test was conducted with one gal polyethylene containers of isopropanol contained in fire retardant cartons. The fire developed slowly with the limited amount of burning isopropanol released from the containers readily controlled by the ceiling sprinklers. Now that the technical feasibility of providing adequate fire resistance for some package/container configurations has been demonstrated, the commercial viability of the packaging is being evaluated by individual manufacturers and users.

There has been one other successful demonstration of sprinkler protection effectiveness for a specific flammable liquid and plastic container combination. In this case (Dean, 1985), the liquid was 80-proof (40%) distilled spirits and the container was a 1.75 L (0.46 gal) bottle of polyethylene terepthalate (PET). Two large scale sprinklered fire tests were conducted at FMRC (sponsored by the Distilled Spirits Council of the glass bottles.

Results from both the solid pile storage test and the palletized storage test indicated that protection established for glass containers of distilled spirits would also be applicable to PET containers. These results confirmed some preliminary single container test observations that the PET bottle was at least as fire resistant as the glass bottles.

Now that there is convincing test data to demonstrate that some flammable liquid in plastic container products warrant isolation and special protection while others do not, tests similar to the one developed in the NFPRF project and the FMRC commodity classification project (Chapter 5) can be utilized to quantify the fire resistance of specific commodities. Fire protection engineers and codes and standards writers will then determine appropriate sprinkler protection on the basis of these tests. Their determination will need to consider appropriate ignition scenarios and exposure fires since these factors have been shown to have a significant influence on the fire resistance and protectability of flammable liquids in small containers, both for metal and plastic containers. As of this 1995, testing has led to substantiated recommendations for Class IIIB liquids in plastic containers, and for various liquids in metal containers, but not for the lower flash point liquids in plastic containers.

References

'Analysis of Tank Fires in the Petroleum Industry', API Safety & Fire Protection Department, American Petroleum Institute, 1988.

Anderson, C., Townsend, W., Zoot, J. and Cogwill, G., "Effects of a Fire Environment on a Rail Tank Car Filled with LPG", US Department of Transportation Federal Railroad Administration Report No. FRA-OR&D 75-31, 1974.

American Petroleum Institute, *Welded Steel Tanks for Oil Storage*, API Standard 650, Eight Edition, November 1988.

American Petroleum Institute, *'Installation of Underground Petroleum Product Storage Systems'*, API Publication 1615, Second Edition, 1987.

American Petroleum Institute, *'Venting Atmospheric and Low-Pressure Storage Tanks'*, API Standard 2000, Second Edition, 1973.

American Petroleum Institute, *'Guides for Fighting Fires in and around Petroleum Storage Tanks'*, API Publication 2021, 1974.

Bainbridge, B.L. and Keltner, N.R., Heat Transfer to Large Objects in Large Pool Fires, *J. Hazardous Materials*, **20**, 21–40, 1988.

Benedetti, R.B., *Flammable and Combustible Liquids Code Handbook*, Third Edition, National Fire Protection Association, 1987.

Capizzani, R.E., An Overview of Flammable Liquid Drum Storage and Protection, *18th AIChE Loss Prevention Symposium*, 1984.

Crowl, D.A. and Louvar, J.F., *Chemical Process Safety Fundamentals with Applications*, PTR Prentice Hall, Englewood Cliffs, NJ, 1990.

Das, A.K., 'Safe Separation Distances Between Hydrocarbon Storage Tanks Based on Radiation due to Fire in One Tank', Worcester Polytechnic Institute M.S. Thesis, December 1983.

Davie, F.M., Mores, S., Nolan, P.F. and Hoban, T.W.S., Evidence of the Oxidation of Deposits in Heated Bitumen Storage Tanks, *J. Loss Prevention in the Process Industries*, **6**, 145–150, 1994.

Dean, R.K., 'Full-Scale Storage Fire Tests of Distilled Spirits in PET Plastic Bottles', FMRC Report Prepared for the Distilled Spirits Council of the United States, Inc, 1985.

Dean, R.K., 'Exploratory Testing of Flammable Liquids in Plastic Containers,' FMRC J.I. ONOE4.RR, 1986.

Delichatsios, M.A., Exposure of Steel Drums to an External Spill Fire, *15th AIChE Loss Prevention Symposium*, 1981.

Dimpfl, L.H., Asphalt Tank Explosions, in *Fire Protection Manual for Hydrocarbon Processing Plants*, C.H. Vervalin, ed., vol 1, Third Edition, Gulf Publishing, 1985.

Duggan, J.J., Gilmour, C.H. and Fisher, P.F., Requirements for Relief of Overpressure in Vessels Exposed to Fire, *Transactions of the ASME*, **66**, 1944 (also in *NFPA Quarterly*, October 1943).

El-Iskandarani, Bilal M. K., 'Emergency Relief Venting Analysis for Flammable Liquid Storage Tanks,' Worcester Polytechnic Institute, Masters Thesis, 1994.

Epstein, M., Fauske, H.F. and Hauser, G.M., The Onset of Two-Phase Venting via Entrainment in Liquid-Filled Storage Vessels Exposed to Fire, *J. Loss Prevention in the Process Industries*, **2**, 45–49, 1989.

'Fact-Finding Report on National Flammable Liquid Container Storage Research Project,' Underwriters Laboratories Inc Report to the National Fire Protection Research Foundation, September 1990.

Fauske, H.K., Epstein, M., Grolmes, M.A. and Leung, J.C., *'Emergency Relief Vent Sizing for Fire Emergencies Involving Liquid-Filled Atmospheric Storage Vessels'*, Plant/Operations Progress, vol 5, No. 4, pp 205–208, 1986.

Fauske, H.K., 'Emergency Relief System Design for Reactive and Non-Reactive Systems: Extension of the DIERS Methodology,' AIChE Loss Prevention Symposium, Minneapolis, 1987.

'Final Report on Explosions in Very Large Tankers', International Chamber of Shipping, London, 1976.

'Fire Spread in Storage Tank Farms', *Contact*, Swiss Reinsurance Company, 1987.

FM Data Sheet 7-29, 'Flammable Liquids in Drums and Smaller Containers,' Factory Mutual Research Corporation, September 1989.

FM Data Sheet 7-88, 'Storage Tanks for Flammable and Combustible Liquids', Factory Mutual Research Corporation, July 1976.

Forrest, H.F., Emergency Relief Vent Sizing for Fire Exposure When Two Phase Flow Must Be Considered, *AIChE 19th Loss Prevention Symposium*, March 1985.

Fritz, R.H. and Jack, G.G., Water in Loss Prevention: Where Do We Go from Here, *Hydrocarbon Processing*, August 1983 (also in *Fire Protection Manual for Hydrocarbon Processing Plants*, C.H. Vervalin, ed., Third Edition, vol 1, 1985).

Garrison, W.G., ed., '100 Large Losses: A Thirty Year Review of Property Damage Losses in the Hydrocarbon-Chemical Industries,' M & M Protection Consultants, 1988.

General Electric, 'Bulging and Rupture Characteristics of 55-Gallon, Steel Closed-Top Drums at Known Pressure,' Company Report No. R-65-Ch-SD-529, Silicone Products Department of General Electric, September 30 1965.

Gordon, G.A., 'Relative Fire Resistance of Shipping Drums,' Sonoco Fibre Drum Inc, Lombard, Illinois, March 1989.

Gregory, J.J., Mata, R. and Keltner, N.R., 'Thermal Measurements in a Series of Large Pool Fires,' Sandia Report SAND85-0196, August 1987.

Grolmes, M.A. and Epstein, M., Plant/Operations Progress, October 1985 (also 'DIERS Appendix C: Externally Heated Vessels,' available from AIChE, June 1983).

Hall, J.R., A Fire Risk Analysis Model for Flammable and Combustible Liquid Products in Storage and Retail Occupancies, *Fire Technology*, **31**(4), November 1995.

Henry, M.F., ed, *Flammable and Combustible Liquids Code Handbook*, Second Edition (based on the 1984 edition of NFPA 30), National Fire Protection Association, 1984.

Herzog, G.A., Recent Major Floating Roof Tank Fires and Their Extinguishment, *Fire Journal*, July 1974.

Hickey, H.E., Foam System Calculations, *The SFPE Handbook of Fire Protection Engineering*, Section 4, Chapter 5, NFPA, 1995.

Hill, J.P., 'International Foam-Water Sprinkler Research Program: Task 3A and Task 3B Tests,' FMRC Report to National Fire Protection Research Foundation, 1991.

Holman, J.P., *Heat Transfer*, Seventh Edition, McGraw-Hill, 1990.

Isner, M.S., $49 Million Loss in Sherwin-Williams Warehouse Fire, *Fire Journal*, **65**, 1988.

Kaura, M.L., 'Aid to Firewater Design,' *Hydrocarbon Processing*, **59**, December 1980, 138–147.

Kitagawa, T., Presumption of Causes of Explosions in Very Large Tankers, *Fourth Intl. Symposium on Transport of Hazardous Cargoes by Sea & Inland Waterways*, 1975.

Lees, F.P., *Loss Prevention in the Process Industries*, Butterworth, 1980.

Leung, J.C. and Nazzario, F.N., Two-Phase Flow Evaluations Based on DIERS, API and ASME Methodologies, *AIChE Loss Prevention Symposium*, 1989.

Lougheed, G.D. and Crampton, G.P., 'Fire Tests of Distilled Spirits Storage Tanks,' National Research Council Canada Report No. CR-5727.1, also available from Association of Canadian Distillers, Ottawa, 1989.

Merritt, R.C., Foam-Water and Automatic Sprinkler Protection for Flammable Liquid Process Structures, *17th Annual AIChE Loss Prevention Symposium*, RC83-TP-7, 1983.

Mudan, K. and Croce, P.A., Fire Hazard Calculations for Large Open Hydrocarbon Fires, *The SFPE Handbook of Fire Protection Engineering*, NFPA, 1995.

Nash, P., The Fire Protection of Flammable Liquid Storage with Water Sprays, *Inst. Chem. Engs. Symp. Series No. 39*, 1974.

National Academy of Sciences, 'Pressure-Relieving Systems for Marine Bulk Liquid Containers,' 1973.

Newman, R.M., Fitzgerald, P.M. and Young, J.R., 'Fire Protection of Drum Storage Using "Light Water" Brand AFFF in a Closed Head Sprinkler System,' FMRC JI 22464, March 1975.

NFPA 11, 'Standard for Low Expansion Foam and Combined Agent Systems,' National Fire Protection Association, 1983.

NFPA 30, 'Flammable and Combustible Liquids Code,' National Fire Protection Association, 1987, 1996.

NFPA 231, 'Standard for General Storage,' National Fire Protection Association, 1987.

Nugent, D., 'Class 1B Flammable Liquid Sales and Storage Fire Tests,' Schirmer Engineering Corporation Final Report, October 1991, (also available from SFPE).

Nugent, D., 'Fire Tests Involving Class II Combustible Liquid in Plastic Containers Using a Fire Retardant Package Assembly and Class IB Flammable Liquid in Metal Containers,' Schirmer Engineering Corporation Final Report, November 1991.

Perzak, F.J., Kubala, T.A. and Lazzara, C.P., 'Fire Tests of Five-Gallon Containers Used for Storage in Underground Coal Mines,' Bureau of Mines RI 8946, 1985.

Pignato, J.A., Storage Tank Protection Using Aqueous Film-Forming Foam, *Fire Journal*, November 1978.

Post, L., Glor, M., Luttgens, G. and Maurer, B., The Avoidance of Ignition Hazards due to Electrostatic Charges During the Spraying f Liquids Under High Pressure, *J. Electrostatics*, **23**, 99–109, 1989.

Przybyla, L. and Ghandi, P., Flammable Liquids in Plastic Containers, *Fire Journal*, May/June 1990.

Redding, D. and O'Brien, A., Large Loss Fire in the United States During 1984, *Fire Journal*, **16**, November 1985.

Richards, R.C. and Munkenback, G.J. Jr., 'Fire Exposure Tests of Polyethylene and 55 Gallon Steel Drums Load with Flammable Liquids-Phase II, Final Report,' CG-D-86-77, August 1977.

Robertson, R.B., Spacing in Chemical Plant Design Against Loss by Fire, *Process Industry Hazards*, 157, 1976.

Rubber Reserve Co., 'Heat Input to Vessels,' Rubber Reserve Co. Safety Memorandum No. 89, 1944.

Rubber Reserve Co., 'Protection of Vessels Exposed to Fire,' Rubber Reserve Co. Safety Memorandum No. 123, 1945.

Russell, L.H. and Canfield, J.A., Experimental Measurement of Heat Transfer to a Cylinder Immersed in a Large Aviation Fuel Fire, *J. Heat Transfer*, August 1973.

Scheffey, J.L., Status Report on Fire Testing of Liquids Stored in Intermediate Bulk Containers, *Proceedings Fire Suppression and Detection Research and Applications Symposium*, 1997.

Shipp, M.P., 'Hydrocarbon Fire Standard–An Assessment of Existing Information,' Fire Research Station, Building Research Establishment, OT/R/8294, 1983.

Solberg, D.M. and Borgnes, O., Thermal Response of Process Equipment to Hydrocarbon Fire, *16th AIChE Loss Prevention Symposium*, 1982.

Southwest Research Institute, 'Fire Tests of Steel and Plastic Containers of Paint Thinner,' Prepared for American Iron and Steel Institute, April 16 1987.

Sumitra, P.S. and Troup, J. M. A., 'Fire Protection Requirements for Six-Barrel High Palletized Storage of Distilled Spirits,' Prepared for Distilled Spirits Council of the United States, Inc, FMRC JI OC2R6.RR February 1979.

Tavares, R., 'Investigation of Solvent-Based Coatings in 55-Gallon Drum Test Apparatus,' FMRC J.I. OG3R1.RA, Prepared for National Paint and Coatings Association, 1982.

Tavares, R. and Delichatsios, M.A., 'Pressure Relief in Flammable Liquid Drums by Pressure-Activated and Plastic Bungs,' FMRC JI OF0R4.RA, prepared for Steel Shipping Container Institute, 1981.

Townsend, W., Anderson, C., Zoot, J. and Cowgill, G., 'Comparison of Thermally Coated and Uninsulated Rail Tank Cars Filled with LPG Subjected to a Fire Environment,' Department of Transportation Federal Railroad Administration Report FRA-OR&D 75-32, 1974.

Underwriters Laboratories Inc., 'Fact-Finding Report on National Flammable Container Storage Research Project,' September 1990.

Underwriters Laboratories Inc. 'International Foam-Water Sprinkler Research Projectm Task 4–Palletized Storage Fire Tests 1 Through 13,' February 1992.

Wachtell, G.P. and Langhaar, J.W. 'Fire Test and Thermal Behavior of 15-Ton Lead-Shielded Cask,' E.I. DuPont de Nemours and Co., DP-1070, Engineering and Equipment TID-4500, 1966.

Wackerlig, H., The Aftermath of the Sandoz Fire, *Fire Prevention*, **199**, 13–20, May 1987.—doubtvol in bold

Wagers, W.D., Capizzani, R.E., Hill, F.C., Lust, M., and Trondel, R.J., 'Deluge Protection for Class I Flammable Liquids Stored Three High in 55 Gallon Drums,' Paper No 63C, August 18 1981 (presented A > I > Ch.E 1981 Summer National Meeting, Detroit, Michigan).

Wang, Z.-X., A Three Layer Model for Oil Tank Fires, *Proceedings of the Second Intl. Symposium on Fire Safety Science*, pp. 209–220, 1991.

Yao, C., 'Flammable Liquid Drum Storage Fire Protection System Development: Phase I,' Prepared for Manufacturing Chemists Association, FMRC J.I. 16425, 1967, 16425.1, 1968.

Yamaguchi, T. and Wakasa, K., Oil Pool Fire Experiment, *Proceedings of the First Intl. Symposium on Fire Safety Science*, pp. 911–918, 1985.

9 ELECTRICAL CABLES AND EQUIPMENT

9.1 Electrical cables: generic description

We start this chapter with a simple generic description of electrical cables, and then proceed to a discussion of representative industrial fire scenarios, relevant cable flammability tests and standards, and cable fire protection options for industrial facilities.

In the broadest possible categorization of cables, we distinguish power cables from communication or signal cable. Figure 9.1 shows a representative cross-section of a single conductor power cable. Surrounding the multiple strand conductor are a layer of insulation, a polymer film, and an outer jacket. The flammability of polymer jacketed power cable is governed in large part by the insulation and jacket layer material composition and their thickness in comparison to the conductor core diameter. Some of the more commonly available insulation materials include polyethylene, cross-linked polyethylene, and Ethylene Propylene Rubber (EPR). The most commonly used jacket is polyvinyl chloride, but many other polymers and co-polymers are available for both the jacket and insulation. Furthermore, there are often fire retardants and other additives in the insulation and jacket materials that influence the cable flammability. Two other key factors are the electrical current passing through the cable and the surrounding environment, i.e. ambient air temperature and the proximity of adjacent cables.

Multi-conductor power cables have multiple insulated conductors within the outer jacket. In some cases filler strips are also inserted inside the jacket to provide special characteristics such as moisture resistance. A label on the cable jacket would designate the number and gauge of the conductors and the cable listing category and listing organization.

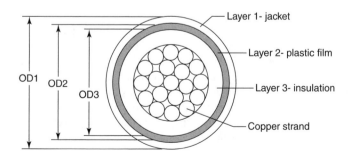

Figure 9.1. Single conductor power cable cross-section. Reproduced by permission of J.R. Reynolds, c/o FTI Consulting, Inc.

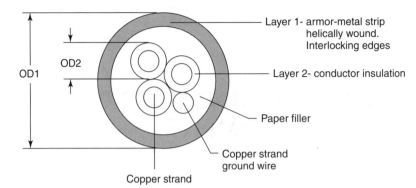

Figure 9.2. Armored cable cross-section. Reproduced by permission of J.R. Reynolds, c/o FTI Consulting, Inc.

Figure 9.2 shows another type of power cable, the armored cable, designated as Type AC cable in the National Electrical Code (NEC), NFPA 70. This cable has multiple insulated conductors embedded in a paper-filled, helically wound flexible metal wrap. Grounding is accomplished via an internal metal bonding strip that should be in contact with the armor for its entire length. Industrial uses for this type of cable include branch circuits (conductors running between the final overcurrent protective device and the electrical outlet) and feeders (conductors between the service equipment and the power supply and the overcurrent device), and in cable trays.

Another common type of power cable is the metal-clad cable designated as Type MC cable in the NEC. The MC cable can have multiple insulated conductors (including optical fibers) enclosed in either armor or corrugated metal sheath. As shown in Figure 9.2, paper fills the space between the conductors and sheath. The MC cable has similar uses as the AC cable.

A third common type of power cable is the Mineral-Insulated, metal-sheathed cable called MI cable in the NEC. It has one or more conductors insulated with compressed refractory mineral insulation and enclosed in a continuous metal sheath. It is often used in lighting circuits as well as power delivery service from the electric utility, and in branch and feeder circuit connections from the power supply. Since it has virtually no combustible materials and is inherently resistant to moisture, the only MI cable use restriction in the NEC is for corrosive atmospheres.

There are several other types of specialized power cable for special applications. One relevant application for industrial facilities is the Power and Control Tray Cable (Type TC cable in the NEC) for use in cable trays. It has two or more insulated conductors under a flame-retardant nonmetallic sheath.

There are numerous types of communication cable. The generic types shown here are unshielded communication cable (Figure 9.3), shielded communication cable (Figure 9.4), optical fiber cable (Figure 9.5), and coaxial cable (Plate 9). Both the shielded and unshielded communication cables have a polymeric jacket that has a major bearing on the cable ignitability and flame spread. The flexible aluminum or copper mesh sheath in the shielded cable provides reduced electromagnetic interference. A non-scientific survey of a few communication cable manufacturers indicates that the most commonly used jacket material for these types of communication cable is PVC, with PTFE and fluorinated copolymers also in abundant supply. However, there are probably hundreds of combinations of insulation and jacket material composition, with many of the newer compositions designed to pass the more stringent fire resistant tests for communication cable as described here in Section 9.3.

The two coaxial cables shown in Figure 9.5 contain a center conductor surrounded by a solid dielectric (often PTFE), a metal sheath, and an outer polymer jacket. As with the previous types of

ELECTRICAL CABLES AND EQUIPMENT

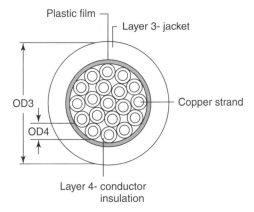

Figure 9.3. Non-shielded communication cable cross-section. Reproduced by permission of J.R. Reynolds, c/o FTI Consulting, Inc.

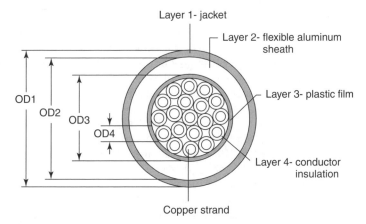

Figure 9.4. Shielded communication cross-section. Reproduced by permission of J.R. Reynolds, c/o FTI Consulting, Inc.

Figure 9.5. Coaxial cable cutaway (from www.belden.com)

cable, coaxial cables are often bundled into one external jacket. The most commonly used jacket material for coaxial cables is polyethylene (flame retardant), but PVC is also readily available.

The fiber optic bundled cable cross-section shown in Plate 9 has numerous coated fiber filaments inside a high-strength aramid lining and a polymer outer jacket. The aramid lining provides structural resistance for the delicate glass fibers; in some designs it is wrapped around individual fibers. As with all the other types of cables, the composition and size of the outer jacket has a major influence on the cable's flammability. The most commonly used jacket material currently used for fiber optic cables is flame retardant polyethylene, available in a variety of densities. PVC, cross-linked polyolefin, and several other jacket materials are also readily available. More than half the cable's weight is usually due to the jacket or series of jackets used (Tomasi, 2000).

Long runs of cables in industrial facilities are often placed in cable trays and ladders. The tray bottoms can have either an open ladder structure or perforated or solid bottom sheet metal. Article 318 of the NEC describes current US requirements for the construction, installation, and grounding of cable trays, and the selection and placement of cables within the trays. Some of the guidelines particularly pertinent to fire protection include decreasing allowable cable ampacity to account for reduced air cooling in the trays, new requirements for separating most high voltage (over 600 V) and low voltage cables, and limitations on the cable fill area (equivalent to fill height limitations for tightly spaced cables and a given tray width).

Cable installations in office and commercial buildings and in many modern light industrial buildings are often in various concealed spaces. Examples of concealed spaces shown in Figure 9.6 include spaces above suspended ceilings, raceways and open spaces under raised floors, and risers. The types of cables that can now be installed in these concealed spaces depend on the flammability ratings and test methods described in Section 9.3.

Cables in Class I, Division 1 hazardous locations and high voltage (>600 V) cable not in cable trays are required by the NEC to be installed in conduit. Most of these installations are in metallic conduit. Although the metal conduit does prevent the cables from being exposed directly to flame, it does allow the cables to be heated by heat conduction if they are in direct contact with the conduit wall. Tests reported by Khan (1984) indicate that PVC/Nylon insulated 12-AWG wires inside 1-inch diameter electrical metal tubing (EMT) can fail electrically after 72–251 seconds immersion of the conduit in a burner flame.

The installation of cables involves pulling the cable through conduit, risers, cable trays, and other types of raceways. If the cable jacket and possibly the insulation are damaged when pulling it through a congested raceway (that may have a fill fraction approaching or even exceeding the limits specified in the NEC), both the electrical performance and fire resistance of the cable can be compromised. To facilitate the cable pulling without damaging cable jacketing, cable lubricants are often used. The residue of these waxes, emulsions, and other lubricants can also significantly influence the propensity for ignition and flame spread in the conduit, cable tray, raceway, etc.

9.2 Cable fire incidents

Fire loss records from large industrial insurers such as IRI (1988) and FM (1998) indicate that grouped cable fires are responsible for many large loss fires, and the most frequent cause of these fires is an electrical fault often due to cable insulation failure. The degraded insulation causes arcing, and the resulting excessive current ignites either the cable insulation/jacketing or some other combustible material that may have inadvertently accumulated in the tray. The subsequent fire often burns for a long time before it is either discovered or effectively suppressed. The three main reasons for the difficulties in locating and suppressing these cable fires are (1) inaccessibility of many cable trays and raceways, (2) problems in seeing or otherwise locating the seat of the fire, and (3) inordinate delays in de-energizing the cables. The following incident accounts, which are from other sources, provide anecdotal accounts of this type of fire scenario.

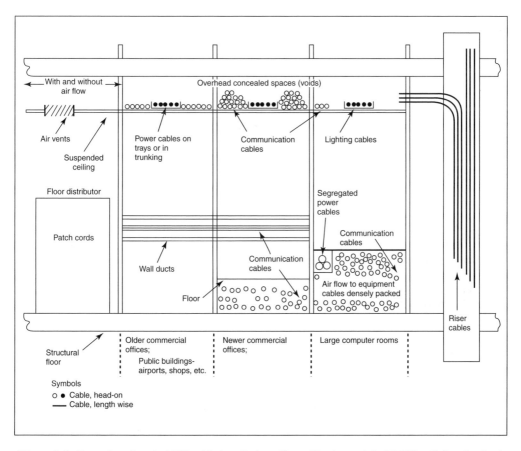

Figure 9.6. Examples of typical UK cable installations (from Chapin *et al.* in LPC Fire Safety Seminar)

New York telephone exchange fire – 1975

The 1975 New York City telephone central office fire was apparently initiated by a short circuit or arcing at an open splice in polyethylene jacketed power cable in a basement cable vault. There was no automatic detection or suppression in the vault. The fire was discovered when smoke propagated up to the main distribution frame on the first floor. Since the phone lines in the building were inoperative by that time, an employee transmitted an alarm from Fire Department call box in the street. Heavy smoke in the cable vault prevented firefighters from finding the actual location burning.

Cables and electrical equipment on the first two floors were destroyed, and approximately 4000 people were needed to assist in the cleanup and replacement of equipment. Most of the damage was due to soot and the corrosive combustion product (HCl) from burning PVC insulation.

Hinsdale telephone central office fire – May 1988

The Hinsdale, Illinois, telephone central office fire started in an overhead cable tray that was overfilled with a combination of power and signal cable. According to the fire investigators report (FTI, 1989), cable jacketing and insulation had been damaged two months earlier during the removal of old cables from the tray. The fire was initiated on Mothers Day 1988 during peak calling hours that would have caused an unusually high electrical load in the cables. Investigators attributed the initial electrical fault to contact between the metal sheath of an armored cable and a damaged dc power cable. Even though smoke detectors were effective in sending an alarm to a remote monitoring station, the response to the alarm at the unattended facility was delayed by at least one hour.

Thick black smoke throughout the first floor of the central office made it very difficult for firefighters to see the area actually burning. After they did observe arcing from the cables, firefighters and company employees had great difficulty in turning off the backup battery power supply energizing the cables because no battery isolation switch or instructions was available. Before the fire could be suppressed, acid fume (HCl) combustion products from the PVC jacketed cables caused virtually total damage to all the electrical switching equipment in the facility. Additional details are provided in Appendix B.

Cable tray fire in coal-fired power plant – Illinois 1998

Arcing in an overhead cable tray at a coal-fired power plant ignited the cable insulation and accumulated coal dust. The fire smoldered for several hours as the plant fire brigade monitored it but was unable to reach it. There was no automatic detection or suppression system in the fire area. Efforts to extinguish the fire with hose lines were unsuccessful. Besides the $5.4 million loss, the fire caused a loss of power to communities as far as 150 miles away (Badger, 1999).

Moscow TV transmission tower fire – August 2000

The 1772 ft high Ostankino television tower in Moscow is the world's second tallest freestanding structure. In August 2000, the major Russian television networks and twelve radio stations transmitted their signals from the tower. At about 3:30 pm on August 27th, a fire broke out, apparently from a short circuit or arcing in cable or electrical equipment at approximately the 1500 ft elevation on the tower. Fire propagated down the electrical cables in the tower central core. As shown in Figure 9.7, the fire eventually propagated down to the 328 ft level, a distance of about 1200 ft. Plate 10 shows the smoke emanating from the tower as the fire intensified and propagated downward.

Three people died when a freight elevator being used to haul extinguishing equipment up the tower became trapped and then fell about 860 ft to the base of the tower. Firefighters had great difficulty trying to extinguish the cable fire in the narrow confines of the tower. They eventually extinguished the fire 26 hours after it was discovered. Damage to the electric circuits and transmission equipment prevented television broadcasts throughout Moscow for three days. Approximately 25% of the steel cables used to support the tower were also damaged and need to be replaced.

Newswire accounts (Associated Press, August 31 2000) quote Moscow Fire Chief Leonid Korotchik saying there was a three hour delay in getting authorization to de-energize electrical circuits in the tower. This delay caused several additional short circuits in the cable trunk at various elevations. Korotchik said "If we had cut the power earlier, we would probably been able to localize the fire." He also said the transmission feeder cables overheated because "there was huge overloading" as additional equipment was installed in recent years. Although the types of cables were not identified, there were accounts of "pieces of burning cable insulation quickly spreading the fire downward." There was vague mention of some type of automatic foam suppression system that became inoperative (perhaps because it ran out of foam concentrate) early in the fire. Other news accounts said the fire burned through makeshift asbestos barriers.

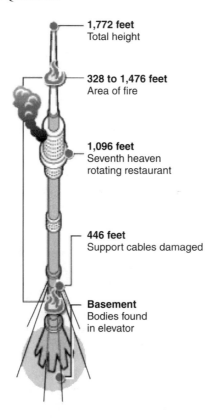

Figure 9.7. Moscow Ostankino TV tower cable fire (from Reuters)

Lead-sheathed cable incidents

MC cables are not immune to arcing induced fires. According to FM Data Sheet 5-31/14-5, the lead sheaths are prone to failure due to lighting, overloading, and holes/cracks allowing moisture penetration. The resulting arcing can melt the lead sheath and ignite the paper filler and cable insulation. The flames can melt the sheath on adjacent cables and cause the fire to escalate.

Even if there is no fire, arcing in MC cables can lead to a large loss. As an example, a phase-to-ground/sheath arc in a three-conductor 750 MCM underground cable melted a section of copper conductor as well as lead sheathing. The arcing was caused by galvanic corrosion of the lead sheathing. The total loss was almost $500,000.

Cutting and welding incidents

Insurance company compilations indicate that cutting/welding is the second most frequent cause of cable tray fires. The cutting/welding may either ignite the cable jacket directly, or it may ignite some other combustible near the cable tray. One such incident occurred in a pulp mill. The inaccessible cable trays spread the fire along the mill walls without firefighters being able to get to them. The result was a $10 million loss with damage throughout the entire mill.

Cable fire spread incidents

There are several well-known fires in which electrical cables were ignited by a small exposure fire, and the cables subsequently spread the fire over a large area. Once such fire was the 1975 Browns

Ferry Nuclear Plant fire in Alabama. The fire was ignited when cable installers attempted to check the seal at a cable wall penetration by lighting a candle near the foam insulation penetration filling. They were looking for evidence of smoke propagation through the penetration. Indeed, fire and smoke burned through the penetration and into a cable vault containing control cable needed to shut down the nuclear reactor. The resulting fire disabled electrical control of the reactor, which had to be shutdown by manually lowering control rods into the reactor core. The estimated damage has been reported to be about $500 million.

Another example of a small flame exposure starting a large cable fire occurred in the Setagaya Telephone Cable Tunnel in Tokyo on November 16 1985 (Sugawa and Handa, 1990). The cables were accidentally ignited by 'gasoline torches' when two maintenance workers left for lunch. Smoke was observed venting from a tunnel manhole within a few minutes after the fire started. However, 17 hours of firefighter water spray application were needed to extinguish the fire in the tunnel. 104 polyethylene-jacketed cables burned out over a length of about 200 m. Telephone service was disrupted for 89,000 subscribers for 7–10 days.

There have been several damaging cable fires in European power plants (Fire Prevention No. 103). Two of those fires involved lubricating oil exposure fires. In the first fire, which occurred at a plant in Gothenburg, Sweden, the fire spread up through a cable chase into an electrical relay room and a control room. The plant was shut down for a year. In the second fire, which occurred at a Swiss power plant, the cables were ignited in the turbine hall and spread through unprotected openings into the adjacent cable spreading rooms. The fire caused about $4 million damage; about one-third of the damage was attributed to HCl produced by the PVC insulated cables. This plant was shut down for seven months. Two other cable fires described in the Fire Prevention article involved power plants under construction. In both these cases, the cable fire did extensive damage to the control room. In one case, extinguishment attempts with carbon dioxide were unsuccessful; the fire was eventually extinguished with water.

The 1975 New York City World Trade Center fire (Lathrop, 1975; N.Y. Board of Fire Underwriters, 1975) started in an 11th floor office area, and spread to PE-insulated/PVC-jacketed cables in a telephone closet. Fire spread along the telephone cables upward to the 16th floor, and downward to the 10th floor. There were no vertical fire stops in the through-floor openings for the cables. Fire may also have spread horizontally via cabling in the plenum utility ducts above the suspended ceiling. The property damage was in excess of $1 million.

Hirschler's 1992 review of electrical cable flammability testing also contains a listing of 24 cable fires in U.S. industrial and commercial facilities between 1953 and 1965. The largest loss, $24 million in 1965 dollars according to the NFPA compilation in Hirschler's paper, occurred at a space tracking station in Florida.

9.3 Cable flammability testing and classifications

A myriad of laboratory-scale and larger-scale cable flammability tests and associated classifications have evolved in various countries for various applications. Table 9.1 lists the designations, and associated flammability test requirements in the 1999 edition of the National Electrical Code. The four major categories of reduced flammability cables in the NEC, in order of decreasing flammability requirements, are: plenum cables; riser cables; general purpose cables, and limited use cables. The various cable designations reflect the cable construction or intended usage, as well as the flammability rating. There is a fifth category, Circuit Integrity cable, intended to certify the ability of fire alarm cable to function with a two-hour fire resistance rating.

The primary pass/fail criterion in the various classification tests listed in Table 9.1 is the length of flame spread along the exposed cables. In the case of the plenum cables and the optional Limited Smoke (LS) rating for UL 1685 tested cables, there is also a limit on the smoke concentration

ELECTRICAL CABLES AND EQUIPMENT 305

Table 9.1. Cable flammability categories in NEC articles 725, 760, 770, 800, 820 and 830

Application	Cable designations	Test designation	Test description	Pass/fail criteria
Plenums and Air Ducts	CMP for Communications Cable; OFNP or OFCP for Optical Fiber cables; CATVP for coaxial cables; PLFP and NPLFP for fire alarm circuits. BLP for Network-Powered Broadband Communication Cable; CL2P and CL3P for remote control and power-limited circuits.	NFPA 262 UL 910	Multiple cables exposed for 20 minutes to 88 kW gas flame in 1 ft wide, 25 ft long horizontal tunnel with 240 ft/min air flow velocity.	Maximum Flame travel distance = 5 ft; Peak Optical Density = 0.5; Average Optical Density = 0.15 as measured in exhaust duct.
Risers	CMR for Communications Cable; OFNR and OFCR for Optical Fiber cables; CATVR for coaxial cables; PLFR and NPLFR for fire alarm circuits; BMR for Network-Powered Broadband Communication Cable; CL2R and CL3R for remote control and power-limited circuits.	ANSI/UL 1666	Multiple cables exposed to 145 kW gas flame for 30 minutes in a 19 ft high concrete shaft divided into two compartments at the 12 ft level.	Flame does not spread to top of tray.

(continued overleaf)

Table 9.1. (*continued*)

Application	Cable designations	Test designation	Test description	Pass/fail criteria
General Purpose except for Risers, Plenums, and Ducts; Nonmetallic Sheathed Cables in Trays	CM for general purpose communications cable; OFN and OFC for Optical fiber cables; CATV for coaxial cables. FPL and NPLF for fire alarm circuits; CL2 and CL3 for remote control and power-limited circuits. BM for Network-Powered Broadband Communication Cable. PLTC: Power-Limited Tray Cable.	UL 1685 formerly ANSI/UL 1581 IEEE 383 Includes CSA FT-4, C22.2 No. 0.3-M	Multiple cables loaded into 8-ft high, 1 ft wide vertical tray and exposed to a 20 kW gas flame for 20 minutes.	Flame (char length) does not spread to the top of the 2.4 m high tray in UL 1581/1685. Maximum vertical flame spread of 1.5 m in CSA C22.2. Optional Limited Smoke (LS) rating requires limited smoke release rates and total smoke generated.
Limited Use "Flame Retardant" Cable	CMX for limited use Communications Cable; CATVX for limited use coaxial cable; BLX for limited use Network-Powered Broadband Communication Cable CL2X and CL3X for raceways and dwellings.	VW-1: Vertical Wire Flame Test in UL 1581	1.5 ft length of single cable is exposed for 15 seconds to small burner flame. Exposure repeated four more times.	No more than 10 inch of cable burns, nor does it drip and ignite a cotton swab on floor, nor does it continue to burn for more than one minute after burner flame is removed.
Fire Alarm Circuit Integrity	CI	UL 2196	Furnace test with standard time-temperature history.	2-hour fire resistance.

(more precisely, optical density) produced in the test. Hirschler (1992) and Babrauskas *et al.* (1991) have reviewed and critiqued these tests. Both reviews criticize the absence of heat release rate data and associated pass/fail criteria in these tests because it is almost impossible to do scenario-specific fire hazard assessments (at other than the tested configuration) without such data. Since then, heat release rate and smoke release rate data are being acquired in UL 1685 tests. Hirschler (1992) and Gandhi *et al.* (1992) showed that cone calorimeter measurements of cable heat release rates could be correlated against the vertical flame spread data from the vertical cable tray tests. Barnes *et al.* (1996) attempted to correlate the heat release rate data from 21 cables tested via UL 1685 to the corresponding cone calorimeter heat release rate data. The correlation was good for the cables that passed UL 1685, but was poor for the cables that failed UL 1685.

International Electrotechnical Commission (IEC) standards categorizing cable flammability use the test configurations and pass/fail criteria summarized in Table 9.2. Some of the tests are analogous to the UL 1581/1685 tests, but there are apparently no international equivalents to the NEC plenum and riser cable tests (UL 910 and UL 1666). Babrauskas *et al.* (1991) have criticized the IEC 1034 test that measures cable smoke generation because the alcohol exposure fire is not a realistic exposure fire scenario.

Barnes *et al.* (1996) have compared the results of the UL 1685/CSA FT-4 vertical cable tray tests to those of IEC 323-3 for 21 sets of cables. Fourteen cables passed both tests; five cables failed both tests; two cables passed IEC 323-3 but failed UL 1685/CSA FT-4. Thus, the UL 1685/CSA FT-4 test is slightly more challenging than the IEC 323-3 test. They also found that there is good correlation in the char lengths measured for the cables that passed both tests.

Table 9.2. International standards for cable flammability and fire resistance

Standard designation	Test description	Pass/fail criterion	Application or comment
IEC 331	Exposure to 750 °C	Electrical Continuity maintained.	May be replaced by variable time-temperature exposure.
IEC 332-1 and IEC 332-2 (now 60332-1 and 60332-2)	0.60 m length of single cable is exposed to small tilted burner flame for at least 60 s depending on cable mass or diameter.	Damaged portion of cable does not extend to within to within 0.05 m of cable clamp.	Several revisions proposed. Similar but less severe than UL 1581 VW-1.
IEC 332-3 (now 60332-3)	Cables in 3.5 m high vertical tray exposed to 20 kW burner flame. Three different categories of cable loading density and flame exposure time.	Maximum char length = 2.5 m above burner.	International version of IEEE 383.
IEC 1034 (BS 6724)	Horizontal 1 m length cables exposed to alcohol pool fire in a 3-m cubical room. Vertical cables used in CEGB version of test. Photometer measures light obscuration.	Limitation on smoke-generated light beam obscuration. (Pass if >60% transmission)	Railway and utility applications. Used in many European countries to determine if cable should be classified as Low Smoke.
IEC 754-1 and IEC 754-2	Cable inserted in 6 cm ID Tube Furnace. Measures pH of combustion gases in aqueous solution, and conductivity of cable.	?	?

The IEC does not specify how the tests should be used to select cables for various applications. One important organization that is currently developing application standards for cables in building construction is the Commission of the European Communities (CEC). The CEC has established six different categories of cable flammability (CEC, 2000). There is a category for noncombustible (e.g. MI) cable, and there are several categories that entail various flame spread and corrosivity tests. Classification criteria will include flame spread, total heat released, heat release rate, fire growth rate, total smoke produced, and smoke production rate, and the occurrence/duration of flaming droplets. The demarcations between categories have not been established/published as of this writing (October 2000).

Factory Mutual has its own cable flammability testing and classification scheme that is based on tests in the laboratory-scale flammability apparatus developed by Tewarson and Khan (1989). Two different types of tests are conducted, ignition tests and flame propagation tests. The time-to-ignition, t_{ig}, tests are used to determine a Thermal Response Parameter, TRP, defined as the inverse slope of a plot of $t_{ig}^{-1/2}$ versus applied heat flux. (Theoretical considerations for thick materials suggest that TRP is related to cable ignition temperature and thermal inertia.) The flame propagation test is conducted by measuring the peak heat release rate, $\dot{Q}$ generated with a 0.61 m vertical length of cable in a tube with an oxygen concentration of 40 v%. Data reported by Tewarson and Khan indicate that the rate of self-sustained vertical flame propagation, v_f, can be calculated from:

$$v_f^{1/2} \approx \frac{\left[\dfrac{0.40\dot{Q}}{n\pi D}\right]^{1/3}}{TRP} \qquad [9.3.1]$$

where n is the number of cables of diameter D in a vertical tray, $\dot{Q}$ is in kW, TRP is in kWs$^{1/2}$/m^2, and v_f is in m/s. Based primarily on this correlation, a Fire Propagation Index, FPI, was defined as

$$FPI = \frac{\left[\dfrac{0.40\dot{Q}}{\pi D}\right]^{1/3} 10^3}{TRP} \qquad [9.3.2]$$

Using the calculated value of FPI, a particular cable is classified into one of three groups. A FM Group 1 cable has a FPI < 10 and is not expected to undergo self-sustained flame propagation. A FM Group 2 cable has a FPI between 10 and 20, and is expected to have limited self-sustained flame propagation rates. A FM Group 3 cable has a FPI larger than 20, and is expected to have accelerating flame propagation rates. FM considers Group 1 cables to not need installed detection/suppression systems unless there is a separate threat of smoke or corrosion, i.e. unless there is a non-thermal damage threat.

Although procedures were established to test and mark cables as FMRC GP-1, GP-2, or GP-3 (FMRC, 1989), most cables do not have such designations. The reasons for the limited adoption of the FM groupings may be a combination of: (1) limited availability of the FM flammability apparatus; (2) criticism by Babrauskas *et al.* (1991) of the unsubstantiated need to test with 40% oxygen; and (3) lack of any established equivalence to the more established UL and NEC tests and categories. Although there are no formally established equivalencies between the FM groupings and the UL/NEC categories, some data were presented by Tewarson and Khan (1989) to explore a possible relationship. Their comparisons indicate that the UL/NEC ratings of CMP and CMR for plenum and riser cables, respectively, may have some equivalency in terms of FM Group 3 cables. On the other hand, cables that passed the UL 1685/1581 vertical tray test for general purpose communication cable fell into all three groupings.

Tewarson and Khan (1989) also presented some data on cable response to electrical current overloads. They increased the current through each cable until sudden vaporization of the cable insulation was observed. Data for six individual cables indicate that the ratio of the critical overload current for insulation vaporization to the cable rated ampacity at 90 °C is in the range 2.9 to 3.3. If an ignition source is present when the cable insulation vaporizes due to current overload, the flame propagation is reported to be virtually instantaneous. This is because the vaporization along the entire cable length produces an extended flammable vapor-air mixture around the cable. Premixed flame propagation through the vapor-air mixture occurs at rates two orders-of-magnitude greater than those for diffusion flame propagation.

Neither the UL tests, nor the IEC tests, nor the FM tests account for the flammability/fire resistance of energized cables. This is unfortunate in view of the propensity of arcing initiated cable fires. Tests conducted in several laboratories indicate that reproducibility is more difficult with the electrical ignition source than with an exposure fire or imposed external heat flux. However, the need for such tests in determining cable fire suppression requirements is particularly important, and standardized test for that purpose have been proposed.

9.4 Vertical cable tray fire test data

Data from vertical cable tray fire tests are reviewed here with a view toward characterizing the extent of flame propagation, the heat release rates generated, the smoke release rates generated, and, in the case of halogenated cable materials, the production of acid gas, i.e. HCl. Results are summarized for cables that have passed the UL 685/1581 fire tests, and those that did not pass.

Barnes et al. (1996) analyzed vertical cable tray fire test results in terms of cable insulation and jacketing material composition. Two of the four best performing cables, in terms of heat release rate and smoke release rate, had halogenated jackets and insulation; the third had a halogenated jacket and a non-halogen insulation, and the fourth had no halogens in either the jacket or insulation. The five cables with non-fire-retarded low-density polyethylene insulation were the poorest performers in that they failed both vertical tray fire tests.

An important consideration in assessing the damage and injury potential of cable fires is the rate and amount of hydrogen chloride generated during combustion of chlorinated jacket and/or insulation. Barnes et al. examined the measurements of HCl generation in the vertical cable tray tests and found that they were roughly correlated with the cable heat release rate. There was no direct correlation with the chlorine content of the insulation/jacket materials since some of the more chlorinated materials had relatively low burning rates. Their overall conclusion is that material composition is not a reliable indication of cable flammability in the absence of fire test data for heat release rates, smoke release rates, and HCl generation rates.

Gandhi et al. (1992) have shown that flame propagation length in the UL 1685/1581 test is roughly linearly proportional to peak heat release rate, and that in order to satisfy the 2.44 m flame height limitation, the peak heat release rate is less than 90 kW. Later data reported by Barnes et al. (1996) are also consistent with this conclusion.

If the UL 1685/1581 data are to be applied to other vertical tray configurations with different tray widths, it is more useful to present the data in terms of heat release rate per unit area of burned cables. The specific definition in this case is:

$$\dot{Q}'' = \frac{\dot{Q}_{max}}{L_f w_c} \qquad [9.4.1]$$

where $\dot{Q}''$ is the peak heat release rate per unit area of burned cables, L_f is the extent of flame propagation up the cable tray, and w_c is the tray width occupied by cables.

Cables failing UL 1685/1581 have a L_f of 2.44 m, while cables passing UL 1685/1581 will have a L_f given in the test data.

Vertical cable tray heat release rate data reported by Barnes *et al*. (1996) and Gandhi *et al*. (1992) have been used to calculate values of $\dot{Q}''$ for cables that have passed and those that have failed UL 1685/1581. Statistical results are shown in Table 9.3. The median for cables passing UL 1685/1581, (208 kW/m^2), is less than half the median for the cables that failed UL 1685/1581 (459 kW/m^2). Cumulative distribution data shown in Figure 9.8 for passing cables can be fit with a normal distribution using the median and standard deviation values shown in Table 9.3.

Smoke release rate data reported by Barnes *et al*. (1996) have also been analyzed to calculate the median values shown in Table 9.3 for: mass burning rate per unit cable burn area ($\dot{m}_b''$); soot yield, and the soot release rate per unit cable burn area. The median value of $\dot{m}_b''$ for cables passing UL 1685/1581 (2.9 g/m^{2}s) is about 20% lower than the median value of $\dot{m}_b''$ (3.6 g/m^{2}s).

Table 9.3. Vertical cable tray heat release rate and smoke release rate data from UL 1685/1581 tests

	Cables passing UL 1685/1581	Cables failing UL 1685/1581
Number in Sample	34	7
Median L_f (m)	1.12	2.44
Median $\dot{Q}''$ (kW/m^2)	208	459
$\dot{Q}''$ Standard Deviation (kW/m^2)	38	172
Median $\dot{m}_b''$ (g/m^{2}s)	2.9	3.6
Median Soot Yield (g/g)	0.038	0.024
Median Soot Release Rate per unit cable burn area (g/m^{2}s)	0.11	0.09

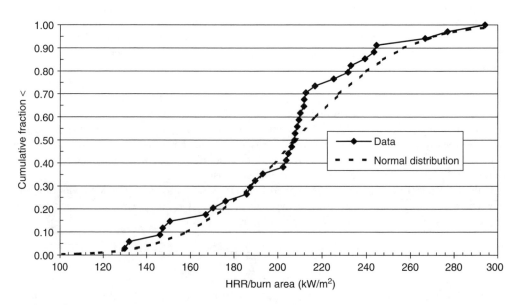

Figure 9.8. Vertical cable tray heat release rates per unit tray burn area for cables passing UL 1685

for failing cables. Since the median heat release rate for passing cables is less than half the median heat release rate for failing cables, the primary difference between the cables must be in their effective heats of combustion, defined as $\dot{Q}''/\dot{m}_b''$.

The median soot yields shown in Table 9.3 have been calculated from the Barnes et al. reported values of average specific extinction area (SEA in m²/g), based on optical density data and air flow rates through the calorimeter. The particular equation used for soot yield is $y_s = \text{SEA}/8\,\text{m}^2/\text{g}$, which is based on the correlation reported by Gandhi (1993) for most building materials tested in the cone calorimeter. Since its applicability to vertical cable tray test data has yet to be determined, the soot data in Table 9.3 should be considered much less reliable than the heat release rate and mass burning rate data. In fact, the median soot yield calculated for passing cables (0.038) is substantially less than the soot yields reported by Tewarson (1995) for a variety of cables based on his laboratory calorimeter tests. The range of cable soot yields reported by Tewarson is 0.076 to 0.16.

After appropriate value of soot yield and $\dot{m}_b''$ are determined, the soot release rate $\dot{m}_s''$ can be calculated from

$$\dot{m}_s'' = y_s \dot{m}_b'' \qquad [9.4.2]$$

Calculated median values of $\dot{m}_s''$ shown in Table 9.3 are approximately 0.1 g/m²s for both passing and failing UL 1685/1581 cables. Thus the primary difference in smoke hazard between passing and failing cables is due to the length of cable burned rather than to the soot generation rate per unit cable burned area.

9.5 Horizontal cable tray fire test data

Horizontal cable trays are used in a wide variety of installations including cable vaults and tunnels, ventilated plenums, telecommunication/switch rooms, and large open facilities. Depending on the installation, cables can include power cables, various communication cables, and, in some cases, mixtures of both types. Sometimes vertical trays are interspersed with the horizontal trays. In many applications, the trays are configured in vertical stacks. Cable loadings in the trays can vary from a single layer to overstuffed trays with cables extending above the tray sidewalls.

Table 9.4 provides a tabulation of six fire test programs devoted to characterizing the free burn fire characteristics of horizontal cable tray fires. The test configurations and cable loadings are summarized along with the ignition source and the range of measured heat release rates and/or horizontal flame spread rates. The ignition source in every test program except the Hinsdale fire simulation (Figure 9.9) was an exposure fire.

Reported cable tray heat release rates listed in Table 9.4 are in the range 200–490 kW per m² of tray horizontal area, with the higher values corresponding to the heavy cable loadings shown in Figure 9.9. Horizontal flame spread rates varied from 0 to 3 m/min, with the latter value measured by Sugawa and Handa (1989) in their simulation of the 1985 Setagaya telephone cable tunnel fire. The Sugawa and Handa tests were conducted with 16 horizontal trays in a 1.7 m (5.6 ft) high ventilated tunnel. In many other tests with fewer trays and larger/taller enclosures, flame spread rates were in the range 4–43 cm/min. A special hazard was observed in the foamed insulation coaxial Radio Frequency cable tests conducted by Riches (1988) and by the Naval Research Laboratory (Alexander et al., 1989).

Cable tray fire heat release rates reported by Newman (1983) are of the form

$$\dot{Q} = \dot{Q}_{\max}[1 - e^{-(t-t_0)/\tau}] \qquad [9.5.1]$$

where

$$\dot{Q}_{\max} = N_{tray} A_{tray} \dot{Q}'' \qquad [9.5.2]$$

Table 9.4. Horizontal cable tray free burn tests

Reference	Tray and room size	Cable loading	Ignition source	Peak heat release rate – per unit tray area (kW) – (kW/m^2)	Flame spread rate	Notes
Sumitra, EPRI NP-1881 1982	12, 8 ft long, 18 in wide ladder trays, 6 tiers high, in FM Test Center	1. No. 12 AWG PE/PVC cable, not qualified per IEEE 383; 2. No. 12 AWG EPR/Hypalon, qualified per IEEE 383; 3. Silicone rubber/asbestos cables, qualified per IEEE 383. 3" Deep trays 40% loaded. Total length = 7000 to 14,000 ft.	Heptane pool fire; 400 kW	2600–4600 kW (200–370 kW/m^2)	9–33 cm/min	Flame spread upward to trays in top tier in 6.5–40 minutes depending on cables used.
FTI Hinsdale Fire Report, 1989	6 ft long × 22 in wide under hood.	Combination of power and communication in four horizontal tray tests. 12 layers per tray	Arcing/short circuit at end of one power cable	450–500 kW (440–490 kW/m^2)	4–12 cm/min	Flame spread along top layer of cables
Hill, EPRI NP-2660	Mixed horizontal and vertical trays; as in Sumitra report Trays in 12.2 m by 12.2 m by 6.1 m (20 ft) high ventilated room	3–4 layers of non-qualified PE/PVC cables in each tray. 2 tests with ceiling tray. 4 tests altogether, tray location and size of exposure fire also varied.	Methanol Pool fires. 185 kW and 286 kW	Projected free burn HRR given in J. Newman's report = 3300–4700 kW	43 cm/min in lower trays before sprinkler actuation	Smoke Concentrations and Smoke Detector alarm times given in Newman's report. Sprinkler actuation in Hill's report
Sugawa and Handa, 1989	16 38 cm (15 in) wide trays in a 2.3 m wide by 1.7 m (5.6 ft) high by 9 m (30 ft) long concrete tunnel	42 PE-sheathed telephone cables, 2 tests with same cable loadings.	Cotton sheet ignited by gasoline torch in center of tunnel.	Not determined	Horizontal: 3 m/min for upper trays Downward to lower trays: = 1 m/min	Wire electrical failure (short) times recorded for about 1200 wires showed most cables failing within 5–6 min.
Riches, 1988	9 in and 12 in wide, 24 ft long trays in a 12 ft diameter, 9 ft high, 65 ft long tunnel	Mix of communication and power cable including several types of coaxial cables. About 12 tests conducted with different cable loadings and varying numbers of trays.	20–40 kW propane burner	Not determined	Fire in most tests self-extinguished after igniter was removed. Self-propagating flame spread rate = 2–4 cm/min	Foamed insulation coaxial cables exploded after several minutes of flame exposure.
Hoover, et al., 1997	0.38 m wide by 7.2 m (24 ft) long tray in ceiling plenum	5 different tests with various communication cable.	150 kg wood crib 1 MW	Not reported	0–1 m/min	No flame spread on CMP cable

Figure 9.9. Cable trays used in Hinsdale fire simulation (from FTI report, 1989. Reproduced by permission of J.R. Reynolds, c/o FTI Consulting, Inc.)

In these equations, t_0 is the time at which self-sustained flame propagation begins, τ is an empirical constant that varies from test-to-test, N_{tray} is the number of trays, and $A_{tray} = w_c L_{tray}$ is the tray horizontal area loaded with cables.

If the flame is assumed to propagate at a velocity v_f along the horizontal cable trays, the time constant, τ, in equation [9.5.1] is given by

$$\tau = \frac{L_{tray} N_{tray}}{v_f N_{burn}} \quad [9.5.3]$$

where N_{burn} is the number of cable trays burning at the time that significant horizontal flame spread begins. In other words, if vertical tray-to-tray flame spread occurs prior to any significant horizontal flame spread (because the trays are spaced very closely), then $N_{burn} = N_{tray}$, and $\tau = L_{tray}/v_f$. At the other extreme, if the trays are spaced far apart and only one tray is burning as horizontal flame spread begins, $\tau = N_{tray} L_{tray}/v_f$. Values of τ reported by Newman (1983) were in the range 60– 560 sec.

Several cable tray fire test programs have examined the effect of ventilation velocity on flame spread rate, v_f. The most prevalent conclusion, as reported by Babrauskas et al. (1991) and Riches (1988), is that ventilation velocities have no apparent effect on flame spread rate, at least for ventilation velocities in the range 0.5–1.6 m/s. Other studies reviewed by Babrauskas et al. indicate that there is a maximum flame speed which occurs at a ventilation velocity of about 1.8 m/s. The maximum flame spread rate for the tests summarized in Table 9.4 occurred in the Sugawa and Handa (1989) tests with a wind induced air velocity of about 1 m/s in the cable tunnel prior to the fire.

Equations [9.5.1]–[9.5.3] together with the heat release rates per unit area given in Tables 9.3 and 9.4, and the flame spread velocities given in Table 9.4, provide a basis for estimating the thermal threat of horizontal cable tray fires. The non-thermal threat associated with smoke generation can be included taking into account the median soot yields and soot specific generation rates shown in Table 9.3. Chloride generation rates are too difficult to generalize, and require test data for the cable in question.

9.6 Cable fire suppression tests

9.6.1 SPRINKLER AND WATER SPRAY SUPPRESSION TESTS

Several fire test programs were conducted in the 1970s and 1980s to determine cable tray fire suppression requirements using water sprays from nozzles and sprinkler heads. The Factory Mutual Research Corporation test program reported by Sumitra (1982) was intended to test the effectiveness of two different spray delivered densities (flow rate per unit horizontal area delivered to the top cable tray) on the stacked tray configuration shown in Figure 9.10(a). The water was discharged from a large nozzle situated 4 ft above the center of the tray stack.

The cables, trays, and exposure fires used in these FMRC tests were the same as those in the free burn tests listed in the top row of Table 9.4. Initiation of water spray occurred at various times corresponding to the burn out of the exposure fire and, in some cases, to the full fire involvement of the cables. Specific water application times for each of the suppression tests are listed in Table 9.5. As indicated in Table 9.5, two different water delivered densities were employed, $0.16\,\text{gpm/ft}^2$ and $0.45\,\text{gpm/ft}^2$.

A delivered density of $0.16\,\text{gpm/ft}^2$ was sufficient to achieve suppression in all the tests with horizontal cable trays. Water spray was able to penetrate through/around the upper trays and reach all the cables. Comparing the results of the first four tests listed in Table 9.4, it is clear that early water application and use of a larger spray density both produce earlier suppression with less fuel consumed during suppression.

Comparison of FMRC Test 13 and Test 6 provides some interesting and perhaps unexpected insights. The tests were identical except for the cable material used. Test 13 was conducted with EPR/CSPE (chloro-sulfonated polyethylene) cable, which is less flammable than the PE/PVC cable used in Test 6, i.e. the EPR/CSPE passed the IEEE 383 test while the PE/PVC did not pass. During the period prior to water application, the EPR/CSPE cable fire had a much slower flame spread than the PE/PVC. Hence, water was applied much sooner in Test 6 than in Test 13. However, the delayed water application made the EPR/CSPE cable tray fire more difficult to extinguish because the fire had apparently burned deeper into the cable when water was eventually applied. Thus, early water spray/sprinkler application is a key factor in achieving efficient suppression in less flammable cables as well as in the more flammable cables.

The only FMRC cable tray test that did not result in complete suppression was Test 15, which was conducted with a 16 ft high vertical cable tray that partially obstructed the water spray discharge such that a portion of the horizontal trays was shielded from the spray. Fire

Table 9.5. Key data from FMRC cable tray sprinklered fire tests (from Sumitra, 1982)

Test #	Cable materials insulation/jacket	Tray configuration	Water delivered density (gpm/ft^2)	Time of application (min)	Fuel consumption after water application (kg)	Time to extinguish (min)
5	PE/PVC	Horizontal	0.45	8:08	5	2
6	PE/PVC	Horizontal	0.16	8:30	9	3
7	PE/PVC	Horizontal	0.45	11:12	12	5
8	PE/PVC	Horizontal	0.16	11:00	28	17
9	Si/As	Horizontal	0.16	58:00	<1	2
13	EPR/CSPE	Horizontal	0.16	44:00	78	31
14	EPR/CSPE	Horizontal	0.45	98:00	7	6
15	PE/PVC	Hor & Vert	0.45	13:10	103	—[a]
16	PE/PVC	Hor & Vert	0.16	9:12	69	23
17	PE/PVC	Hor & Vert	0.16	15:25	73	21

[a]Fire was never extinguished in the shielded portion of cable trays

ELECTRICAL CABLES AND EQUIPMENT

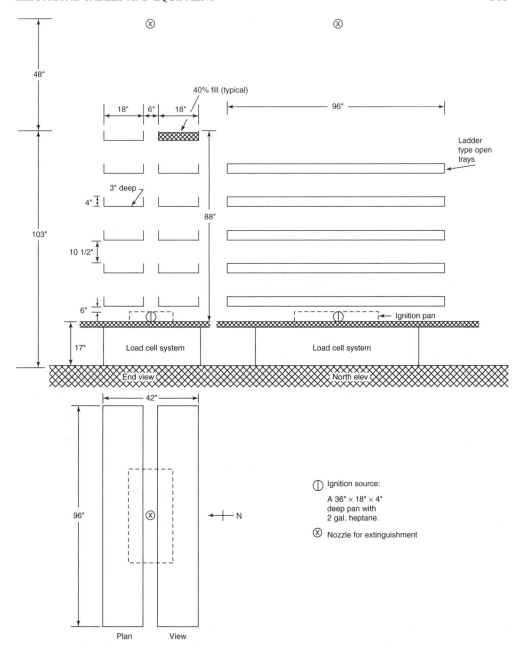

Figure 9.10(a). FMRC horizontal cable tray sprinklered fire tests

in the unshielded portions of the tray was extinguished within two minutes of water application. However the shielded section of the tray continued to burn until virtually the entire cable jacket/insulation was consumed, i.e. that portion of the fire was never extinguished. In the last two tests listed in Table 9.5, a shorter vertical tray was used such that there was no obstruction or shielding of the water spray. The short vertical tray did increase the fire suppression challenge beyond that of the horizontal trays alone (as evidenced by the more fuel consumed), but the suppression was eventually achieved after 21–23 minutes of water application.

FMRC conducted four sprinklered cable tray fire suppression tests (Hill, 1982) using various versions of the mixed horizontal and vertical tray configuration with PE/PVC cables. These tests were conducted in a 40 ft by 40 ft by 20 ft high, ventilated enclosure equipped with closed head ceiling sprinklers spaced 10 ft on center. The $\frac{1}{2}$-inch orifice sprinklers had 160 °F links and discharged water at a density of 0.30 gpm/ft^2. In the first test, the fire heat release rate in the ventilated enclosure was not sufficient to actuate the sprinklers. In the other three tests, suppression was rapidly achieved with two or three sprinklers operating. Apparently the multiple sprinkler actuations and the 4 ft elevation above the 16 ft high vertical trays allowed the spray to reach the entire array of cable trays without any significant shielding.

Gustafson (1976) described a series of three sprinklered cable fire tests conducted in Finland in a 20 m (66 ft) long structure with a 2 m by 2 m (6.6 ft by 6.6 ft) cross-section. There were six levels of 60 cm (2 ft) wide cable trays on each side of the test structure, and an 80 cm (2.6 ft) wide corridor down the center. The trays were filled with a combination of power and signal cables, primarily PVC insulated. Sprinklers with 57 °C (135 °F) links were installed at 4 m (13 ft) spacing on the ceiling and pressurized to each discharge about 260 lpm (69 gpm), corresponding to a discharge density of 0.70 gpm/ft^2. The cables were ignited by a small pan fire in a different location in each test. Only the two sprinklers nearest the ignition site actuated in each test. The sprinkler discharge did extinguish the cable fire before it spread across the corridor or downward from the first-ignited tray. Even the presence of a steel plate on one of the trays in the third test did not prevent water spray from reaching the cables burning under the plate.

Sandia National Laboratories (Chavez and Lambert, 1986) conducted both water spray nozzle and automatic sprinkler fire tests with arrays of horizontal and vertical cable trays. In some tests, IEEE 383 qualified cable was used (cross-linked polyethylene insulation and jacket), while non-qualified PE/PVC cables were used in other tests. The trays were situated in an enclosure with a 24 ft by 25.5 ft floor area. In the sprinklered tests, two pendant sprinklers were located over the periphery of the approximately 9 ft wide by 12 ft long array of trays. In the water spray tests, ten nozzles were located around the periphery with the sprays discharging directly onto the trays at a density in accord with NFPA 15 (0.15 gpm/ft^2). Both the spray nozzles and the sprinklers were able to achieve complete suppression. The ceiling sprinklers required a slightly longer time to achieve suppression because the sprinkler discharge had to gradually penetrate the blockages represented by the upper trays. In all cases, the 5 minute spray duration was able to cool the cables sufficiently to prevent re-ignition.

9.6.2 GASEOUS SUPPRESSION SYSTEM TESTS

The gaseous agents that have been used to suppress cable fires include Halon 1301 (CF_3Br), carbon dioxide, and various fluorinated hydrocarbons offered as halon replacements. The Sandia National Laboratories test program (Chavez and Lambert, 1986) included total flooding tests with Halon 1301 and CO_2. Each test involved either two or five horizontal or vertical cable trays with the same XPE/XPE or PE/PVC used in the water spray/sprinkler tests. A Halon 1301 concentration of 6% was used with the enclosure ventilation closed during discharge and the following 10 or 15 minute hold period. The fires were suppressed by the Halon, but in some cases the cables continued to smolder throughout the hold period. However, the smoldering cables did not re-ignite when ventilation was resumed. The Sandia researchers recommended a minimum hold time of 15 minutes to prevent re-ignition of IEEE-383 qualified cable. They recommended a similar hold time for a carbon dioxide suppression system based on their tests. The NFPA 12 standard for CO_2 systems calls for a 50% concentration to be maintained for a minimum period of 20 minutes for electrical fires.

ELECTRICAL CABLES AND EQUIPMENT

More recent fire suppression testing of electrical cables has been on a smaller scale and has involved the various halon replacement gaseous agents, primarily fluorinated hydrocarbons. Decomposition of the fluorinated agents during suppression has produced large concentrations of HF when the cable remains energized and the gas is exposed to an electric arc during the hold period. On the other hand, early actuation of the agent discharge de-energizing the cables has successfully prevented toxic/corrosive concentrations of HF during suppression and the following hold/cooling period.

NFPA 2001 (Clean Agent Fire Extinguishing Systems, 1999 Edition) says "If electrical equipment cannot be de-energized [prior to agent discharge], consideration should be given to the use of extended discharge, the use of a higher initial concentration, and the formation of ... decomposition products." The design concentrations for the various new gaseous agents are established in tests conducted/monitored by the listing agencies.

9.7 Passive protection: coatings and wraps

The fire resistance of cable trays can be substantially increased through the use of different forms of thermal insulation. Protective coatings can be sprayed or brushed on to the cables. Thermal blankets can be wrapped around the trays. Insulated metal shields are also available.

Early versions of these thermal insulation methods did provide at least some increased fire resistance but had the drawback of significantly de-rating the ampacity of power cables. One such design, shown in Figure 9.10, consisted of ceramic-refractory blankets placed in and around the cable trays. The blanket reduced heat transfer from the cables during normal current-carrying and allowed them to increase heat up and possibly begin degrading (Schultz, 1985).

Many modern cable tray insulation products do not cause as much ampacity de-rating because either they do not require blanketing to be inserted into the tray itself (Figure 9.11) or because

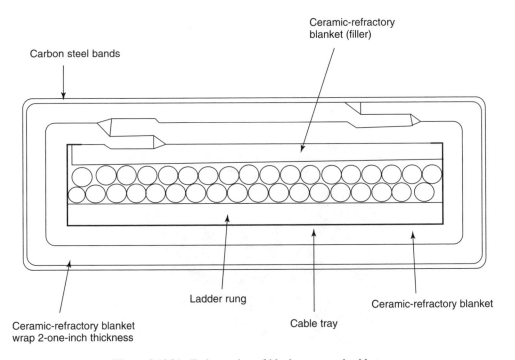

Figure 9.10(b). Early version of blanket-wrapped cable tray

Figure 9.11. Wrapped cable tray installation

they utilize intumescing or hydrated insulations that achieve their insulation effectiveness by swelling and/or endothermic changes in composition at high temperatures. Figure 9.12 shows an installation consisting of a composite shield (intumescing layer between metal sheets) that can be used both under the tray and in the wall penetration.

The Factory Mutual Approval Guide lists many different cable coatings that do not require ampacity de-ratings when applied per the manufacturers' instructions to achieve a coating thickness in the range 1 to 3 mm depending on the specific product. These coatings are tested to verify

Figure 9.12. Composite shield under cable tray and in wall penetration

ELECTRICAL CABLES AND EQUIPMENT

they will prevent flame spread in cables exposed to a small fire (37 kW) directly under the tray, or to sparks/arcs in the tray.

Nowlen (1989) has reviewed the tests conducted at Sandia National Laboratories on the effectiveness of various cable coatings and cable tray insulations/shields. Their tests involved two different size exposure fires and three different types of cables. Both cable ignition and loss of electrical integrity were used as cable failure criteria. Results indicated that most of the coatings prevented electrical failure for the small exposure fire (estimated to be about 40 kW), but that all six coatings tested allowed cable failures for the larger diesel fuel exposure fire in the 14 ft by 25 ft by 10 ft high enclosure. Failure times for coatings applied to a non-rated PE/PVC 3-conductor cable were in the range 3–19 minutes. Tray covers and ceramic wool blankets also failed to prevent eventual failures in the PE/PVC cables exposed to the diesel fuel pool fire, but did prevent failures in cables that had passed the IEEE-383 (UL 1685/1581) fire test.

A wide variety of cable and tray wall and floor penetration seals have been tested and listed by UL, FM, and other certifying organizations. Their fire resistance ratings vary from one hour up to four hours per the standardized tests described in Chapter 3.

9.8 Protection guidelines and practices

Factory Mutual Loss Prevention Data Sheet 5-31 provides protection options for cables and trays. As shown in Table 9.6, these options depend on the cable flammability designation and, in some cases, on the presence of detection/alarm or of minimum cable tray separation distances. If the cable is either FM Group 1 or Group 2 (or UL 910 qualified) with installed detection and cable access, there is no requirement for either a suppression system or cable/tray insulation. No allowance is given for cables passing UL 1686 (Riser) or UL 1685 (General Purpose).

The tray spacing requirements alluded to in the last row of Table 9.6, and illustrated in Figure 9.13, are intended to prevent flame spread between trays. According to the explanation in Data Sheet 5-31, the vertical and horizontal distances, V and H, were calculated using heat release rate, flame height, and critical heat flux data obtained in the FMRC Cable Flammability test apparatus for cables with the highest flammability rating in each FM Group. Since the length of cable burning was assumed to be much greater than the tray width, the calculated separation

Table 9.6. FM requirements for grouped cables (FM Data Sheet 5.31)

Cable flammability	Installed detection & alarm	Cable tray spacing	FM required installed suppression or thermal insulation
FM Group 1	N.A.	N.A.	None
FM Group 2 or UL-910 qualified (plenum cable)	Detection system with constantly monitored alarm and cables accessible to manual firefighting.	N.A.	None
Neither FM Group 1 or 2 nor UL 910 qualified	Requires fire detection for 'early detection.'	N.A.	Automatic Sprinklers or Gaseous Suppression System or Approved Coating or Tray Wrap
FM Group 2 or 3	N.A.	Minimum Vertical and Horizontal Separation per Table in Data Sheet	None

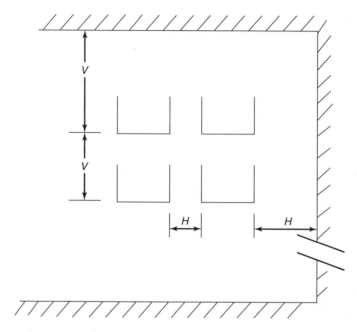

Figure 9.13. Cable tray separation distances. © 2002 Factory Mutual Insurance Company, with permission

distances in each category depend only on the tray width. Calculated values of V shown in Data Sheet 5-31 vary from $1\frac{1}{2}$ to 3 tray widths for FM Group 2 cables and from $2\frac{1}{2}$ to $4\frac{1}{2}$ tray widths for Group 3 cables. Horizontal separations, H, are in the range $\frac{1}{2}$ to $1\frac{1}{2}$ tray-widths.

Industrial Risk Insurers (1988) has developed fire protection for 'high priority cable trays,' i.e. cable trays "that are essential to production, a critical process, or an emergency power system." There are two broad cable flammability categories in the guidelines. The less flammable, or 'IRI-approved' category includes AC and MI cable and NEC-designated Power and Control Tray Cable (Type TC). Cable coatings are required for the more flammable, unapproved, cables installed in two or three-tier trays, even if the trays are in a sprinklered building. The guidelines recommend dedicated water spray protection for four or more tiers of any cable in an un-sprinklered building, and for four or more tiers of unapproved cable even in a sprinklered building. Heat detectors installed in the trays are also recommended when there are four or more tiers of trays.

Although many of the provisions of the National Electrical Code (NFPA 70) pertain to fire prevention, there are no specified requirements for fire detection or suppression, or for installing thermal insulation to increase fire resistance. Several other NFPA standards have detection or suppression recommendations/requirements for cables installed in specific facility areas. For example, the NFPA 75 Standard for the Protection of Electronic Computer/Data Processing Equipment (1999 Edition) requires an installed smoke detection under a raised floor containing cables. The NFPA 850 Recommended Practice for Electric Generating Plants and High Voltage Direct Current Converter Stations (1996 Edition) has the following provisions for cables:

- 'Cable spreading rooms and cable tunnels should be protected with automatic sprinkler [at a design density of $0.30\,\text{gpm/ft}^2$], water spray, or gaseous extinguishing systems,' as well as with an 'early warning fire detection system.'

- 'Cable trays subject to accumulation of coal dust and the spread of an oil spill should be covered by sheet metal.'

- A Fire Risk Evaluation considering plant-specific design, layout, and operations 'should consider the provision of fire suppression systems or fire retardant coatings or both for protection of cable concentrations from exposure fires.'

The plant-specific fire risk evaluation recommended in NFPA 850 is consistent with the engineering approach to fire protection presented in this textbook. In this case, the engineering approach would entail considerations of the flammability of current and anticipated future cable installations, and the potential impact of a scenario-specific cable fire on facility operations and personnel safety. Fire spread both along the cable tray, as well as potential spread to adjacent rays would be part of the analysis.

Various criteria can be utilized to help decide where and when installed fire suppression systems and/or passive protection are warranted. If a quantitative approach is utilized, one possible criterion might be to establish a maximum acceptable heat release rate, or smoke generation rate. For a given cable (or possibly cable flammability category) the heat/smoke generation rate is the product of the specific rate per unit tray surface area multiplied by the tray width and the maximum allowable flame spread length. Thus, the use of a maximum allowable heat/smoke release rate is tantamount to a maximum allowable flame spread length, or, for a given flame spread rate, to a maximum, allowable burn time. These considerations, in conjunction, with anticipated fire detection and personnel response, lead to a determination of where/when installed suppression or passive protection is needed. The possible spread of fire to adjacent cable trays can also be considered using methodology similar to that used by FM in establishing their minimum tray separation distances. Data on heat/smoke release rates for the plant-specific cable is preferable for these evaluations, but the data in Sections 9.4 and 9.5 may suffice when that data is not available or when it is not a reliable indication of future facility modifications and usage.

In establishing maximum acceptable heat release rates or smoke generation rates or maximum allowable burn times, the potential thermal and non-thermal damage to critical facility equipment and circuits are often key considerations. Reports of test data for thermal damage to cables (Lee, 1981) indicate that "electrical integrity failure of cables is shown to be dependent on ... the nature of the products formed upon heat flux exposure ... not entirely dependent on insulation/jacket degradation." Lee and others observe failures due to corrosive combustion products (primarily HCl) even when the insulation/jacket is not significantly degraded. Nowlen (1989) reports that electrical failure is highly dependent upon the cable termination exposure. This suggests that a fire risk evaluation to determine when suppression or passive protection is needed should carefully consider these details; if a reliable quantitative determination cannot be made, prudence would suggest installation of suppression or passive protection as described in Sections 9.6 and 9.7 and the relevant system installation standards.

One indication of the complications inherent in predictions of adequate cable separation and circuit redundancy is the twenty-foot cable tray separation tests reported initially by Cline *et al.* (1983) and reviewed by Nowlen (1989). These tests were intended to investigate the adequacy of separating cable trays by a horizontal distance of 20 ft to prevent loss of electrical integrity of the cables in both trays. The indirectly exposed second cable tray was situated within two feet of the 10 ft high ceiling in the relatively small test enclosure. The direct exposure fire caused the failure of the cables in the first tray, and the ceiling layer that developed eventually caused the failure of the non-qualified (per IEEE-383, UL 1685) cables in the second tray, even when the cables had a fire retardant coating or a ceramic blanket. The coating and the ceramic blanket did prevent cable failure in tests involving IEEE-383 qualified cables. One implication of these results is that trade-off decisions between active and passive protection should explicitly compare the cable surroundings used in the testing versus the facility.

Industry practices in using the various guidelines recognize the need for plant-specific special considerations. For example, there are a wide variety of practices for the specific case addressed in NFPA 75, i.e. cables under a raised floor. Some new facilities have decided to only utilize CMP qualified cable, and do not install any detection or suppression systems under the floor. Other facilities that cannot remove older, more combustible cables under the floor, utilize installed suppression and detection systems. Of course, many facilities install smoke detection (as required by NFPA 75), and anticipate suppression will be achieved by a combination of de-energizing circuits and using portable fire extinguishers.

One large facility with cable runs in several tunnels has decided to rely on use of low flammability cable, early detection, and manual response by a full-time professional plant fire brigade equipped with smoke ejection apparatus. Many other facilities have adopted the FM guideline that manual firefighting is not feasible in a congested cable tunnel, and have installed automatic sprinklers.

9.9 Electronic equipment flammability and vulnerability

9.9.1 ELECTRONIC COMPONENT FLAMMABILITY

Statistics reported by Keski-Rahkonen et al. (1999) on electrical fires in European and US nuclear plants indicate that the failed electrical components most often involved in these fires are as follows: switches and circuit breakers: 26%, transformers 26%, contact terminals 18%, wire/cable 10%. The most common first ignited components are: transformer oil 28%, cable insulation 21%, switches/breakers 10%. The failed switches and circuit breakers allow shorts and arcing to ignite combustible cable insulation and electronic components often installed on printed circuit boards. Since cable flammability has already been covered in this chapter, and transformer fires are discussed in Section 9.10, this section will focus on the flammability of small electronic components on printed circuit boards. The larger-scale issue of electronic cabinet flammability is discussed in Section 9.9.2.

Keski-Rahkonen et al. (1999) conducted a series of electrically ignited fire tests on small printed circuit boards containing various capacitors, resistors, semiconductors, transistors, and diodes. The circuit board substrate was a fiberglass laminate. The circuit boards were subjected to over-currents in some tests, excess voltage in other tests, and reverse polarities is still other tests. Most tests resulted in some smoke generation followed by the mechanical failure of the electronic component or connector before flame broke out. Two tests with power transistors on 4.5×15 cm circuit boards did result in brief periods (25–70 seconds) of flaming. The peak heat release rates were about 450 W, corresponding to a peak heat release rate per unit surface area of $67 \, kW/m^2$. The latter value is quite small in comparison to heat release rates per unit area measured in cone calorimeter tests with more combustible polymers. It is also smaller than the lowest values shown in Figure 9.8 for cables that have passed the UL 1685 vertical tray test, and those in Table 9.4 for horizontal cable tray tests. Therefore, it seems that the primary hazard of these small electronic components and printed circuit boards is that an electrical fault could initiate a small fire that may in turn ignite either the more combustible wiring and cables or the plastic housing of the electronic equipment.

The plastics used for electronic equipment housings include Fire-Retarded Acrylonitrile-Butadiene-Styrene (FRABS), polycarbonate, PVC, and fire-retarded high-density polystyrene. Underwriters Laboratories standard 94 classifies these and other plastics with respect to the results of flammability tests on small laboratory size samples. Krock and Smith (1996) reported results of both UL 94 fire tests and calorimetry tests with computer monitors constructed from some of these plastics. The peak heat release rates for the computer monitors were in the range 12–40 kW. Based on an average surface area of $7 \, ft^2$ ($0.65 \, m^2$), the corresponding peak heat release rates per unit area are $18–62 \, kW/m^2$. Since these values are also lower than the heat release rates for most wires and cables, and since there can be extensive runs of wiring in electronic enclosures, it is clear that the wiring is usually the primary source of heat released in electronic equipment fires.

9.9.2 ELECTRONIC CABINET FLAMMABILITY

Electronic cabinets are common in control rooms, computer rooms, motor control centers, telecommunications centers, and other electrical areas. Most cabinets are made of sheet metal. Some of these cabinets are free-standing while others are placed on tables, control consoles, shelves, etc. Some of the cabinets have ventilation fans, while others have openings for natural ventilation.

Sandia National Laboratories (Nowlen, 1989) conducted a series of fire tests on ventilated electronic cabinets loaded with cables and ignited electrically by simulating a faulty terminal block connection generating 200 watts. Two of the three tests in the 0.38 m³ cabinets involved cables that had not passed the IEEE 383 (UL 1685) vertical tray test, while the third test involved qualified cables. Cabinet ventilation rates were 1 air change per hour in the two tests with non-qualified cables, and eight air changes per hour in the test with qualified cable. The fires developed slowly, but eventually resulted in the complete involvement of all the cable in the cabinets. Peak heat release rates of 800 kW to 1300 kW occurred at times ranging from 10 minutes after ignition to about 27 minutes after ignition. Smoke generation rates were not measured, but the report noted that the smoke accumulation under the 5.5 m (18 ft) high test enclosure ceiling totally obscured visibility within 6–15 minutes.

The VTT Building Technology group in Finland has conducted two series of fire tests in naturally ventilated 0.77 m³ electronic cabinets loaded with circuit boards, relays, connectors, wiring and cables. The first series of cabinet fire tests (Mangs and Keski-Rahkonen, 1994) was conducted with varying combustible loads (from 30–91 kg). The second series of tests (Mangs and Keski-Rahkonen, 1996) was conducted with a constant combustible loading (66 kg) to determine how heat release rates varied with the size of the ventilation openings, and the propensity for fire spread to adjacent cabinets. A propane burner was used to ignite the cables in the cabinets in both test series.

The VTT cabinet fires developed slowly and reached peak heat release rates (ranging from 50–380 kW) that depended on the cabinet effective ventilation area and the amount of electrical components that actually burned. The effective ventilation area was equal to the nominal open area of the cabinets plus the openings that developed because of burn damage and distortion during the tests. In some tests, gaps developed around the cabinet doors, and in at least one test the door opened because the locking mechanism failed. The percentage total mass burned varied from 10–56%, and the effective heats of combustion varied from 9–20 kJ/g.

Figure 9.14 shows the heat release rate smoothed data from a VTT electrical cabinet test with a door ventilation area, A_i, of 0.04 m², and a cabinet ceiling ventilation area, A_e, of 0.079 m². During the first 15 minutes of the fire, the heat release rate increases steadily as the fire spreads to the various electrical components. From 15 minutes to about 95 minutes, the heat release rate is limited by the air flow into the cabinet. The peak heat release rate of 180 kW lasts about 12 minutes in this test.

Mangs and Keski-Rahkonen (1994) derived an equation for the oxygen-limited peak heat release rate at flashover. Their derivation is based on the following assumptions. The cabinet has two openings; the lower opening has area A_i, and the upper opening has area, A_e. The vertical distance between the midpoints of the two openings is designated as H. The combustibles in the cabinet have a heat of combustion of 2.97 kJ/g-air, which is representative of many hydrocarbons. The combustion efficiency is about 50%, and the average temperature of combustion products in the cabinet is about 660 °C. Using these assumptions, and using SI units for H and the two areas, the equation for the peak heat release rate, $\dot{Q}_{max}$, is

$$\dot{Q}_{max} \approx 4.3 MW \sqrt{\frac{H}{\frac{3.3}{A_e^2} + \frac{1}{A_i^2}}} \qquad [9.9.1]$$

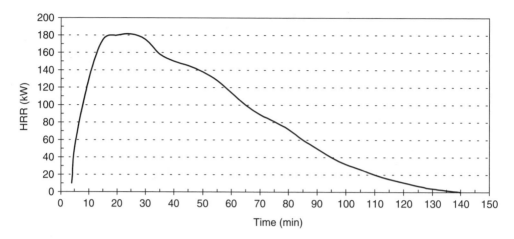

Figure 9.14. Heat release rate smoothed data from VTT electrical cabinet Test 3

The calculated value of $\dot{Q}_{max}$ based on equation [9.9.1] is in accord with the 180 kW shown in Figure 9.14, and is in approximate agreement with other VTT data with known ventilation areas. However, the reader is cautioned that equation [9.9.1] does not account for burning outside the cabinet, nor does it account for the higher heat release rates that result if/when the cabinet effective ventilation area increases during the fire. It also does not account for the significantly higher heat release rates that were observed with the fan-ventilated cabinets in the Sandia cabinet fire tests.

VTT also measured smoke release rates in their second round of cabinet fire tests. Peak smoke release rates varied from $10-34 \, m^2/s$. The corresponding peak heat release rates were in the range 100–180 kW. VTT determined that the cabinet wall temperatures were sufficiently high to ignite cable against the wall in an adjacent cabinet. They concluded that fire spread to the adjacent cabinet would occur during the time interval from 11–16 minutes after fire initiation in the first cabinet. They did not assess the impact of the smoke from the cabinet fire on any exposed electronic equipment.

9.9.3 ELECTRONIC EQUIPMENT VULNERABILITY

Some electronic equipment is vulnerable to failures caused by exposure to smoke or acid gases produced during electrical fires. Smoke particles suspended in air are attracted to electrically charged surfaces and tend to build conductive bridges between connectors. Therefore, exposed connectors represent potential failure sites for leakage currents and associated faulty signals. Sandia National Laboratories conducted tests to monitor the operation of various electronic components during exposure to smoke generated from heated fuels. Representative results reported by Tanaka *et al.* (2000) for components commonly used in digital instrumentation and control systems are summarized in Table 9.7.

One of the methods for preventing smoke contamination of circuit boards is to apply conformal coatings to the boards. These coatings are either sprayed, brushed, or vacuum deposited onto the boards. Tanaka *et al.* tested the effectiveness of the following coatings: acrylic, epoxy, parylene, polyurethane, and silicone. In general, the coatings did reduce the occurrence of smoke-induced leakage currents. The best results were obtained with parylene and polyurethane. It is not clear whether the application of these coatings significantly increases the combustibility of the circuit boards.

ELECTRICAL CABLES AND EQUIPMENT

Table 9.7. Results of smoke exposures to operating electronic components

Electronic component	Failure mode	Test results (Tanaka *et al.*, 2000)
Circuit Boards	Leakage Currents	Board resistance decreased upon exposure to smoke from high fuel loadings (200 g/m^3) and high humidity (75%).
Digital Connectors: series, parallel, Ethernet.	None	Smoke conductivity is too low ($<10^{-6}$ Siemens); to cause failures, even at high smoke concentrations.
Memory Chips: (SRAM, DRAM, & EPROM)	Leakage current, voltage drift.	Low voltage (3.3 V) SRAM chips experienced intermittent failures when exposed to a combination of high smoke concentration and humidity.
Hard Disks	None	No disc failures during two hours exposure at fuel loadings as high as 30 g/m^3 and humidity of 75%

When electronic components are exposed to halogenated combustion products, there is a threat of corrosion induced failures of exposed connectors and insulators. Corrosion of the contacts can cause an increase in circuit resistance. On the other hand, Chapin and Gandhi (1997) maintain that the predominant threat to electronic components is shorts and arcing caused by metal migration, electrolytically corroded conductors, and other electrochemical degradation processes. According to Chapin and Gandhi, these processes increase exponentially with relative humidity. Hence, there is a need to decrease the humidity to accelerate the recovery of electronic equipment damaged by corrosive smoke.

Reagor (1992) has reported on the experiences of the telecommunications companies in dealing with corrosive combustion products generated in central office fires involving PVC jacketed cables. The major problem is the formation of chloride layer (often zinc chloride formed from the reaction of HCl with the zinc in transistor chip coatings) of the surface of exposed switches. Table 9.8 shows how the level of damage increases with increasing chloride deposits, until equipment restoration is no longer feasible.

Thermal damage to telecommunications equipment depends upon the exposure temperature and duration. According to the NFPA 76 draft standard, most equipment can remain operational at long-term exposures up to 70 °C (158 °F). Permanent damage occurs when the wire insulation, circuit boards, etc. are heated to their polymer softening temperatures. This is far less likely to occur than smoke damage in telecommunications facility fires. In fact, NFPA 76 estimates that 95% of the fire damage in Telephone Central Office fires is due to smoke products, whereas only 5% is thermal damage.

Table 9.8. Damage induced on electronic switches by chloride contamination

Chloride deposit on equipment surface (micrograms/sq-inch)	Effect on electronic switch (from Reagor, 1992)
30–60	Normal accumulation over 20 year operating life
<200	Equipment can easily be restored to service
200–600	Can be restored to service providing a dry, smoke-free environment was established shortly after fire and maintained.
>600	Cleaning is no longer cost effective, so equipment should be replaced

9.9.4 DETECTION AND SUPPRESSION OF ELECTRONIC EQUIPMENT FIRES

Both fire protection engineers and personnel responsible for the management and operation of electronic equipment facilities agree on the importance of early, reliable fire detection systems. NFPA 76 stipulates that Very Early Warning Fire Detection (VEWFD) systems be used for the telecommunications equipment in a large facility (over 2500 ft^2), and that Early Warning Fire Detection be used in small facilities. The characteristics of both types of detection system are summarized in Table 9.9. Both the detector spacings and the sensor sensitivity settings are lower than the usual smoke detector sensitivity specifications for general use; thus providing earlier warning. The in-situ performance tests are taken from British Standard 6266 (1992).

Many telecommunications facilities use sensitive sampling smoke detection systems. The intent is to detect small concentrations of pyrolysis products at an incipient stage of a fire such that a buoyant plume has not yet developed. The smoldering fire specified for performance testing of VEWFD systems represents the type of smoldering fire that these systems should be able to detect.

Another approach to early detection of electrical fires involving PVC insulation is the use of a hydrogen chloride detector. HCl is generated at an early stage of PVC pyrolysis, when the PVC temperature is about 250 °C (Beyler, 1995). One type of commercial HCl detector marketed for this purpose utilizes a vibrating quartz crystal coated with a special zinc alloy. Upon exposure to HCl, a salt precipitate forms, increasing the crystal mass, and decreasing its vibration frequency. A microcontroller senses the change in vibration frequency and triggers an alarm. Although the same principle should be applicable to HF detection, apparently no commercial HF detector is currently being marketed for detecting fires of fluorinated insulation of fluorinated suppression agent decomposition.

Although there is widespread agreement on the need for early detection of electronic equipment fires, there is no such agreement on suppressing these fires. The prescriptive portion of the NFPA 76 draft standard does not require automatic suppression for telecommunications equipment. In the performance-based approach to choosing fire protection for such equipment, the draft standard emphasizes the importance of preventing fire spread to adjacent equipment. The Appendix of NFPA 76 describes fire tests in which automatic sprinklers successfully prevented fire spread from one switch cabinet to the cabinets in adjacent rows across a wiring aisle. However, there were copious quantities of smoke in the test enclosure, and the expectation is that this smoke would have entered the cabinets if they had been equipped with ventilation fans.

Gaseous suppression systems tripped by smoke detectors have the advantage of earlier system activation and fire extinguishment, and thereby far less smoke damage. These systems have been

Table 9.9. Early warning and very early warning fire detection systems per NFPA 76

Parameter	Early warning fire detection	Very early warning fire detection
Maximum Sensor or Port Spacing per prescriptive approach	400 ft^2	200 ft^2
Minimum Sensitivity Setting per prescriptive approach	Alarm: 1.5 % per ft obscuration	Alert: 0.2 % per ft obscuration Alarm: 1.0 % per ft obscuration
In-Situ Performance Test	Respond to a crucible fire of a mixture lactose and potassium chlorate within two minutes.	Respond to heated PVC wire (15 amp current) within 5 minutes; Fire size is <1 kW

used in many telecommunications facilities and other facilities with large quantities of electronic equipment. The primary hesitations in using these gaseous systems have been their expense, and their reliability in preventing re-ignition. The latter concern is that the gaseous agents do not cool the burning materials sufficiently to prevent re-ignition after the agent concentration has been reduced; e.g. after personnel are allowed to re-enter the room. Of course, at that time there would probably be personnel on site to manually suppress any re-ignited fire.

One common concern for both water-based and gaseous suppression systems is the threat of contaminating vulnerable electronic equipment with scale and residue in the suppression system piping. Hence, NFPA 76 requires that suppression system piping be cleaned and checked more rigorously than it would be for other applications. The general bias against automatic suppression systems on the part of management and operating personnel is reflected in the NFPA 76 Appendix comment "The high record of reliability [at telecommunications facilities] has been achieved mostly without the use of automatic suppression systems." A careful risk analysis would be needed to establish the accuracy and applicability of that assertion to specific facilities, considering the other modes of fire protection in place at that facility.

9.10 Transformer fire protection

9.10.1 TRANSFORMER GENERIC DESCRIPTION

The transformers of concern here are power and distribution transformers used either to increase voltage for transmission from a power plant, or to decrease voltage at a substation or industrial or commercial facility. These transformers are usually categorized in terms of their power rating in kVA, their voltage rating (usually the primary circuit voltage), and the dielectric material around the core. Power transmission transformers usually have power ratings from about 2500 kVA to as much as 100,000 kVA (100 MVA). Distribution transformers are usually rated from 10 kVA to 5000 kVA. Figures 9.15a and 9.15b are photographs of a substation transformer, and a smaller distribution transformer, respectively.

The major components of a transformer are the core, the windings, the tank/casing, the radiator, and the bushings. The transformer core is a ferromagnetic material that provides a path of high magnetic permeability from the primary circuit to the secondary circuit. The windings allow a secondary voltage to be induced from the a.c. voltage in the primary circuit. The tank/casing, which is usually a reinforced rectangular structure in these transformers, contains the dielectric fluid, the core and the windings. The radiator (finned structure in photographs) provides a heat

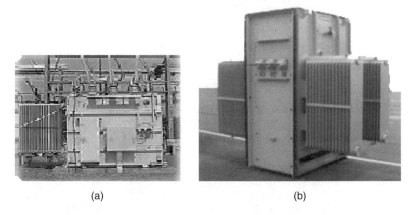

Figure 9.15. (a) Substation transformer, (b) distribution transformer

transfer path to dissipate the internal heat generated in the transformer. The bushings (located on top of the casing in Figure 9.15a and on the front face in Figure 9.15b) provide for the electrical connections to the primary and secondary circuits.

The types of dielectric materials used in transformers include air and insulating solids (dry-type transformer), mineral oil, a less-flammable liquid (fire point > 300 °C), or a nonflammable liquid, including askarels (generic name for chlorinated hydrocarbon insulating liquids). Dry-type transformers are often smaller and hotter than the liquid filled transformers, and are used more frequently in or near commercial buildings. Askarel transformers were used in many commercial buildings until environmental and public health regulations curtailed their production and use in large transformers. Askarels are generally composed of 60–70% polychlorinated biphenyls (PCB), which are now considered highly toxic and environmentally hazardous (they bio-accumulate in a variety of species). Since the phase out of askarels, other nonflammable and less flammable liquids have been developed for use in large transformers. The less flammable liquids include silicones, which have lower pool fire burning rates because of the production of silicon dioxide which forms a protective layer over the burning liquid. Mineral oil-filled transformers are usually the most flammable, and are primarily used in outdoor locations.

9.10.2 TRANSFORMER FIRE SCENARIOS

The most common fire scenario for liquid-filled transformers starts with degradation and breakdown of the high voltage windings insulation. A high-impedance, low-current fault develops in the windings as a result of the insulation breakdown. The turn-to-turn or layer-to-layer windings fault grows rapidly, and the coil impedance decreases. If the electrical fault protection does not rapidly isolate the transformer from the power grid, the fault current increases and causes arcing. The arc decomposes the dielectric fluid and generates a growing gas bubble as illustrated in Figure 9.16a. The bubble pressure and volume causes the liquid pressure to increase to the point that the rectangular tank deforms and begins to fail, often at a bottom corner (Figure 9.16b). Upward displacement of the transformer oil by the growing bubble causes increased stresses and deformation and rupture of the tank, violently ejecting transformer fluid and gaseous decomposition product. Ejection of hot windings conductor or insulation debris can ignite the released transformer oil and vapor. A spray fire and flowing liquid fire result.

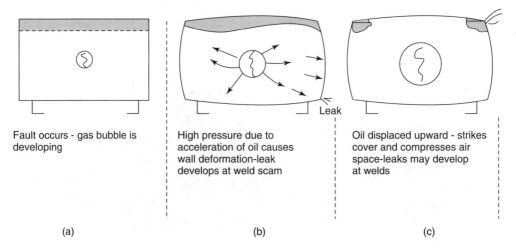

Figure 9.16. Arc initiated transformer tank failure sequence. Reprinted from NEMA, 1980 by permission of the National Electrical Manufacturers Association

A similar worst-case fire scenario involves an overload current that is sustained if the circuit breakers and/or fuses are inoperative for some reason. The sustained overload current gradually raises the temperature of the transformer oil. Heat transfer calculated results presented in the NEMA 1980 report indicate that a sustained current of 150% of the rated load, can cause the oil temperature to increase from about 50 °C to about 135 °C over a period of 3–4 hours. This elevated temperature causes insulation deterioration, possibly in the form of charring or tracking. Eventually a fault to ground develops along the charred insulation surface. This tracking fault can flash into a long, high current arc and generate a large high-pressure bubble as shown in Figure 9.16a. Sustained lower energy arcing in mineral oil will produce bubbles consisting primarily of hydrogen and ethylene (NEMA, 1980). Once these gases mix with any residual air in the tank ullage, the arc can also initiate a tank deflagration that will also readily cause tank rupture.

The NEMA 1980 report includes test data on the arc energies and associated pressures needed to cause tank failure. In the case of three-phase pad-mounted tanks rated for 25–100 kVA (i.e. small distribution transformer), tank ruptures occurred at spark energies of 800–1400 kW-sec discharged over about three cycles at 60 Hz. Arc energies in the range 100–750 kJ produced tank deformation, but not rupture, except in one case when a bushing blew out. Hydrostatic tests showed a yield strength of about 30 psig, and that tank weld seam failure occurred at a fluid pressure of 90 psig. The associated tank deformation at tank rupture was about 25% of the original 9.5 ft^3 tank volume. Report authors project that the spark energy needed to rupture larger tanks of similar strength should increase roughly as tank volume, e.g. a 20 cubic foot tank would rupture at spark energies of 1600–2800 kJ.

Another fire scenario involves a leaky indoor transformer casing (perhaps due to corrosion) that eventually produces a growing pool of transformer fluid on the floor. If the liquid or vapor reaches an external ignition source (perhaps a pilot light on a gas-fired appliance in an adjacent area), a pool fire will result. Even if there isn't any nearby ignition source, the loss of dielectric fluid can cause an internal fault to develop in the transformer, and cause arcing at either the bushings or the circuit breakers protecting the transformer. According to the NEMA 1980 report, there have been many transformer failures of this nature, and there would have been many more fires if it were not for the widespread use of askarel transformers for almost fifty years.

9.10.3 TRANSFORMER FIRE INCIDENTS

There are several incident reports consistent with the first (internal arcing) fire scenario. One such incident in 1988 occurred at a 715 MVA main generator power transformer at a utility power plant. On the day prior to the fire, the transformer high temperature alarm actuated. Upon checking, a winding temperature of 110 °C, an oil temperature of 88 °C, and an oil pressure of 8 psig were noted. The normal values at full load would be about 95 °C winding temperature, about 80 °C oil temperature, and about 2 psig oil pressure. The transformer was taken out of service for the usual nightly reduction in utility load, and was re-energized the next morning. About three hours after being restarted, the transformer tank ruptured around its base and became engulfed in flames.

The immediate cause of the transformer rupture was a phase-to-ground fault that resulted from the prior overheating and insulation breakdown of the windings. The transformer circuit breaker tripped, but not in time to prevent the massive arcing that occurred in the transformer tank. A water deluge system for the transformer was manually tripped about two minutes after tank rupture, but the water supply pipe for the deluge system had cracked due to the movement of the transformer when it ruptured. Therefore, the deluge system was ineffective, and the fire had to be extinguished by the responding firefighters approximately thirty minutes after their arrival. Since a replacement transformer had to be shipped and installed, electric service from that plant unit was down for about a month.

In a 1978 incident, time delay fuses at a 1000 kVA (1 MVA) utility distribution transformer failed to prevent arcing and tank rupture. Burning oil from the ruptured transformer flowed from the transformer concrete pad through an open gutter into the switchgear room of an adjacent building. Automatic sprinklers in the switchgear room extinguished the resulting fire.

Several transformer fire incidents were initiated by bushing failures. One series of such incidents occurred at a Canadian hydroelectric plant described by Kennedy (1974). The powerhouse had 19 MVA transformers in individual three-sided vaults. Each vault was protected by a heat detector activated water deluge system as illustrated in Figure 9.17. There were four detectors and three spray nozzles in each vault. The nozzles discharge 95 gpm at a pressure of 75 psig.

The transformer bushings are porcelain and contain a dielectric oil similar to mineral oil. Defects in the ceramic can cause leakage currents that produce carbon tracks on the bushings. Tracking faults on the bushing cause further degradation and lead to arcing. When this happened on one of the 19 MVA transformers, the tank cover weld seam ruptured and allowed oil to discharge and be ignited by the bushing tracking current. The heat detectors tripped the deluge system and the fire was extinguished by the time utility personnel reached the unattended plant. However, the combination of fire exposure and water spray cooling apparently caused a hairline crack on another bushing. When the generators and transformers were put back in service, the second bushing cracked and released oil. One day later, a third bushing failed and caused another arc-induced tank rupture and fire. The deluge system also extinguished this fire in the vault, but the burning oil flowed out of the vault into the adjacent corridor where it was extinguished by firefighters about 90 minutes after ignition. There was extensive smoke damage throughout the powerhouse.

Sometimes manual firefighting for transformer vaults is much more difficult. Eisner (1983) described a transformer fire in a substation in a sub-basement of a Manhattan office building. The transformer fire was started by a short circuit caused by water flooding from a ruptured water main. When the mineral oil was released from the ruptured transformer casing, it floated and burned on the water. For some reason, the transformers in the vault could not be de-energized remotely for several hours. Several attempts to extinguish the fire with fog lines were unsuccessful. Eventually foam was pumped into the vault through holes drilled in the roof. After applying almost 3000 gallons of foam over a period of several hours, the mineral oil fire was extinguished.

Firefighter fear of applying water spray to a possibly energized transformer was a key factor in a 1997 fire involving a 138–69 kV step-down transformer at a Texas utility. Although the transformer was de-energized within 10 minutes of ignition, there was a nearly two hours delay to confirm the de-energizing of every circuit around the transformer. After receiving this confirmation, firefighters applied water spray to cool the transformer and foam to the burning transformer oil in the containment area. The fire was extinguished after 27 minutes of water/foam application, but it re-ignited within the transformer about 10 hours later, and was not completely extinguished until the mineral oil was completely drained and the transformer tank was flooded with water and AFFF. The transformer was a $2 million total loss.

Transformer fire smoke damage is sometimes exacerbated by PCB contamination. NIOSH Publication 86-111 lists seven transformer fires between 1981 and 1984 involving PCB contamination. The most well known incident occurred in the basement of a state office building in Binghamton, New York. A switchgear fire adjacent to an askarel filled transformer cracked one of the transformer bushings and released approximately 180 gallons of askarel. Askarel decomposition products were carried by the smoke through a vertical shaft and ducting, and contaminated the entire building. Despite extensive and expensive cleanup, the building remains unoccupied.

Most of the large askarel transformers have been either replaced or converted to other dielectric fluids. If the replacement reduces PCB levels to less than 50 ppm, environmental regulators in most countries allow the transformers to remain in service. Furthermore, small, sealed transformers

ELECTRICAL CABLES AND EQUIPMENT 331

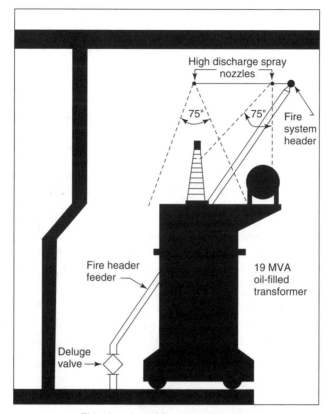

Elevation view of fire protection system

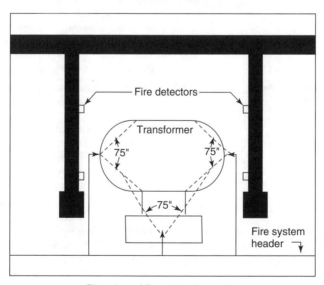

Plan view of fire protection system

Figure 9.17. Transformer vault fire protection layout (from Kennedy, 1974)

containing askarels can remain in service indefinitely as long as they are inspected for leakage and emergency responders are informed.

9.10.4 INSTALLATION AND FIRE PROTECTION GUIDELINES

Transformer installation and fire protection guidelines are given in the National Electrical Code (NFPA 70) and NFPA 850. The NEC requirements are summarized in Table 9.10. Depending on the dielectric fluid and voltage rating, there are various separation, liquid confinement, and/or fire resistance requirements. In many indoor applications, the transformer must be installed in a vault. NEC requirements for the transformer vault include a minimum three hour fire resistance, automatic closing fire dampers on all ventilation openings, and drainage provisions for a combined transformer rated capacity greater than 100 kVA.

Most outdoor installations require either a fire-resistant barrier or a trench filled with crushed stone. The crushed stone significantly reduces the burning rate of an ignited transformer oil release. Laboratory tests described by Zalosh and Lin (1995) showed that the presence of a gravel bed reduced the average burning rate of transformer oil to 1/8th of its value without the gravel. This effect becomes more pronounced as the oil level approaches the top of the gravel.

NFPA 850 contains minimum separation guidelines for outdoor oil-filled transformers at power plants. Reductions in minimum required separations due to firewalls are also offered. Minimum fire resistance ratings for indoor installations are also provided.

9.10.5 WATER SPRAY PROTECTION OF TRANSFORMERS

Section 4-5.4 of NFPA 15 describes water spray protection designs for transformers. The minimum required flow rates are 0.25 gpm/ft^2 (10.2 l/min-m^2) of projected rectangular prism area

Table 9.10. National electric code fire protection requirements for transformers (from NEC Article 450)

Dielectric material	Power rating (kVA)	Voltage rating	Indoors or outdoors	NEC requirements
Dry-Type	<112.5	>600 V	Indoors	>1 ft separation from combustibles
	>112.5	Any		Install in a room with a minimum fire rating of 1 hour
	Any	>35,000 V		Install in a room with a minimum fire rating of 1 hour
	>112.5	Any	Outdoors	>1 ft separation from combustibles
Less Flammable Liquid (fire point > 300 °C)	Any	<35,000 V	Indoors	Liquid-confinement area or Automatic suppression system Or in a vault
		>35,000 V	Indoors	In a vault
	Any	Any	Outdoors	Need either spatial separation, a fire resistant barrier (a dike, curb, or stone-filled trench), or an automatic water spray system.
Nonflammable Liquid	Any	<35,000 V	Indoors	Liquid-confinement area
		>35,000 V		In a vault
Askarel	Any	>35,000 V	Indoors	In a vault
Oil	<10	<600	Indoors	Fire barrier or suppression system
	Any	>600	Indoors	In a vault
	Any	Any	Outdoors	Need either spatial separation, a fire resistant barrier (a dike, curb, or stone-filled trench), or an automatic water spray system

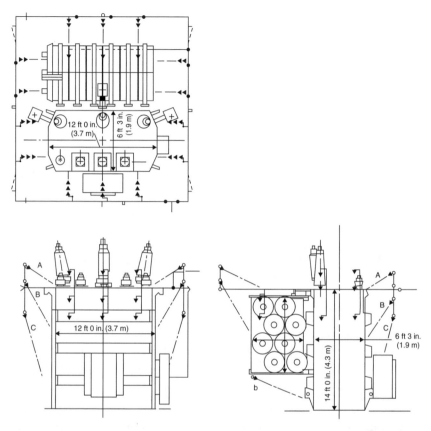

Figure 9.18. Water spray nozzle layout for transformer protection. Reprinted with permission from NFPA 15 *Water Spray Fixed Systems for Fire Protection*. Copyright © 1996 National Fire Protection Association, Quincy, MA 02269. This reprinted material is not the complete and official position of the National Fire Protection Association, on the referenced subject which is represented only by the standard in its entirety

enveloping the transformer, and 0.15 gpm/ft^2 (6.1 l/min-m^2) of non-absorbing ground surface area of exposure. The standard also specifies that water piping not be run directly over the top of the transformer, and that specified clearances be maintained from energized un-insulated components. It also states that the water spray should not impinge directly on energized bushings or lightning arrestors. Figure 9.18 shows a sample nozzle layout consistent with these requirements.

Detailed design of nozzle layout requires knowledge of how the spray coverage pattern varies with nozzle orientation angle (relative to vertically down) and with wind speed. Nozzle manufacturers usually provide data on spray coverage patterns and distances as a function of nozzle orientation. Lin (1991) has investigated how the recommended spray projection distances might be affected by wind speed and fire size. The affects of fire plumes and wind speeds are quite sensitive to spray droplet size and momentum. For a representative case of a 0.8 mm diameter drop from a 40 psig nozzle, Lin calculated a reduction in projected spray distance of 55% when the wind speed increased from 5–15 m/s.

A member of the NFPA Committee responsible for transformer fire protection recommendations in NFPA 850 has provided the following perspective on the use of water spray protection for transformers (Hathaway, 1997). "The benefit of such systems under catastrophic conditions has been debated. Some contend that the violent nature of a transformer failure renders automatic water spray protection useless. Although this may have occurred, in many cases the automatic water spray systems did survive explosions and were credited with controlling fire, limiting damage, and minimizing plant downtime." The incident narratives in Section 9.10.3 of this chapter are consistent with Hathaway's conclusions. In one incident the transformer explosion did rupture the water spray piping, but in the Canadian hydroelectric plant incidents the water spray system was indeed effective.

The General Services Administration (1980) conducted tests to measure the electrical shock hazard associated with the use of water spray on energized transformers in a room or vault. A 225 kVA 3-phase dry transformer was installed in a 300 ft^2 test room equipped with two ceiling sprinklers. The transformer was connected to a 375 volt phase-to-phase, 210 volt phase-to-ground power supply with 300 ampere rated circuit breakers. Three ammeters were set up at different levels in the room to measure the current through 475 ohm resistors simulating the electrical resistance through wet people in the room. Various faults were induced in the transformer and water flows of 32 gpm were supplied to each of the two sprinklers. In eight of the nine tests, the measured currents through the people-simulating resistors were all less than the 10 milliamps considered to represent a serious shock hazard because at least some people can no longer let go of an energized object at that current. In the last test, the fault current was sustained for a longer period of time and two of the resistors did measure currents of 10 and 100 milliamps before the power was shut down.

GSA concluded "an accidental sprinkler discharge in conjunction with various electrical faults in the electrical equipment poses minimal shock potential to personnel within the room. A phase-to-ground fault in a transformer will induce large current flows to ground which in themselves will be dangerous to personnel coming in contact with it irregardless of a sprinkler discharge." GSA does recommend using automatic sprinkler protection for transformer rooms and for switchgear rooms.

Another personnel hazard associated with the use of water sprays on transformer oil fires is the potential for a temporary flare-up and splattering of hot oil. This is primarily a concern for manual firefighting using water spray nozzles. Heard (1980) investigated the effectiveness of various manually applied extinguishing agents on fires involving two high flash point oils. One liquid was a high molecular weight hydrocarbon with a fire point of 312 °C (RTEmp), and the other was a polydimethylsiloxane with a fire point of 343 °C. When a water spray nozzle attached to a hose line was used to extinguish the 4 ft diameter pool fires, there were flare-ups when the spray reached the burning liquids. Both fires were successfully extinguished within 3–7 seconds of water spray application. No flare-ups or splattering occurred when the fires were extinguished with a dry chemical (Purple K) portable extinguisher. The dry chemical application does leave a residue that requires cleanup.

One additional consideration in the use of water spray on transformer fires is the need for drainage and containment of the oil-contaminated water runoff. The expenses associated with provisions for drainage/containment and cleanup as well as water spray system installation need to be compared to the expenses associated with the increased fire damage to the transformer and surrounding equipment if no fixed fire suppression system is installed. In the case of the 1997 Texas utility transformer fire, the absence of an installed water spray system, together with the delay in manual application of water spray, caused the complete loss of a very expensive transformer.

References

Alexander, J., Bogan, D., Eaton, H., Greene, J., Paholski, J., Sheinson, R., Smith, W., Tatem, P., Wang, H., Williams, F., Henderson, C., Polito, P., Stone, J. and Mc Lain, W., 'Navy Electrical Cable Full-Scale Fire Tests of 1985 at Mobile, Alabama,' Naval Research Laboratory Memorandum Report 6395, February 1989.

ANSI/UL 1581-1991, 'Reference Standard for Electrical Wires, Cables, and Flexible Cords,' Underwriters Laboratories, 1991.

ANSI/UL 1666-1997, 'Standard Test for Flame Propagation Height of Electrical and Optical Fiber Cable Installed Vertically in Shafts,' Underwriters Laboratories, 1997.
Babrauskas, V., Peacock, R., Braun, E., Bukowski, R. and Jones, W., 'Fire Performance of Wire and Cable: Reaction-to-Fire Tests – A Critical Review of the Existing Methods and of New Concepts,' NIST Technical Note 1291, 1991.
Badger, S., Large-Loss Incidents, *NFPA Journal*, November/December 1999.
Barnes, M., Briggs, P., Hirschler, M., Matheson, A. and O'Neil, T., A Comparative Study of the Performance of Halogenated and Non-Halogenated Materials for Cable Applications. Part 11 Tests on Cables, *Fire and Materials*, **20**, 17–37, 1996.
British Standard 6266, 'Fire Protection for Electronic Data Processing Installations,' 1992.
Chapin, J., Tan, C., Willis, A., Pye, K., Hoover, J. and Caudil, L., Full-Scale Fire Tests of LAN Data Communication Cables Used in Concealed Space Applications, *The Loss Prevention Council Fire Safety Seminar*, 1999. (See also same title paper by J. Hoover et al. in *Fire Risk and Hazard Assessment Research Application Symposium*, National Fire Protection Research Foundation, June 1997.)
Chapin, J.T. and Gandhi, P., Comparison of LAN Cable Smoke Corrosivity by U.S. and IEC Test Methods, *Proceedings National Fire Protection Research Foundation Fire Risk & Hazard Assessment Research Application Symposium*, 1997.
Chavez, J. and Lambert, L., 'Evaluation of Suppression Methods for Electrical Cable Fires,' NUREG/CR-3656, SAND83-2664, 1986.
Cline, D., von Riesemann, W. and Chavez, J., 'Investigation of Twenty-Foot Separation Distance as a Fire Protection Method as Specified in 10 CFR, Appendix R,' Sandia Report SAND83-0306, NUREG/CR-3192, 1983.
Commission of the European Communities, 'Classes of Reaction to Fire Performance for Power, Control, and Communication Cables,' Commission Decision RG N208, Annex, September 2000.
Eisner, H., Fire Beneath the Streets, *Firehouse*, November 1983.
Factory Mutual Loss Prevention Data Sheet 5-31/14-5, 'Cables and Bus Bars,' Revised 1998.
Factory Mutual Research Specification Test Standard 3972, 'Cable Fire Propagation,' 1989.
Fustich, C., 'Transformer Room Fire Tests,' General Services Administration Technical Report under project No. PB 79-08, 1980.
Gandhi, P., Przybyla, L. and Grayson, S., Electric Cable Applications, Chapter 16 in *Heat Release in Fires*, Elseveir, 1992.
Gandhi, P., Using Calorimeter Data and a Zone Model to Predict Smoke Obscuration in Room Fires – A Parametric Study, *Fire Safety Journal*, **20**, 115–133, 1993.
Gustafson, N.-E., Sprinkler Tests in a Cable Duct in Rautaruuki Oy's Factory in Brahestad 23-24.9.1975, *Industribrand Meddelanden*, 18–24, 3/1976.
Hathaway, L., Electric Generating Plants, *NFPA Fire Protection Handbook*, Eighteenth Edition, p. 9–213, 1997.
Heard, D., 'Study of Fire Extinguishment of Transformer Fluids,' FMRC J.I. 0F0N1.RG, prepared for Department of Transportation, 1980.
Hill, J., 'Fire Tests in Ventilated Rooms Extinguishment of Fires in Grouped Cable Trays,' EPRI NP-2660, Project 1165-1, December 1982.
'Hinsdale Central Office Fire,' Joint Report of the State Fire Marshall and Illinois Commerce Commission Staff, prepared by Forensics Technologies International Corporation, March 1989.
Hshieh, F.Y., Shielding Effects of Silicon-Ash Layer on the Combustion of Silicone and Their Possible Applications in the Fire Retardancy of Organic Polymers, *Fire and Materials*, **22**, 69–76, 1998.
IEC 332, 'Tests on Electric Cables under Fire Conditions,' International Electrotechnical Commission. 1989.
IEC 1034, 'Measurement of Smoke Density of Electric Cables Burning under Defined Conditions,' International Electrotechnical Commission. 1991.
Industrial Risk Insurers, Cable Trays: Confining the Cables and Losses, *The Sentinel*, Second Quarter, 1988.
Kennedy, K., Fire-Fighting Experience with Indoor Transformers, *Canadian Electrical Association*, 1974.
Khan, M., 'Electrical Failure of Wires Inside 1-Inch Conduit Under Simulated Fire Conditions,' FMRC J.I. 0H4R4.RC, 1984.
Krock, R. and Smith, G., Performance Based Fire Classifications for Plastics Used in Electronic Enclosures, *Proceedings of the 1997 Fire Risk and Hazard Assessment Research Application Symposium*, National Fire Protection Research Foundation, 1997.
Lathrop, J., World Trade Center Fire, *Fire Journal*, July 1975.
Lin, W.-H., 'Analysis of Fire Detection and Suppression Systems for Outdoor Oil-Filled Transformers,' M.S. Thesis, WPI, 1991.
Mangs, J. and Keski-Rahkonen, O., 'Full-Scale Fire Experiments on Electronic Cabinets,' VTT Publication 186, 1994.
Mangs, J. and Keski-Rahkonen, O., 'Full-Scale Fire Experiments on Electronic Cabinets, II,' VTT Publication 269, 1996.
NFPA 15, 'Standard for Water Spray Fixed Systems for Fire Protection,' 1996 Edition.
NFPA 70, 'National Electrical Code,' 1999 Edition, National Fire Protection Association, 1999.
NFPA 75, 'Standard for the Protection of Electronic Computer/Data Processing Equipment' (1999 Edition).

NFPA 76, 'Draft Standard for the Protection of Telecommunications Facilities,' 1999.
NFPA 262, 'Standard Method for Test for Fire and Smoke Characteristics of Wires and Cables,' National Fire Protection Association, 1994.
NFPA 850, 'Recommended Practice for Fire Protection for Electric Generating Plants and High Voltage Direct Current Converter Stations,' National Fire Protection Association, 1996.
NEMA Fluid-Filled Transformer Flammability Study Committee, 'A Report on the Fire Safety Test Methods and Performance Criteria for Transformers Containing PCB Replacement Fluids,' National Electrical Manufacturers Association Report for National Bureau of Standards and Department of Energy, 1980.
New York Board of Fire Underwriters Report, 'One World Trade Center Fire,' 1975.
Newman, J. 'Fire Tests in Ventilated Rooms, Detection of Cable Tray and Exposure Fires,' EPRI NP2751, February 1983.
NIOSH Publication No. 86-111, 'Polychlorinated Biphenyls (PCBs): Potential Health Hazards from Electrical Equipment Fire or Failures,' National Institute of Occupational Safety and Health, February 1986 (available on Web at www.cdc.gov/niosh/86111_45.html).
Nowlen, S., 'A Summary of Nuclear Power Plant Fire Safety Research at Sandia National Laboratories, 1975–1987,' NUREG/CR-5384, SAND89-1359, 1989.
Reagor, B., Smoke Corrosivity: General, Impact, Detection, and Protection, *J. Fire Sciences*, **10**, 169–179, 1992.
Riches, W., 'Full Scale Horizontal Cable Tray Fire Tests,' Fermi National Accelerator Laboratory Report September 1988.
Robinson, P., Copper versus Fiber, *Cabling Systems*, February 2000.
Schultz, N., Fire Protection for Cable Trays in Petrochemical Facilities, *1985 AIChE Loss Prevention Symposium*, Paper 59E, 1985.
Skjordahl, J. and Metes, W., 'Fact-Finding Report on Comparative Flame Propagation and Smoke Development Tests on Communication Cables in Various Test Geometries,' Underwriters Laboratories Project 76NK8188, 1978.
Sugawa, O. and Handa, T., Experiments on the Behavior of Telephone Cables in Fires, *Proceedings of the 2nd Intl. Symposium on Fire Safety Science*, pp. 781–790, Hemisphere Publishing, 1989.
Sumitra, P., 'Categorization of Cable Flammability Intermediate-Scale Fire Tests of Cable Tray Installations,' EPRI NP-1881, Project 1165-1, August 1982.
Tanaka, T., Nowlen, S., Korsah, K., Wood, R. and Antonescu, C., Technical Findings on the Impact of Smoke Exposure for Digital I & C, *Proceedings International Topical Meeting on Nuclear Plant Instrumentation, Controls, and Human-Machine Interface Technologies*, November, 2000.
Tewarson, A. and Khan, M., 'Electrical Cables – Evaluation of Fire Propagation Behavior and Development of Small-Scale Test Protocol,' FMRC J.I. 0M2E1.RC, January 1989.
Tewarson, A., *SFPE Handbook of Fire Protection Engineering*, 2nd Edition, p. 3–80, 1995.
Thomasi, G., The Basics of Fiber Optic Cable Design, *Cabling Systems*, March/April 2000.
UL Subject 2196, 'Standard for Test of Fire Resistive Cables.'
Zalosh, R. and Lin, W.-H., Effects of a Gravel Bed on the Burning Rate and Extinguishment of High Flash Point Hydrocarbon Pool Fires, *Proceedings of the 29th Annual Loss Prevention Symposium*, Paper 4E, 1995.

APPENDIX A: FLAME RADIATION REVIEW

The following is offered as a review of the essential equations and data required for engineering estimates of flame radiation effects on surrounding materials and property. Simplifying assumptions associated with the equations and data are summarized, but it is assumed that the reader has previously seen, or has access to, cited references containing equation derivations and descriptions of the experiments conducted to obtain the data.

A common fire protection engineering problem is the calculation of the radiant heat flux impinging on a target at some distance from a flame. The problem is composed of the following three parts: (1) flame radiant emissive power calculations; (2) flame height calculations; and (3) radiation configuration factor determinations.

A.1 Flame emissive power

The emissive power, E, of a flame is the radiant energy emitted per unit time per unit flame surface area. Several alternative approaches have been used to calculate flame emissive powers.

If the flame can be approximated as a radiant grey body (radiant intensity independent of wave length), E is given by

$$E = \varepsilon \sigma T_f^4 \qquad [A.1]$$

where ε is the flame emissivity, σ is the Stefan–Boltzmann constant (5.67×10^{-11} kW/m$^2 \cdot$ K^4), T_f is the flame radiation temperature (K), and E is the flame emissive power (kW/m^2).

Strictly speaking, flames are not grey bodies since gaseous combustion products emit radiation at discrete spectral bands. However, equivalent grey body emissivities of volumes of hot gaseous combustion products have been determined by Hottel and can be found, for example in Figure 2.32 of Drysdale (1985). Emissivities for luminous, sooty flames are often approximated by the equation

$$\varepsilon = 1 - \exp(-kL) \qquad [A.2]$$

where k is the effective emission/absorption coefficient (m^{-1}), and L is the mean equivalent beam length of the flame (m).

Mean equivalent beam lengths depend slightly on flame geometry (Drysdale, 1985, Table 2.8), but are approximately equal to the flame radius. Published data on diffusion flame emission/absorption coefficients are listed in Table A.1. Reported values for nominally the same material differ by as much as a factor of four. These differences are important for flames with

Table A.1. Effective flame emission/absorption coefficients, k, and radiation flame temperatures, T_f

Material	$k(m^{-1})$	$T_f(K)$	Reference[a]
Benzene	2.6	1190	Mudan (1984), Drysdale (1985)
Butane	2.7	–	Mudan (1984)
Diesel Oil	0.43	–	Drysdale (1985)
Ethyl Alcohol	0.37	–	Mudan (1971)
Furniture (assorted)	1.13	–	Drysdale (1985)
Gasoline	2.0	1240	Mudan (1984)
Gasoline	–	1300	Drysdale (1985)
Hexane	1.9	–	Mudan (1984)
Hydrogen (liquid)	7.0	–	Mudan (1984)
JP-4	–	1200	Mudan (1984)
Kerosene	2.6	1600	Mudan (1984)
Kerosene	–	1260	Drysdale (1985)
LNG	3.0	1500	Mudan (1984)
Methanol	4.6	1500	Mudan (1984), Drysdale (1985)
PMMA	0.5	–	Drysdale (1985)
PMMA	1.3	1400	deRis (1979)
PMMA	1.5	1260	Orloff (1981)
Polypropylene	1.8	1350	deRis (1979)
Polystyrene	1.2	–	Drysdale (1985)
Polystyrene	5.23	1190	deRis (1979)
Wood Cribs	0.5, 0.8	–	Drysdale (1985)
Xylene	1.2	–	Mudan (1984)

[a]Data attributed to Mudan (1984) were compiled primarily by Attalah and Allan (1971). Data attributed to Drysdale (1985) were compiled from a variety of sources cited in his text

beam lengths on the order of one meter or less, but are insignificant for larger flames since the emissivity is effectively equal to one.

Radiation flame temperatures needed in equation [A.1] are also material dependent. As indicated in Table A.1, flammable liquid pool fire diffusion flame temperatures are reported to be in the range 1190–1600 °K (1680–2420 °F), while solid polymer diffusion flame temperatures are in the narrower range 1190–1400 °K (1680–2060 °F). The corresponding emissive powers for polymers are 120–220 kW/m² (10–20 Btu/sec-ft²). Under-ventilated fires would be expected to produce lower effective flame temperatures because of heat absorption by soot.

In situations where reliable flame temperature or emission/absorption coefficient data are not available, flame emissive powers may be calculated from flame heat release rate data and flame surface area estimates. In these cases E is

$$E = Q_r/A_f \qquad [A.3]$$

where Q_r is the radiant heat release rate (kW), and A_f is the flame surface area (m²).

The radiant heat release rate, Q_r, is often expressed as

$$Q_r = \chi_{rad} Q = \chi_{rad} M H_c \qquad [A.4]$$

where χ_{rad} is the ratio of the radiative heat release rate to the theoretical heat release rate, M is the fuel mass burning rate (g/s), H_c is the theoretical heat of combustion (kJ/g), and Q is

Table A.2. Chemical, convective and radiative combustion efficiencies (data from Chapter 3-4 of *SFPE Handbook*, 1995)

Material	H_c	χ	χ_{conv}	χ_{rad}
Gases				
Ethane	47.5	0.99	0.79	0.20
Propane	46.4	0.95	0.68	0.27
Butane	45.7	0.95	0.68	0.27
Ethylene	47.2	0.91	0.59	0.32
Propylene	45.8	0.89	0.50	0.39
1,3-Butadiene	44.6	0.74	0.34	0.40
Acetylene	48.2	0.76	0.37	0.39
Liquids				
Heptane	44.6	0.93	0.59	0.34
Octane	44.4	0.92	0.61	0.31
Benzene	40.1	0.69	0.28	0.41
Styrene	40.5	0.67	0.27	0.40
Methanol	20.0	0.95	0.81	0.15
Ethanol	26.8	0.97	0.73	0.24
Isopropanol	30.2	0.97	0.73	0.24
Acetone	28.6	0.97	0.73	0.24
Methyl Ethyl Ketone	31.5	0.97	0.67	0.24
Polydimethyl Siloxane	25.1	0.61	0.51	0.10
High MW Hydrocarbons	43.9	0.84	0.56	0.28
Solids				
Red Oak	17.7	0.70	0.44	0.26
Douglas Fir	16.4	0.79	0.49	0.30
Pine	17.9	0.69	0.49	0.21
Polyoxymethyene	15.4	0.94	0.73	0.21
Polymethylmethacrylate	25.2	0.96	0.66	0.30
Polyethylene	43.6	0.88	0.50	0.38
Polypropylene	43.4	0.89	0.52	0.37
Polystyrene	39.2	0.69	0.28	0.41
Silicone	21.7	0.49	0.34	0.15
Polyester	32.5	0.63	0.33	0.30
Epoxy	28.8	0.59	0.30	0.30
Nylon	30.8	0.88	0.53	0.35
Polyethylene-25%-Cl	31.6	0.72	0.32	0.40
Polyethylene-36%-Cl	26.3	0.40	0.24	0.16
Polyethylene-48%-Cl	20.6	0.35	0.19	0.16
Polyvinyl chloride	16.4	0.35	0.19	0.16
Fluoropolymers	5.3	0.32	0.17	0.15
Flexible Polyurethane Foams				
GM 21 (29 kg/m^3)	26.2	0.68	0.33	0.35
GM 23 (FR 29 kg/m^3)	27.2	0.70	0.38	0.32
GM 25 (44 kg/m^3)	24.6	0.69	0.29	0.40
GM 27 (FR 44 kg/m^3)	23.2	0.71	0.33	0.38

(*continued overleaf*)

Table A.2. (*continued*)

Material	H_c	χ	χ_{conv}	χ_{rad}
Rigid Polyurethane Foams				
GM 29 (35 kg/m^3)	26.0	0.63	0.26	0.37
GM 31 (FR 32 kg/m^3)	25.0	0.63	0.28	0.35
GM 35 (64 kg/m^3)	28.0	0.63	0.28	0.35
GM 37 (320 kg/m^3)	28.0	0.64	0.31	0.33
GM 41 (36 kg/m^3)	26.2	0.60	0.22	0.38
GM 43 (33 kg/m^3)	22.2	0.67	0.29	0.38
Polystyrene Foams				
GM 47 (16 kg/m^3)	38.1	0.68	0.30	0.38
GM 49 (FR 16 kg/m^3)	38.2	0.67	0.26	0.41
GM 51 (34 kg/m^3)	35.6	0.69	0.29	0.40
GM 53 (29 kg/m^3)	37.6	0.69	0.30	0.39

Table A.3. Theoretical unit heat release rates for fuels burning in the open

Commodity	Heat release rate (Btu/sec)
Flammable Liquid Pool	290/ft^2 of surface
Flammable Liquid Spray	2000/gpm of flow
Wood Pallets (Single Stack)	1000/ft of height
Wood or PMMA (Vertical)	
2 ft Height Burning	30/ft of width
4 ft Height Burning	70/ft of width
8 ft Height Burning	180/ft of width
12 ft Height Burning	300/ft of width
Wood of PMMA	
Top of Horizontal Surface	65/ft^2 of surface
Solid Polystyrene (Vertical)	
2 ft Height Burning	65/ft of width
4 ft Height Burning	150/ft of width
8 ft Height Burning	400/ft of width
12 ft Height Burning	680/ft of width
Solid Polystyrene (Horizontal)	120/ft^2 of surface
Solid Polystyrene (Vertical)	
2 ft Height Burning	45/ft of width
4 ft Height Burning	100/ft of width
8 ft Height Burning	280/ft of width
12 ft Height Burning	470/ft of width
Solid Polypropylene (Horizontal)	70/ft^2 of surface

Table A.4. Theoretical unit heat release rate for commodities burning in the open (compiled by Dr G. Heskestad)

Commodity	Heat Release Rate (Btu/sec per ft^2 of floor area)
Wood Pallets	
Stack 1-1/2 ft High (6-12% Moisture)	125
Stack 5 ft High (6-12% Moisture)	460
Stack 10 ft High (6-12% Moisture)	940
Stack 16 ft High (6-12% Moisture)	1500
Mail Bags, Filled, Stored 0.5 ft High	35
Cartons, Compartmented, Stacked 15 ft High	150
PE Letter Trays, Filled, Stacked 5 ft High on Cart	750
PE Trash Barrels in Cartons, Stacked 15 ft High	175
PE-Fiberglass Shower Stalls in Cartons, Stacked 15 ft High	125
PE Bottles in Compartmented Cartons, Stacked 15 ft High	550
PE Bottles in Cartons, Stacked 15 ft High	175
PU insulation Board, Rigid Foam, Stacked 15 ft High	170
PS Jars in Compartmented Cartons, Stacked 15 ft High	1250
PS Tubs nested in Cartons, Stacked 14 ft High	475
PS Toy Parts in Cartons, Stacked 15 ft High	180
PS Insulation Baord, Rigid Foam, Stacked 14 ft High	290
PVC Bottles in Compartmented Cartons, Stacked 15 ft High	300
PP Tubs in Compartmented Cartons, Stacked 15 ft High	390
PP and PE Film in Rolls, Stacked 14 ft High	550
Methyl Alcohol	65
Gasoline	290
Kerosene	290
Diesel Oil	175

the theoretical heat release rate based on the mass burning rate and H_c, i.e. assuming 100% combustion efficiency.

The combustion efficiency, χ, is the sum of the convective and radiative combustion efficiencies:

$$x = \chi_{conv} + \chi_{rad} \qquad [A.5]$$

Experimental values of these three combustion efficiencies as obtained by Tewarson (1995) are listed in Table A.2, along with values of H_c for a variety of combustible materials. The data in Table A.2 were obtained at well-ventilated conditions in Tewarson's laboratory apparatus with imposed external radiant heat flux impinging on the fuel surface. Large external heat fluxes presumably are representative of large fires or intense exposure fires.

Theoretical heat release rates for an assortment of industrial commodities have been compiled by Alpert and Ward (1984) and are reproduced here as Tables A.3 and A.4. These heat release rates represent steady-state values after the fires are fully developed.

A.2 Flame height

Flame height calculations are needed to determine the flame surface area (in equation [A.3] and the configuration factors described in Section A.3. Buoyant diffusion flame height data have been correlated by Heskestad (1983) for cylindrical and conical flames with base diameter, D. The simplified Heskestad correlation is

$$H_f = 0.23 Q^{2/5} - 1.02 D \qquad [A.6]$$

where H_f is the flame height (m) above the top of the fuel, Q is the actual heat release rate (kW), and $7 < Q^{2/5}/D < 700 \, \text{kW}^{2/5}/\text{m}$.

Flame heights given by equation [A.6] are not applicable to flames exposed to ambient crosswinds. Flame height and tilt correlations for this situation have been reviewed by Mudan (1984) and Drysdale (1985).

According to Delichatsios (1984), flame heights for rectangular wall fires can be correlated by

$$H_f = 0.050 Q_w^{2/3} \qquad [A.7]$$

where Q_w is the heat release rate per unit wall width (kW/m), and H_f is the flame height (m) above the wall base.

Values of Q_w for several wall materials are given in Table A.3.

A.3 Configuration factor

The configuration factor, sometimes called the view factor, for an elemental target area is the geometric factor relating the emissive power of the flame to the radiant heat flux impinging on the elemental target area. Thus,

$$q'' = E \varphi \tau \qquad [A.8]$$

where q'' is the radiant heat flux (kW/m^2) at the remote elemental target, φ is the flame-target configuration factor, and τ is the atmospheric transmissivity accounting for atmospheric absorption.

Values of the configuration factor have been mathematically derived for a variety of flame-target geometries. Table A.5 shows the mathematical representations for eight simple geometries. Figure A.1 shows a graphical representation of φ for an elemental rectangle parallel to, and opposite one corner from, a rectangular flame. Configuration factors for contiguous flames exposing the same elemental target are additive, as shown by Drysdale (1985).

Figure A.2 shows plots of φ for an elemental target area parallel to a cylindrical flame. Configuration factors for a vertical cylindrical flame and a target oriented at the angle that would produce the maximum view factor to the flame are shown in Figure A.3, reproduced from Mudan (1984). Configuration factors for wind-tilted flames have also been computed by Mudan. Formulas and some tables for well over one hundred configurations involving combinations of differential areas and finite areas of varying geometry are available in Howell's (1985) compilation. In general, choosing a differential target closest to the flame will result in a larger configuration factor than for the finite target surface area containing that differential element.

A.4 Atmospheric transmissivity

Radiation attenuation due to atmospheric absorption can be significant over long path lengths if the flame emits in the same spectral bands as the atmospheric gas (water vapor and carbon dioxide) absorption bands. This effect is usually negligible for sooty, grey body flames.

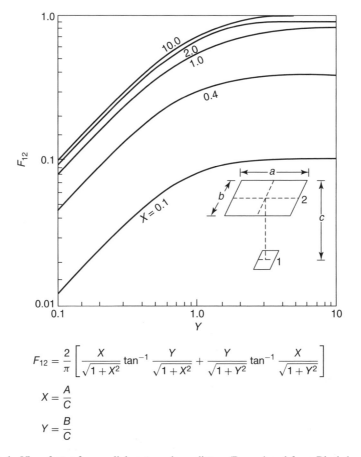

Figure A.1. View factor for parallel rectangular radiator. (Reproduced from Blackshear, 1974)

Atmospheric absorption effects are also important when there is a water spray curtain between the flame and the target. The transmissivity through the spray curtain is given by Heseldon and Hinkley (1965) as

$$\tau = \exp[-3Q''d/4vr] \quad [A.9]$$

where Q'' is the water flow rate per unit horizontal area of curtain (ft^3/sec-ft^2), d is spray curtain depth (ft), v is the average droplet fall velocity (ft/s), and r is the average drop radius (ft).

The data used by Heseldon and Hinkley in deriving equation [A.9] is shown in Figure A.4. Values for v and r depend upon the nozzle diameter and pressure. Small diameter nozzles operated at high pressure produce small drops which also have small terminal fall velocities, thus leading to small values of τ.

A.5 Point source approximation

If atmospheric absorption effects are negligible, and if the target distance is large compared to flame height, the radiant point source approximation provides an attractive simplification of the

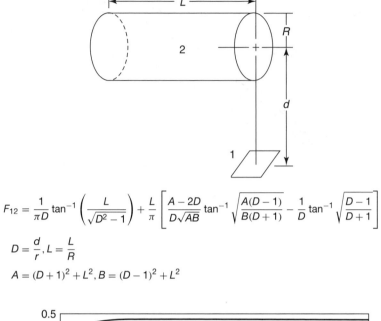

$$F_{12} = \frac{1}{\pi D}\tan^{-1}\left(\frac{L}{\sqrt{D^2-1}}\right) + \frac{L}{\pi}\left[\frac{A-2D}{D\sqrt{AB}}\tan^{-1}\sqrt{\frac{A(D-1)}{B(D+1)}} - \frac{1}{D}\tan^{-1}\sqrt{\frac{D-1}{D+1}}\right]$$

$$D = \frac{d}{r}, L = \frac{L}{R}$$

$$A = (D+1)^2 + L^2, B = (D-1)^2 + L^2$$

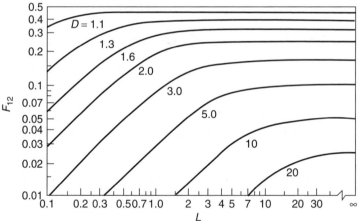

Figure A.2. Cylindrical radiator to parallel received. (Reproduced from Blackshear, 1974)

preceding equations. The radiant heat flux, q'', at a target at a distance x (m) from a point source of radiation is

$$q'' = \frac{Q_r \cos\theta}{4\pi x^2} \quad [A.10]$$

Equation [A.10] is useful for far field estimates of radiant heat fluxes. However, it significantly over-estimates heat fluxes in the immediate vicinity of the flame, i.e. when the distance to the target is small compared to the flame height, and it under-estimates heat fluxes at target distances that are comparable to the flame height.

APPENDIX A: FLAME RADIATION REVIEW

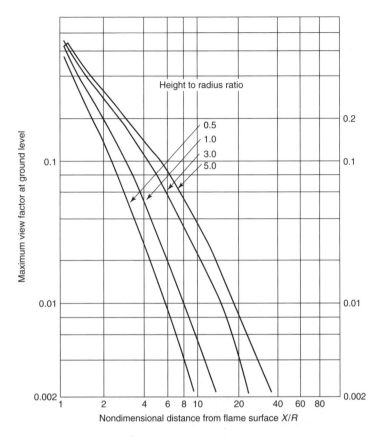

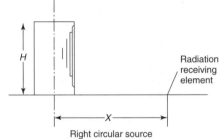

Right circular source

$$F_H = \frac{1}{x}\left[\frac{(B-1/S)}{\sqrt{B^2-1}}\tan^{-1}\sqrt{\frac{(B+1)(S-1)}{(B-1)(S+1)}} - \frac{(A-1/S)}{\sqrt{A^2-1}}\tan^{-1}\sqrt{\frac{(A+1)(S-1)}{(A-1)(S+1)}}\right]$$

$$F_V = \frac{1}{\pi}\left[\frac{1}{S}\tan^{-1}\frac{h}{\sqrt{S^2-1}} + \frac{h}{S}\left\{\tan^{-1}\sqrt{\frac{S-1}{S+1}} - \frac{A}{\sqrt{A^2-1}}\tan^{-1}\sqrt{\frac{(A+1)(S-1)}{(A-1)(S+1)}}\right\}\right]$$

where h = flame height or length/flame radius, S = distance to observer from axis/flame radius,

$$A = \frac{h^2 + S^2 + 1}{2S} \quad \text{and} \quad B = \frac{1 + S^2}{2S}$$

Figure A.3. Configuration factors for a vertical cylindrical flame and a target oriented at the angle that would produce the maximum view factor to the flame. (Reproduced from Mudan, 1984)

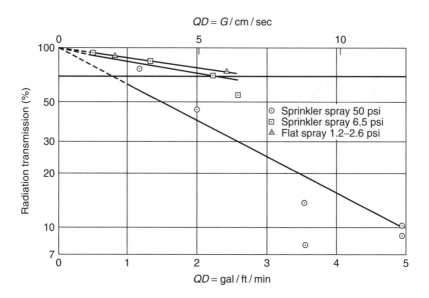

Figure A.4. Effect of factor QD on radiant transmission through water sprays

References

Alpert, R.L. and Ward, E.J., Evaluation of Unsprinklered Fire Hazards, *Fire Safety Journal*, **7**, 127–143, 1984.
Atallah, S. and Allan, D.S., Safe Separation Distances from Liquid Fuel Fires, *Fire Technology*, **7**, 47–56, 1971.
Blackshear, P. (ed.), Heat Transfer in Fires, Scripta Book Co., Washington, DC, 1974.
Delichatsios, M., 'Modeling of Aircraft Cabin Fires,' Factory Mutual Research Corporation J.I. 0H5N5.BU, prepared for Federal Aviation Administration, 1984.
de Ris, J.N., Fire Radiation – a Review, *17th Symposium (International) on Combustion*, pp. 1003–1016, 1979.
DiNenno, P., SFPE TP 82-9, 1982.
Drysdale, D., *An Introduction to Fire Dynamics*, John Wiley & Sons, 1985.
Heselden, A.J.M. and Hinkley, P.L., Measurement of the Transmission of Radiation through Water Sprays, *Fire Technology*, **1**, 130–137, 1965.
Heskestad, G., Luminous Heights of Turbulent Diffusion Flames, *Fire Safety Journal*, **5**, 103–108, 1983.
Howell, J.R., *A Catalog of Radiation Configuration Factors*, McGraw-Hill, 1985.
Mudan, K.S., Thermal Radiation Hazards from Hydrocarbon Pool Fires, *Progress in Energy and Combustion Science*, **10**, 59–80, 1984.
Orloff, L., 'Simplified Radiation Modeling of Pool Fires,' *18th Symposium (International) on Combustion*, The Combustion Institute, 1981, pp. 549–562.
Tewarson, A., Generation of Heat and Chemical Compounds in Fires, Chapter 3-4 in *SFPE Handbook of Fire Protection Engineering*, S.F.P.E. and NFPA, 1995.

APPENDIX B: HISTORIC INDUSTRIAL FIRES

Fire protection "engineering disasters are much more likely to be avoided if future designers, individually, develop a habit of looking back and questioning how each concept grew."[1] In this spirit, brief descriptions of some of the historic industrial fires are presented here.

B.1 General Motors Livonia fire – August 12 1953

According to the Executive Editor of *Business Week* (1953), the General Motors Corporation automobile transmission plant fire in Livonia, Michigan was the biggest business news story of 1953 (*Business Week*, 1953). It demonstrated the importance of fire protection engineering to the entire business community including the upper echelon of management. It also led to some new insights into building fire resistance, and dramatically revealed the disaster potential inherent in omitting fundamental fire protection design features. The following account of the fire is based primarily on the NFPA report ('The General Motors Fire,' 1953).

The plant was located in a 1.5 million ft^2 (140,000 m^2) building laid out as shown in Figure B.1. Manufacturing operations, distributed throughout the main plant area, were separated by a fire wall from the administrative offices at the south end of the plant. The one story steel frame manufacturing building had exterior brick apron walls with large areas of plain glass windows. The roof, which was a major factor in spreading the fire, consisted of steel deck plates covered with asphalt-saturated felt, fiberglass insulation, and a tar and gravel outer surface. There were over three pounds of asphalt per square foot of roof, which was 6.4 m (21 ft) high throughout most of the plant. Steel trusses and columns supporting the roof had no protective coating.

The only interior wall in the manufacturing building was a 4.6 m (15 ft) high, 36.6 m (120 ft) long hollow concrete block wall separating transmission manufacturing operations from a 12,400 m^2 (133,000 ft^2) area occupied by another G.M. division. Automatic transmission manufacturing operations included metalworking, (lathes and presses with lubricating and cooling oil), heat treating (equipment included numerous small quench tanks), and rustproofing (dip tanks). There was no automatic sprinkler protection in these areas despite the presence of large quantities of flammable quantities of flammable liquids. The only sprinklered areas were the receiving, storage, and shipping areas around the periphery of the building.

The fire started in a rustproofing area near the south end of the plant. As indicated in Figure B.2, the area contained a dip tank and an overhead conveyor for transporting metal parts into and out

---

[1] Petroski (*American Scientist*, **80**, 523, 1992) attributes this quote to R.J.M. Sutherland.

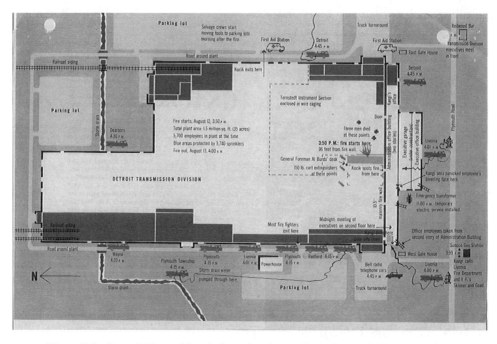

Figure B.1. General Motors Livonia fire; plant layout (from 'Fortune' November 1953)

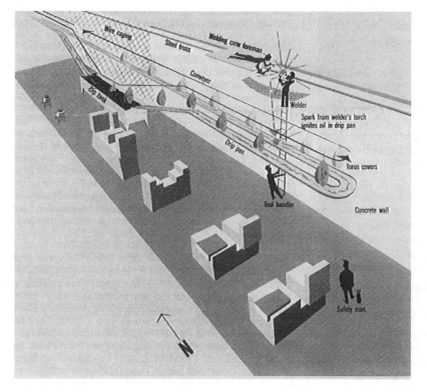

Figure B.2. Rustproofing area of GM Livonia plant where the fire started (from 'Fortune', November 1953)

of the tank. A 36.6 m (120 ft) long, 0.6 m (2 ft) wide drip pan under the conveyor was situated 3.3 m (10.7 ft) above the floor. The rustproofing fluid had a 37 °C (98 °F) flash point. The 1700 L (450 gal) dip tank was protected with an automatic carbon dioxide extinguishing system, but there was no installed protection for the drip pan.

A welding crew was working on some piping above the conveyor and drip pan when welding splatter ignited a 6 mm (1/4 in) layer of liquid in the pan. Two crew members at floor level spotted the fire, and passed two portable carbon dioxide fire extinguishers up to the men on the ladder, as depicted in Figure B.2. Apparently, the fire was almost extinguished in the drip pan when the portable extinguishers were depleted. While employees were getting other manual extinguishers, flames spread over the entire length of the drip pan. Fighting a fire this size from the top of the welder's ladder proved futile. Flames ignited oily condensate on the steel roof truss. At about the same time, the drip pan warped and spilled burning liquid onto the floor and the dip tank. A major escalation of the fire occurred when hot roof tar and asphalt flowed down through joints in the heat warped roof deck, ignited, and dripped onto equipment throughout the plant.

Fire spread along and under the roof eventually caused the roof to collapse and destroy the entire plant except for the office building (Figure B.3). There were six fatalities and property damage of approximately $35 million. Production of the majority of G.M. cars were suspended, since this was the only automatic transmission plant serving seven G.M. automobile divisions.

The NFPA account of the fire ('The General Motors Fire,' 1953) cited the following major deficiencies in plant fire protection:

1. The absence of fire walls and roof vents allowed uncontrolled fire and smoke spread throughout the plant and rapidly prohibited access for manual fire fighting within the building.

2. Inadequate sprinkler protection (only 20% of the plant was sprinklered) prevented fire control and suppression.

3. The automatic carbon dioxide system on the dip tank should have been extended to the drip pan, where manual fire fighting was difficult because of the pan elevation and length.

4. Unprotected and uninsulated structural steel columns, trusses, and decking promoted early roof collapse and fire spread via molten asphalt.

Figure B.3. General Motors Livonia plant was almost completely destroyed when the roof collapsed

A later development prompted largely by this fire included large-scale and small-scale fire tests for insulated steel deck. The large-scale tests demonstrated the contribution of asphalt in various loadings to the propensity for self-sustained fire spread along the deck. The small-scale tests allowed for a measurement of the heat release rate of a sample of the deck and for the classification of the deck fire threat based on test data.

Upon realizing the fire resistance deficiencies in the Livonia plant, General Motors notified (*Engineering News Record*, 1954) the construction industry that new G.M. manufacturing facilities were to include fire barriers in the form of fire walls, parapets, and draft curtains, and virtually 100% sprinkler protection supplemented by spray nozzles for dip tanks, quench tanks, and drip pans. Many other industrial organizations also adopted these new construction requirements for their new facilities.

B.2 McCormick Place fire – January 16 1967

The McCormick Place exhibition hall in Chicago was the largest convention hall in the United States and a prototype for modern assembly building construction. The 330×105 m (1080×345 ft) concrete and steel building had three levels separated from a large theater extending the full height of the building. The lower level was used for exhibition space, commercial kitchens, mechanical equipment rooms, and storage and locker rooms. The second level contained restaurants, meeting rooms, and utility, equipment, and storage rooms. The upper level contained 30,000 m^2 (320,000 ft^2) of undivided exhibition area, with a floor to roof height of 14.3 m (47 ft). The building roof was a layered covering with a gypsum board base supported by steel trusses sitting on two rows of interior columns. Portions of the building were sprinklered, but not the upper and lower level exhibition areas. Standpipe, hose, and portable extinguishers were intended for firefighting in the exhibition areas. Water was supplied from city mains connected to a 37,800 L/min (10,000 gpm) pumping station about 610 m (2000 ft) from the building.

The fire occurred on the eve of a major trade show opening. About 1300 exhibit booths in the exhibition areas were constructed of wood frame, plastic, and fabric. The combustible loading is estimated as about 98 kg/m^2 (20 psf) (Juillerat and Gaudet, 1967). Another major factor in the fire was the presence of folded aluminum wall partitions available for subdividing the exhibit hall.

The ignition source is believed to be faulty electrical wiring temporarily installed in an exhibit booth on the upper level. Maintenance men in the building discovered the fire at approximately 2:00 AM on January 16th. They tried to extinguish the fire with portable extinguishers before calling the Chicago Fire Department at 2:11 AM. There was no attempt to use the standpipe and hose connections. Flames rapidly propagated to adjacent exhibit booths, and heat and smoke accumulated under the roof. Chicago firefighters entering the exhibition area at about 2:15 AM were soon forced out by heat and smoke. External fire fighting was hampered (although it probably wouldn't have affected the fire outcome) by three inoperable yard hydrants and two faulty pumps in the pumping station.

Fire spread downward from the upper to lower exhibit areas when the aluminum folding partitions melted, burned caulking in floor expansion joints, and flowed down two levels. The molten aluminum ignited display materials in the lower exhibit area. Most of the upper and lower exhibit areas and the roof were destroyed, as can be seen from Figure B.4. The fire did not spread to the main theater or any of the sprinklered rooms. There was one fatality and property damage of about $53 million.

NFPA investigators concluded (Juillerat and Gaudet, 1967) that the absence of automatic sprinklers, draft curtains, and a heat and smoke venting system in the main exhibit area were primary deficiencies that probably would have controlled the fire and minimized damage. The Chicago investigating committee report pointed out the need for more reliable public water supplies for fire protection, better communications between building managers and the fire department, and of

APPENDIX B: HISTORIC INDUSTRIAL FIRES   351

Figure B.4. The McCormick Place fire, where most of the upper and lower exhibit areas and roof were destroyed

course more public awareness of fire emergency response procedures. Another noteworthy ramification from the McCormick Place fire was widespread appreciation of the damage vulnerability of unprotected structural steel and concrete, and the total inadequacy of aluminum walls as fire barriers.

B.3 K MART fire – June 21 1982

Contrary to the facilities in the two previous examples, the K Mart Distribution Center in Falls Township, Pennsylvania was a fully sprinklered structure containing fire walls to limit fire spread. Total destruction of a modern industrial facility with these two essential fire protection features provided a dramatic lesson on the need for appropriate fire wall and sprinkler system design, and the need for special protection measures to cope with especially severe fire hazards.

The 331 × 360 m (1085 × 1180 ft) K Mart warehouse was layed out as illustrated in Figure B.5. Fire walls subdivided the warehouse into four quadrants designated as A, B, C, and D. Quadrants B and C were used for storing and retrieving a multitude of consumer products. Quadrant A was used for major appliance storage and product shipping, and Quadrant D was used for receiving and product repacking. The storage configuration consisted of mixed palletized and rack storage to a maximum height of 4.6 m (15 ft). Conveyors and lift trucks were used for product transport, placement, and retrieval.

A 9.1 m (30 ft) high steel deck roof on the one-story warehouse was supported by steel trusses on steel columns. There were no protective coatings on any of the structural steel. Exterior walls were concrete with insulated metal panels. Fire walls separating the four quadrants consisted

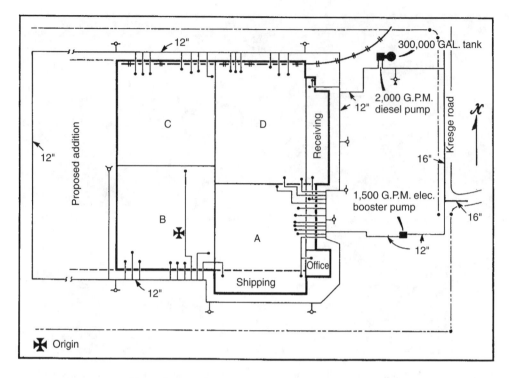

Figure B.5. K MART warehouse layout (from NFPA)

of concrete block tied to steel columns on both sides, and tied to steel trusses at roof level. Figure B.6 shows a wall cross-section.

Ceiling sprinklers were capable of delivering a water density of 16.3 L/min/m^2 (0.40 gpm/ft^2) over the most remote 279 m (3000 ft^2) of the warehouse. Water was supplied from a 7560 L/min (2000 gpm) at 900,000 Pa (128 psig) (diesel engine fire pump connected and adjacent to a 1134 m^3 (300,000 gal) tank. There was also a 5670 L/min (1500 gpm) electric motor booster pump designed to pump from the public water main into the plant yard loop distribution network (Figure B.5). Sprinkler heads were 17/32 in orifice with 141 °C (286 °F) links. Roof heat and smoke vents of 2.2 m^2 (24 ft^2) area in each quadrant provided a vent ratio of 1:96.

The fire started at 12:30 PM in a palletized storage area in Quadrant B. A forklift operator apparently dislodged a carton of aerosol cans containing carburetor and choke cleaner fluid. One of the cans was punctured and escaping liquid and propellant vapors were ignited by the forklift electric motor. Flames rapidly spread up the palletized storage and cans started popping and generating large fireballs. Sprinklers began operating within five minutes from ignition, but were not effective in suppressing or controlling the growing fire.

Plant fire brigade members responding to the fire were soon forced out by flame, smoke, and rocketing aerosol cans. Before they evacuated they observed that sprinklers and deluge curtains at fire wall openings were operating properly. However, exploding cans rocketed through the roof and fire wall openings and in some cases trailed flaming liquid. Weakened structural steel allowed the roof to sag and sprinkler piping to break about 40 minutes after the fire started. Municipal firefighters evacuated the building at this time. The electric booster pump had been shut down 17 minutes after ignition when power was disconnected at the main switchgear. Fire spread throughout the entire warehouse by about 1:30 PM, i.e. one hour after ignition.

APPENDIX B: HISTORIC INDUSTRIAL FIRES

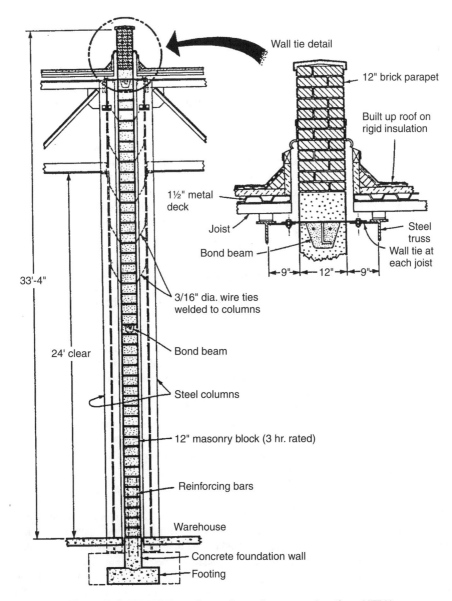

Figure B.6. K MART warehouse fire wall cross-section (from NFPA)

The warehouse was totally destroyed. Property damage costs exceeded $100 million making this fire the most costly industrial fire ever, until another warehouse fire involving aerosol cans in 1985 produced even more damage. Major factors contributing to the extensive damage in the K Mart fire are as follows (Best, March 1983, May 1983; Factory Mutual, 1983):

1. The large quantities of aerosol cans containing flammable liquids and propellants completely overwhelmed the ceiling level sprinklers. There were more than 1 million aerosol cans in the warehouse along with propane cylinders, butane disposable lighters, cans of lighter fluid, and numerous cans of low vapor pressure flammable liquids. This type of high hazard storage

requires more stringent protection in the form of some combination of cut-off walls for hazard isolation, in-rack and/or large drop fast-response sprinklers, and either restricted storage height (NFPA 30B) or fire barriers in the storage racks.

2. Fire walls were deficient in that the ties to unprotected steel on both sides of the wall allowed excessive structural loads to be transmitted to the fire wall when the steel trusses and columns started to sag and collapse. In addition, deluge curtains in the fire wall openings were a poor substitute for fire doors as fire barriers for this type of hazard.

A much less critical factor in this fire was the sprinkler system component layout in the yard (Best, March 1983). The electric pump should have been installed on the utility side of the main plant switchgear, and the post indicator valves should have been further than 14.3 m (14 ft) from the building wall. Ramifications from this loss included the development of improved aerosol fire protection including the classification of aerosol products into deferent flammability levels. Fire wall capabilities were also reevaluated and new guidelines issued addressing proper design to achieve wall stability during a high challenge fire.

B.4 New York Telephone Exchange fire – February 27 1975

The New York City telephone exchange fire on February 27 1975 was the most damaging of three nationally publicized electrical cable fires within a span of 37 days. The other two cable fires were the World Trade Center fire on February 13th (Lathrop, 1975) and the Browns Ferry Nuclear Plant fire on March 22nd (Pryor, 1977). Collectively, these three fires provided dramatic evidence of the fire spread potential of grouped electrical cable, and of the difficulties and dangers of manual firefighting in congested cable areas. Furthermore, the telephone exchange fire demonstrated the potential for remote equipment damage caused by smoke and corrosive combustion products.

The New York Telephone Company building at 13th Street and Second Avenue in Manhattan is used as a central office, i.e. a main switching center connecting other telephone offices via trunk lines. It also served 173,000 individual telephones in lower Manhattan. The 11 storey high building exterior walls are brick and the floors and columns are constructed of reinforced concrete. Windows were made of wired glass sealed closed in metal frames, and on the first two stories shielded on the outside by metal screens.

The distribution of telephone equipment in the building is shown in Figure B.7. Cable entered and exited the building through a $6.1 \times 6.1 \times 84.8$ m ($20 \times 20 \times 278$ ft) long basement cable vault shown in Figure B.8. The trunk main distribution frame, the major input-output connecting point for trunk lines to other central offices, was situated on the first floor directly above the cable vault. The subscriber line main distribution frame was located on the fifth floor. Two different types of electro-mechanical switching equipment (panel switches and the newer crossbar switches) were located on most of the upper floors. The third floor was vacant at the time of the fire. The fourth floor contained administrative offices and a cafeteria as well as some crossbar switching equipment.

Cables in the cable vault were primarily lead jacketed or polyethylene jacketed over paper pulp or polyvinyl chloride insulation (NYC Fire Department, 1975). Polyvinyl chloride (PVC) insulated tip cables ran from the upper areas of the vault through a 15 cm (6 in) wide 82.3 m (270 ft) long slot in the first floor (shown in Figure B.8) to the trunk main distribution frame. Asbestos-and-cement cover plates were fitted around the cables in the open slot. PVC insulated, PVC jacketed cables were also used extensively in cableways throughout the building. The power cable shaft on the south wall and the cable pulling chases in the northwest wall (Figure B.8) apparently were not firestopped, but smaller openings in the floors and ceiling were protected by filling in around the cables with small, fire retardant cotton bags filled with mineral wool (Lathrop, 1975).

APPENDIX B: HISTORIC INDUSTRIAL FIRES

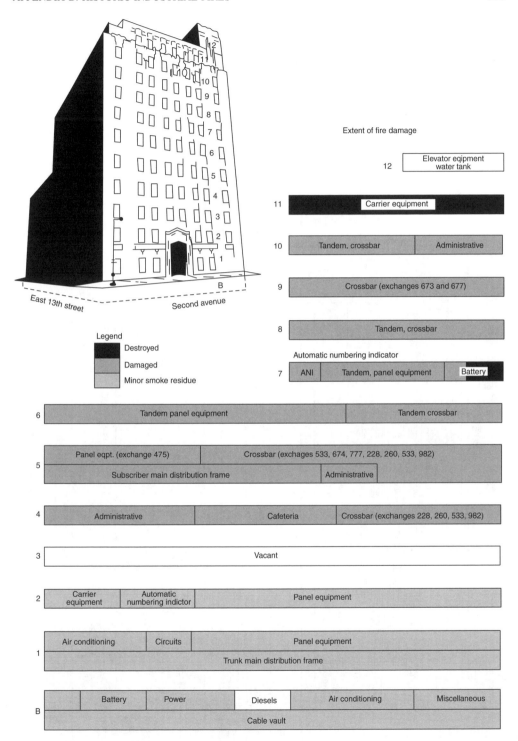

Figure B.7. New York Telephone Exchange telephone equipment distribution and fire damage. Reproduced with permission from IEEE Spectrum, June 1975 (© 1975 IEEE)

356   INDUSTRIAL FIRE PROTECTION ENGINEERING

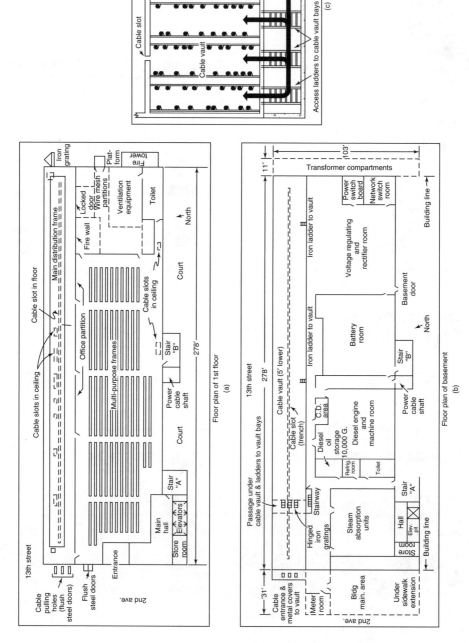

Figure B.8. Cable entries and exits to the New York Telephone Exchange building (from FDNY)

Ionization type smoke detectors were installed in the trunk main distribution frame room on the first floor and in some of the basement standby power rooms, but not in the cable vault. Heat sensitive detection wire was installed within distribution and switching equipment throughout most of the building (Lathrop, 1975). There was no automatic suppression system in the fire area.

The fire apparently was started by a short circuit or arc at an open splice in the polyethylene insulated cable in the cable vault. Shortly after midnight on February 27th, employees discovered smoke in the subscriber main distribution frame area. After a futile attempt to call the Fire Department from a fire damaged line in the building, an employee transmitted an alarm from a Fire Department call box in the street. The alarm was received at 12:25 A.M. It is not clear when the smoke detectors actuated, but they did not transmit the alarm.

When the Fire Department arrived, they found heavy smoke in the basement and light smoke throughout the rest of the building. The smoke and difficult accessibility of the cable vault (Figure B.8) prevented them from finding the exact location of the fire.

The fire spread upward from the cable vault through the 15 cm (6 in) wide slot to the trunk main distribution frame which was separated from the rest of the first floor by interior office partitions. Fire also spread to the second floor and to portions of the upper floors by way of burning cable in the cable chases and floor openings. The absence of equipment and cables on the third floor prevented more fire damage on the upper floors.

After one hour of difficulty in getting to the seat of the fire, the Fire Department attempted to use high expansion foam to extinguish the cable vault fire. Unfortunately, this attempt was not successful because the foam drained out into the sub-basement passageway beneath the cable vault (Figure B.8). They resumed their hose line attack and temporarily brought the cable vault fire under control at 3:40 A.M. However accumulating pyrolysis gases from the hot cable reignited within about 20 minutes. The burning polyvinyl chloride cable generated acrid gases and acidic smoke that drove the firefighters out of the building below the fourth floor, despite their use of oxygen masks. Firefighting efforts (including water application to the cable vault via a jerry rigged cellar pipe) resumed intermittently with over 300 firefighters simultaneously working on the building interior and exterior until the fire was declared under control at 4:46 P.M.

Equipment fire damage is indicated in Figure B.7. Most cables and equipment in the basement and first two floors were destroyed, with cable insulation on the first two floors almost totally consumed. There was scattered smoke and corrosion damage on the upper floors, including the 11th floor. Eventually, 66 miles (105.6 km) of cable were replaced and 10 million electromechanical relays were cleaned manually (Allan, 1975) with hand brushes to remove soot and a special cleaning fluid to remove the chlorides formed from the burning PVC insulation. Spalled and cracked concrete floor bays on the first and second floors also were replaced. Approximately 4000 people were involved in the cleanup and replacement of equipment, which was miraculously completed in one month.

The fire motivated New York Telephone and the New York Fire Department to jointly formulate a plan for upgrading fire protection at telephone exchanges in New York City (Chapman, 1978). Some of the improvements to be implemented included:

1. Installing cementitious vermiculite with a two hour fire resistance rating in the cable slots above cable vaults.
2. Installing smoke detectors throughout the buildings (including cable vaults) and wiring the alarm to a constantly attended supervisory service.
3. Designing and installing a smoke control system in new buildings with more than 2300 m^2 (25,000 ft^2) of floor area; the system should confine the spread of smoke and vent it out of the building.

4. Equipping cable vaults with a special non-automatic sprinkler system supplied with water from a street level Fire Department connection. Sprinkler heads with 177 °C (350 °F) links should be located 3 m (10 ft) on centers in the aisles of the vault.

Similar recommendations were issued by an ad hoc committee representing the entire Bell System prior to divestiture. (Hession, 1983). However, the Bell System continued to rely on manual suppression with portable extinguishers. Since the 1975 New York Telephone Company fire, there have been two more large telephone exchange fires that received national attention, and have stimulated new evaluations of fire protection practices at telephone operating companies.

B.5 Ford Cologne, Germany Warehouse fire – October 20 1977

At the time of its occurrence, the fire at the Ford Motor Company automobile parts and accessories warehouse in Cologne, West Germany was the largest and financially costliest single structure fire ever. It is comparable in many respects to the K Mart warehouse fire. Both fires involved massive destruction to fully sprinklered warehouses that stored commodities that were substantially more combustible than those at the time the warehouse was constructed. They demonstrated the danger of designing fire protection without an assessment or periodic reassessment of the evolving hazards during the operating lifetime of the warehouse.

The Ford Cologne warehouse was built in 1962 and expanded in 1967. The floor plan layout is shown in Figure B.9. The total floor area in the 9.1 m (30 ft) high building was 130,000 m^2 (1.4 million ft^2). Construction features included steel columns and beams supporting steel purlins and two different types of roofing; steel deck and concrete slabs. The steel deck roof had heat and smoke vents with an area of 0.09 m^2 (1 ft^2) per 2.8 m^2 (30 ft^2) of roof area. A steel framed double brick wall with an 8 cm (3.2 in) cavity between the bricks, and with automatic closing fire doors in the wall openings, separated the building at column line 26 into a 75,600 m^2 (814,000 ft^2) area west of the wall and a 51,000 m^2 (550,000 ft^2) area east of the wall.

Most of the automobile parts were stored in baskets $2.2 \times 1.2 \times 1.04$ m ($89 \times 48 \times 41$ in) high stacked as portable racks five or six high, i.e. 5.2–6.1 m high. There were also some fixed racks with slatted shelves. Ordinarily there were 2.4 m (8 ft) wide aisles between the stacks, but on the day of the fire there was temporary (palletized and basket) storage in several of the aisles. As indicated in Figure B.10, storage in the area of fire origin included: plastic consoles, (ABS plastic in cartons), steering wheels (steel core with polyurethane foam padding and PVC outer surface, windshield wipers (rubber blade on a metal arm and packed in a PVC blister box, motor oil (212–227 °C flash point) in metal cans packed in paperboard cartons, and air filters (plastic and flame retardant paper in cartons).

Ceiling level automatic sprinkler were stored throughout the warehouse. The 1/2 in orifice, 74 °C (163 °F) sprinkler heads were supplied by a combination of pressure tanks, fire pumps, booster pumps, and public water mains. The water supply curve recreated after the fire is shown in Figure B.11. It starts at a design density of 0.8 gpm/ft^2 (32.5 L/min/m^2) for the first four flowing sprinklers and decays rapidly to 0.40 gpm/ft^2 (16.3 L/min/m^2) for 716 m^2 (2350 ft^2) (25 sprinklers) and 0.33 gpm/ft^2 (13.4 L/min/m^2) for 1460 m^2 (4800 ft^2) (51 sprinklers). This supply curve satisfied the NFPA 231 guidelines for ordinary combustibles and Class IV commodity 6.1 m (20 ft) high storage, which presumably was the predominant storage anticipated when the building was constructed. Most current sprinkler design guidelines for plastic storage call for higher water densities for 6.1 m (20 ft) high storage with 3 m (10 ft) clearance to the ceiling sprinklers.

The suspected cause of the fire is a discarded cigarette or match in the vicinity of the plastic console aisle storage just south of column K21. Several employees noticed burning cartons and plastic and notified the Plant Fire Brigade. A hose stream was applied to the burning storage approximately four minutes after the fire was discovered, but the fire continued to grow. The first

APPENDIX B: HISTORIC INDUSTRIAL FIRES

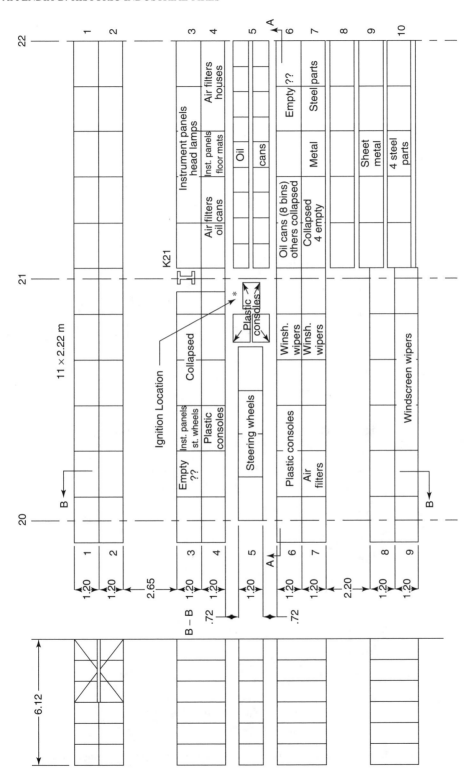

Figure B.9. Ford Cologne warehouse layout

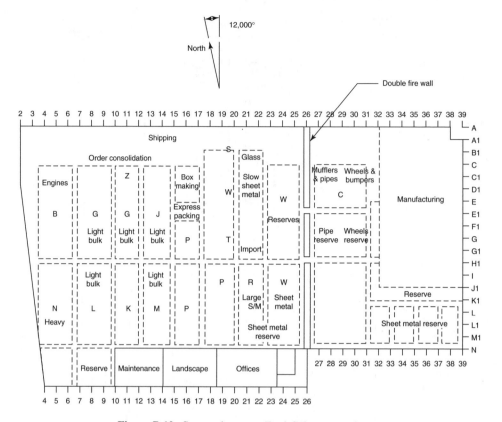

Figure B.10. Storage layout at Ford Cologne warehouse

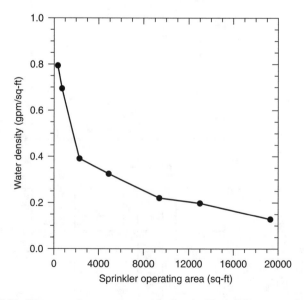

Figure B.11. Water supply curve recreated after the Ford Cologne warehouse fire

sprinklers opened five minutes after the fire was discovered, and about 24 heads were flowing seven minutes after fire discovery. Ford Fire Brigade trucks from a nearby fire station arrived about 10 minutes after the discovery of the fire, and attempted to fight the fire with both dry chemical and water hose streams. Neither the firefighters nor the automatic sprinklers were successful in containing the fire. Electrical cables powering a fire pump, roof smoke ventilators, and building lights burned out twenty minutes after fire discovery. At that point plant and public fire brigade firefighting became problematic and primarily consisted of confining the fire to the west side of the double brick fire wall.

Most of the building structure and contents west of the fire wall suffered heavy fire damage. There was some damage to the fire wall, but sparse combustible loading (mostly metal parts) near the wall and effective firefighting prevented fire spread east of the wall. The total property damage was approximately $100 million in 1978 dollars.

Key factors responsible for the heavy damage include:

1. Temporary aisle storage promoting rapid fire spread and obstructing manual firefighting efforts.
2. Inadequate sprinkler protection for the increased challenge represented by the high piled storage of plastic commodities and combustible liquids (the role of the motor oil in metal cans in this fire was not described in the loss report).
3. Only one fire wall (albeit an effective wall) in a 130,000 m^2 (1.4 million ft^2) warehouse.

B.6 Triangle Shirtwaist Company fire, N.Y.C. – March 25 1911

The Triangle Shirtwaist fire is historic in a number of ways. The tragic loss of life and resulting investigations and 'media' coverage led to the writing of the first edition of the NFPA 101 Life Safety Code, to upgrading of building codes and fire protection regulations throughout the US, and to a myriad of worker safety reforms expedited by the growing labor union movement. The following brief account of the fire is based primarily on information in the book by Leon Stein (1962).

The Triangle Shirtwaist Company operated a ladies garment factory on the top three floors of a 10 story building at the intersection of Greene Street and Washington Place in lower Manhattan. The building, which was completed in 1901, had 929 m^2 (10,000 ft^2) of floor area per floor with relatively open, loft type, wood floor construction. There was no automatic sprinkler system even though approved sprinkler systems were available. According to Stein, decisions on installing automatic sprinklers were governed by insurance premium considerations, and the dominant property insurance brokers in N.Y.C. did not provide the premium reductions offered by other companies in several other cities and towns.

There were two stairways and elevators at opposite ends of each floor as shown in the ninth floor layout in Figure B.12. However, the stairway doors opened in toward the work area because of limited space on each staircase landing. Furthermore, only the Greene St. stairway had an exit on the roof; the other staircase dead-ended at the tenth floor. There was also an external fire escape adjacent to windows but it ended at the second floor and was situated directly over a basement skylight.

A shirtwaist is a female bodice garment popular at that time. Production of the Triangle shirtwaists began on cutting tables located on the eighth floor. The tables were 1.02 m (40 in) wide, 1.06 m (42 in) high, and separated by aisles 0.76 m (30 in) wide. Cutters would pile cotton fabric layers on the tables, and cut the fabric following patterns placed on the fabric piles. After cutting, the fabric was hung on wires stretched over the length of the tables. Scraps were placed in bins located under the tables.

Sewing of the fabric pieces was performed on the ninth floor. About 240 sewing machines were situated in 23 m (75 ft) long, 1.2 m (4 ft) wide tables, as illustrated in Figure B.12. The

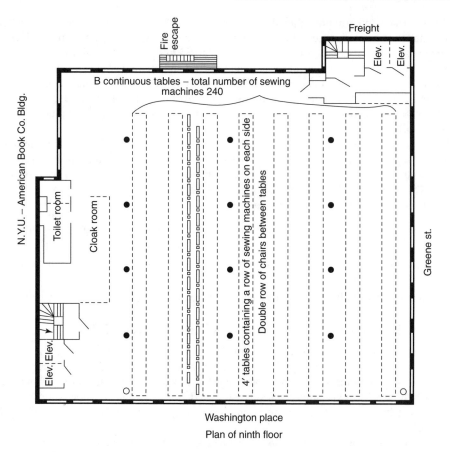

Figure B.12. Triangle Shirtwaist Co. floor layout. Reprinted from Leon Stein, Introduction by William Greider: The Triangle Fire. Copyright © 1962 2001 renewed by Barbara S. Ismail & Walter Stein. Introduction copyright © 2001 Cornell University. Used by permission of the publisher, Cornell University Press

machines were driven by leather belts connected to a flywheel on a rotating shaft in a trough between two tables. The shaft was powered by electric motors located at the Washington Place (south) end of the tables such that there was no aisle at the south end of the floor. The machines and motors were lubricated with oil stored in a barrel near the Greene St. exit door.

The tenth floor was used for pressing and packing garments, for storage, and for executive offices. Only about 50 of the 500 Triangle employees worked on the 10th floor. About 250 worked on the ninth floor, and the other 200 on the eighth floor. Most of the employees were young women; many were recent immigrants to the US.

Despite the recommendations of a fire inspector hired by the insurance company, Triangle never performed fire drills. There were standpipes and hose lines, but there was no water in the lines at the time of the fire.

The fire started at 4:45 PM in a bin under the cutting tables on the eighth floor. When the fire was noticed by cutters, they threw pails of water onto it. However the fabric scraps and the table continued to burn unabated. Within a minute or two after discovery, the fire spread to the hanging fabric pieces and patterns above the tables. Burning fabric brands began flying around the room and landing on other tables where they ignited additional fabric. Smoke accumulated under the ceiling, and the heat and smoke buildup caused a sufficient pressure to burst windows.

APPENDIX B: HISTORIC INDUSTRIAL FIRES 363

Pedestrians on the street below heard 'a big puff' as the windows popped. Soon after, flames spread through the broken windows to the ninth floor above.

Employees began fleeing through the Greene St. stairway, the elevators, and the fire escape. However, the exit door to the Washington Place stairway was apparently locked such that it was not available for escape. Access to the Greene St. exit was limited by the 0.84 m (2 ft 9 in) wide stairway and by the inward opening narrow door separating the vestibule from the work area. (A security person was normally stationed at this door to search employees for stolen goods.) The fire escape weakened from the load of escaping workers and the emerging flames; it collapsed during the fire.

Without any viable egress path, workers panicked and began to leap from the eighth and ninth floor windows. Firefighters on the street below held out nets and blankets, but the impact of the falling bodies rendered the nets useless. Many of the 145 fatalities resulted from trapped workers who jumped. Ladders on Fire Department vehicles only reached the sixth floor, at least 6.1 m (20 ft) below the windows where girls were waiting futilely to be rescued.

Once firefighters reached the eighth and ninth floors via the overcrowded staircase, they were able to effectively control the fire with their hose streams. In fact, the fire was brought under control in 18 minutes and extinguished in about 30 minutes.

The building structure withstood the fire with negligible damage. The fabric, wood tables, wood trim on the walls and windows, the furniture, and the lubricating oil were consumed, as were 145 young lives.

B.7 Hinsdale, Illinois Telephone Central Office Fire

The Hinsdale Central Office (HCO) is a major telecommunications center known as a main hub. Besides serving approximately 42,000 local lines in suburban Chicago, it operates 118,000 trunk lines for local and long-distance routing (FTI Report, 1989). There are links to several long-distance carriers and data transmission lines, and to several cell phone carriers. HCO is operated by the Illinois subsidiary of Ameritech.

In May 1988, most of the switching equipment and the Main Distribution Frame were situated on the first floor of the HCO. The layout of the 100 × 140 ft floor area is shown in Figure B.13. The dashed lines in the figure designate equipment boundaries rather than physical partitions. Most of the first floor was a large open area. The 1AES equipment area on the west side of the floor was used for local calls. The toll call switching equipment on the north end of the floor was situated in 11 ft high racks, called bays, positioned side-by-side from east to west. Overhead cables were situated in trays in the aisles between the bays, and in the east-west direction perpendicular to the aisles. The cable trays were suspended approximately 2 ft below the 14 ft high ceiling.

Telecommunication input and output cables entered and exited a cable vault in the basement of the building. Electrical power lines also entered the building via a basement level transformer vault. After the voltage was stepped down to 208/120 volts, the ac power was converted to 48-volt dc and 24-volt dc for the switches, and to 130-volt dc for other equipment. Since the uninterrupted operation of the switches was a primary concern, there were two backup power supplies. If the commercial power failed, two diesel generators in the basement would be initiated to take up the load. In the short interval between loss of commercial power and generator full operation, a standby battery system in the basement would provide the power needed. The batteries had the capacity to provide full load power for a period of at least four hours. The main AC power supply and the diesel generators had disconnect/shutdown switches in the basement. However, there was no disconnect provision for the batteries.

The two-story high HCO building has a concrete column and concrete beam frame, with 8 in thick concrete floor slabs. There are non-load bearing masonry external walls, and a composite-type roof above a concrete deck. Cable penetrations in the concrete floor slabs are filled with

364 INDUSTRIAL FIRE PROTECTION ENGINEERING

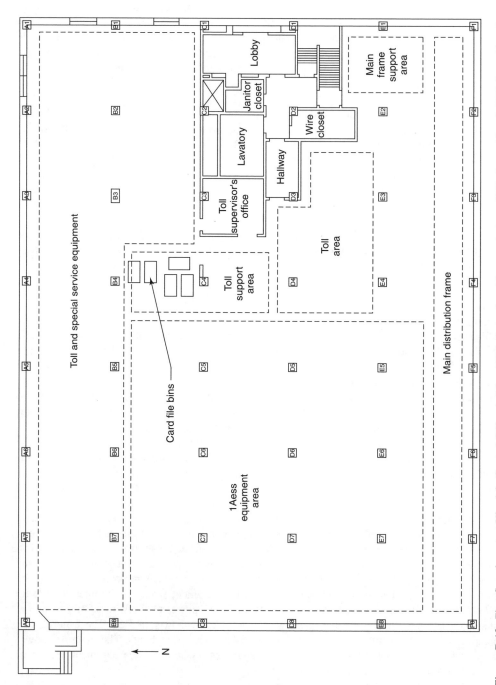

Figure B.13. First floor layout at Hinsdale Central Office in May 1988. Reproduced by permission of J.R. Reynolds, c/o FTI Consulting, Inc.

mineral wool fire stops. The second floor was primarily one large open area with some 1AES equipment.

The HCO had ionization type smoke detectors installed with 20 × 20 ft spacing on the ceiling, and additional detectors in the HVAC ducts. Detector actuation would initiate an audible alarm, illuminate an annunciation panel in the lobby, shut down the air-circulation system, and send an alarm to the Division Alarm Reporting Center (DARC) in Springfield, Illinois, about 200 miles from Hinsdale.

There was no automatic suppression system for the building. Standpipes supplied water to three hose cabinets, and there were Halon 1211 portable extinguishers at various locations in the building. Hinsdale Fire Department personnel had pre-fire planning discussions with telephone company personnel and performed annual fire safety inspections. On the last inspection two months prior to the fire, the only violation cited was for a tied open fire door.

On Sunday, May 8, 1988, there were high winds in the Hinsdale area and several local power outages. As usual for a weekend, the HCO was nominally unoccupied, but had several short-duration visits by telephone company personnel for a variety of reasons. The last employee left the building about one hour prior to the fire, and did not notice anything unusual. The fire timeline listed in Table B.1 begins with this event.

The series of alarms received at DARC beginning at 3:50 PM were initially interpreted as being due to commercial power failure. Power interruption alarms at DARC occur frequently, and were normally accompanied by fire alarms. If there is no fire, the fire alarms should clear once power is restored. As indicated in the timeline, the HCO alarms did clear temporarily when the diesel generators started, and when the batteries began discharging. When the fire alarm re-actuated without clearing, Ameritech instructions were for the DARC technician to call the duty supervisor and the local fire department. However, the DARC technician did not call the Hinsdale Fire Department.

The first telephone company employee to respond to the alarm did not arrive at the HCO until 4:52 PM, approximately one hour after the first series of alarms. Upon seeing a great deal of smoke in the building, he tried to call the Fire Department, but the phone lines were already out. A motorist sent by the telephone company technician notified the Fire Department at 4:58 PM. The first Fire Department personnel arrived at 5:02 PM.

Copious smoke on the first floor of the HCO made it very difficult for firefighters to see where the fire was. They eventually determined that the primary fire was in the overhead cables located in the northeast corner of the first floor, as shown in Figure B.14. Their attempts to extinguish the fire with water spray and with dry chemical discharges were unsuccessful, because the energized cables would rapidly re-ignite. Numerous attempts to de-energize the cables were unsuccessful because of the absence of a battery disconnect switch. At 7:15 PM, a telephone company finally removed the last fuses in the battery power supply circuit. This allowed the firefighters to readily extinguish the primary fire and various residual fires.

When the firefighters first saw the cable tray fire, it had a diameter of about 6 ft. The final fire size, shown in Figure B.14, was about 30 × 40 ft. Most of the cable insulation in the trays in this area was burned away. The extensive damage to the cables can be seen in Figure B.15, which is a photograph of some of the trays after they were lowered to allow viewing.

There was an assortment of power and communication cables in the trays. The power cables varied in size from 6 AWG to 750 MCM (FTI Report, 1989, pp. 3–20), and included some metal-armored cable. Communication cable included shielded and unshielded multi-conductor cable and coaxial cable. Cable insulation and jacketing included polyvinyl chloride (PVC), EPDM rubber, low density polyethylene, and cross-linked polyethylene. The cables had been installed over a period of about 30 years.

Table B1. Hinsdale fire timeline

May 8, 1988 Time	Event
2:46 PM	Central Office Technician leaves HCO after making a minor adjustment on the Main Frame in response to a customer service problem. He did not notice anything unusual at this time.
3:50 PM	The Division Alarm Reporting Center (DARC) in Springfield receives trouble signals from HCO. The signals indicated commercial power failure, a fire alarm trouble alarm, a fire alarm, an air dryer alarm, and a battery discharge alarm.
3:53 PM	HCO alarms cleared, and diesel generator startup indicated.
3:59 PM	DARC receives fire alarm along with diesel generator failure, another fire alarm trouble signal, and another air dyer alarm.
4:00 PM	DARC receives battery discharge alarm, and fire alarm signals are cleared.
4:16 PM	DARC technician calls HCO duty supervisor and reports power failures and battery discharge, which he verified were still active. Duty supervisor unsuccessfully attempts to contact a HCO technician to respond.
4:20 PM	DARC receives fire alarm and the technician informs HCO duty supervisor of alarm.
4:23 PM	DARC receives HCO alarms indicating various equipment was malfunctioning.
4:24 PM	HCO duty supervisor calls Downers Grove Central Office Technician and directs him to investigate the alarm conditions at the HCO.
4:30 PM	Duty supervisor attempts to call both the Hinsdale and Downers Grove Fire Departments. However, both lines were out of order.
4:52 PM	The Downers Grove technician arrives at HCO and observes smoke being emitted from building vent. Upon entering the building, he observes smoke at ceiling level in the equipment room. After unsuccessfully trying to call the Hinsdale Fire Department (because the phones were not working), he leaves the building and stops a passing motorist. He asks motorist to go to fire station and report the fire.
4:56 PM	DARC loses all signals from HCO.
4:58 PM	Hinsdale Fire Department receives notice of HCO fire.
5:02 PM	Hinsdale Fire Department engine arrives at HCO.
~5:03 to 5:30 PM	Firefighters observe heavy black smoke in HCO lobby and other areas on first floor. After donning SCBA, they crawl into toll support area where they observe a small fire in some card bins under the cable trays. After extinguishing the card bin fire, they directed a hose stream onto the overhead cable trays. At this point they had to exit because of a diminished SCBA air supply.
~5:15 to 5:30 PM	Responding Mutual Aid firefighters open a scuttle hatch on the roof. Heavy black smoke is emitted from the open hatch.
~5:30 to 6:00 PM	Two other firefighters entering the fire area observe electrical arcing at ceiling level. They make repeated attempts to extinguish the fire with water sprays. Each time, the fire appeared to be temporarily extinguished, but then re-ignited and seemed to spread. Extinguishment attempts using 20-pound dry chemical extinguishers were also unsuccessful.
~5:30 to 6:35 PM	Various attempts by firefighters and telephone company personnel are made to shut off the electrical power. These attempts included shutting off power entering the building and shutting down the diesel generators. However, electrical arcing and burning continues in the overhead cable trays.
7:15 PM	A telephone company employee finally removes the remaining fuses in the basement. This successfully shuts off battery power to the cables, and virtually extinguishes primary fire.
11:30 PM	Fire is declared out after four hours of extinguishing some residual burning in cable trays.

APPENDIX B: HISTORIC INDUSTRIAL FIRES 367

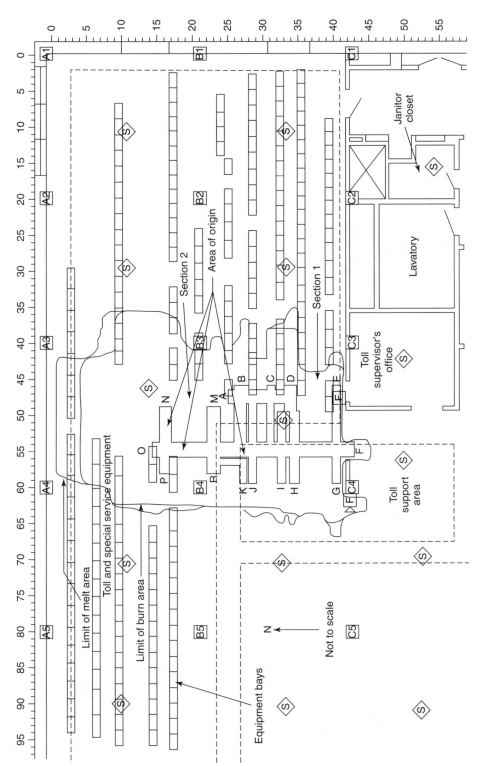

Figure B.14. Fire area in northeast corner of the first floor of Hinsdale Central Office. Reproduced by permission of J.R. Reynolds, c/o FTI Consulting, Inc.

Figure B.15. Burned cable trays lowered for inspection. Reproduced by permission of J.R. Reynolds, c/o FTI Consulting, Inc.

Detailed inspection of the burned cables showed numerous locations with evidence of arcing damage. This evidence included power cables that were severed in two by severe heat, and at least one void space extending deep into the cable pile. There were also several places in which the armored cable was welded to adjacent cable conductors, and other locations where there was arc damage to the aluminum sheets on the bottom of the trays, and to the steel frame of the trays. Photographs in the FTI report show this damage.

Based on the physical evidence and various simulation tests, FTI concluded that the cables were probably ignited by a short circuit due to contact between an insulation-damaged power cable and the grounded metal sheath of an armored cable. They hypothesize that the short was not of sufficient current to actuate a power cable fuse, but generated enough heat to ignite the insulation on adjacent cables.

The FTI report suggests that the cable insulation was damaged two months prior to the fire when some obsolete cables were removed from the trays. In fact, there were some observations of electrical arcing during this cable removal process. The damaged power cable did not short until the current load in the cable tray increased to its peak level due to the increased phone line usage on the day of the fire, which was Mother's Day.

Most of the equipment damage in this fire was due to the smoke and corrosive hydrochloric acid vapors generated by the burning cable insulation. The smoke and gases traveled unimpeded throughout the first floor. The open design of the telephone switching equipment allowed the soot and HCl to contaminate sensitive electronic components and terminals. The 1AES switch was temporarily decontaminated, but eventually all the equipment in the HCO had to be replaced. The property loss was reported by Isner (1989) to be in the range $40–$60 million. The extensive and highly publicized interruption in telephone service was not completely restored until May 20th, twelve days after the fire.

The Hinsdale Central Office fire stimulated several important improvements in fire protection for facilities with large quantities of electrical cable and electronic switches. One improvement was the accelerated development of less flammable cable insulation and jacketing. Another improvement was the widespread adoption of redundant telecommunication pathways for critical applications. A related improvement was the installation of provisions to completely de-energize cables and switches to allow for easier fire suppression. A fourth improvement has been the extensive use of more sensitive smoke detectors to allow earlier detection and notification of incipient fires.

B.8 Sandoz Basel fire

Basel, Switzerland, is situated on the Rhine River where the borders of France, Germany and Switzerland intersect. Several large chemical companies have production and storage facilities in a location known as Schweizerhalle, about 6 km upstream of Basel. In 1987 the Sandoz plant in Schweizerhalle manufactured a variety of chemicals for the textile and agrochemical industries. There are also several warehouses on the site.

Building 956 at the Sandoz Schweizerhalle plant was constructed in 1967 for the storage of machinery. It is 90 m long, 50 m wide, and has two peaks 12 m high. It had a structural steel frame, and was divided into two sections by a 12 cm thick brick wall with sliding fire doors. In 1979 Sandoz converted the warehouse to a flammable liquids warehouse. The conversion involved installing explosion-proof electrical fixtures, sealing the drains from the sewer system, and installing three water spray curtains in the southern half of the warehouse, as shown in Figure B.16. The water spray curtains were to be operated by the plant fire brigade.

There were no automatic sprinklers or smoke detectors in the warehouse. There was a temperature monitoring system that was supposed to trigger flashing lights and associated extra precautions when the temperature exceeded 25 °C. The use of the warehouse for storage of flammable liquids in containers to a height of 10 m was approved by Swiss/Basel government agencies and insurance companies. In fact, a routine inspection was conducted four days before the fire.

On October 31, 1987, the following chemicals were stored in bags and plastic and steel drums in the warehouse:

- 859 metric tons of organophosphate insecticide,
- 12 metric tons of a phenyl-urea derivative used for weed control,

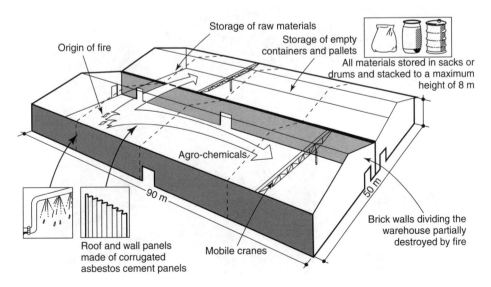

Figure B.16. Sandoz warehouse 956

- 73 metric tons of dinitrocresol derivative herbicide used for weed control,
- 26 metric tons of a fungicide,
- 11 metric tons of water soluble organic mercury compounds,
- 5.6 metric tons of miscellaneous agrochemicals including zinc phosphide,
- 364 metric tons of various formulating agents.

Most of these chemicals are flammable liquids with flash points of 30 °C and higher. Some are powders or pellets. The drums were stored on pallets stacked up to five high in the southern half of the warehouse, as shown in Figure B.17. The northern half contained mostly empty plastic and steel drums and idle pallets.

On October 31st, pallet loads of product in the warehouse were shrink wrapped. The shrink wrap process entailed covering the paper sacks of product with a plastic sheet and then shrinking the sheet by passing a blowtorch back and forth at a distance of approximately 1 ft (30 cm). However, if the blowtorch is not continuously moved, the flame can rapidly burn a whole through the plastic sheet and heat the product. During the fire investigation, the Scientific Branch of the Zurich city police conducted tests and discovered that one particular product, the pigment Prussian blue, is readily ignited by this heat exposure. Moreover, according to the police report, sacks of Prussian blue burn with "a flameless, smokeless and slowly progressing glowing." In fact, it "continues to glow smokelessly for hours without giving off a burning smell." Thus, it is not surprising that the Sandoz employees did not realize that they had inadvertently ignited the Prussian blue as they performed their shrink wrapping.

At 12:19 AM on November 1, 1986, a fire in Warehouse 956 was discovered simultaneously by a Sandoz employee and a Basel police officer. They both observed flames shooting the roof of the warehouse. The Schweizerhalle fire brigade was immediately notified and the brigade chief arrived at the warehouse at 12:22 AM. Minutes later the Sandoz and Ciba-Geigy fire brigades and the public fire department all were on the scene.

At first, the fire chief decided that the fire was spreading too rapidly for it to be extinguished, and that firefighters should concentrate on protecting the adjacent buildings which contained large

Figure B.17. Remains at Sandoz warehouse on the morning after the fire

Figure B.18. Rubble being hosed at the Sandoz warehouse the morning after the fire (from the Swiss Institute for the Promotion of Safety and Security)

amounts of organic solvents and sodium. However, the situation became much more dangerous as steel drums from Building 956 were propelled through the roof and landed in and around the adjacent buildings. An attempt was made to extinguish the primary fire using foam, but that was unsuccessful. Therefore, it became necessary to apply large quantities of water to cool the fire exposed steel drums as well as the adjacent buildings. The estimated water application rate during this period is 30 m^3 per minute (8000 gpm). The fire was declared out shortly before 5 AM. Figure B.18 shows the smoldering rubble being cooled by hosestreams a few hours later that morning.

The torrential water application rate soon overflowed the plant's storm water sewer system and catch basins. Approximately 10,000 m^3 of water runoff contaminated with several tons of toxic chemicals (primary concern was the organophosphates and ethyoxyethyl mercuric hydroxide) flowed into the Rhine River. The mercury content in these compounds is estimated to be 100–200 kg (Matter, 1987). This was the start of an environmental disaster that temporarily destroyed most of the fish in an approximately 250 km section of the Rhine downstream of Basel.

Smoke generated during the fire contained malodorous mercaptans (sulfurous combustion products of thiophosphoric esters). Since the wind was blowing the smoke plume toward the residential areas of Basel, police warned residents to close all windows and remain indoors until morning. There were numerous complaints of eye and respiratory tract discomfort, and several people sought medical treatment for nausea. However, there were many more cases of psychological distress than occurrences of physical distress. Follow-up medical examinations of more than 1500 exposed people indicated that there should not be any long-term health effects.

Extensive decontamination was needed on the impacted section of the Rhine. This included dredging chemical deposits from the riverbed, and restocking the river with various species of fish. Fortunately, the river's vegetation and bacteriological regenerative capacity withstood the chemical impact, and were able to provide nourishment for the restocked fish.

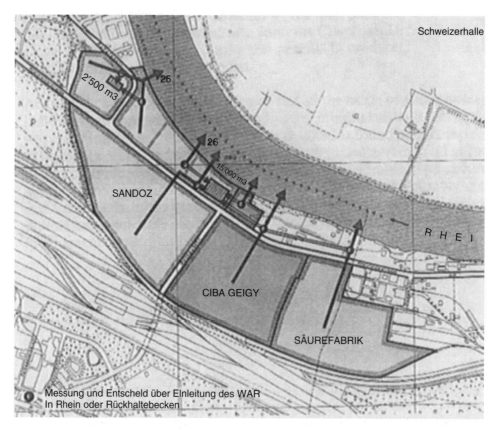

Figure B.19. Locations of the catch basins and treatment facility between the Thine and the Schweizerhalle plant sites of Sandoz, Ciba-Heigy and Saurefabrik (from the Swiss Institute for the Promotion of Safety and Security).

There was also significant concern for the contaminated water runoff that may have entered the groundwater under the warehouse. Subsoil water wells near the site were closed and subsoil water levels below the site were lowered by pumping. Ground water monitoring confirmed that these measures, together with the natural hydrogeologic conditions, did prevent contaminated water from entering the municipal water supply.

Besides the extensive cleanup costs, Sandoz was liable for approximately $60 million (100 million Swiss francs) in claims by people and government agencies affected by the polluted section of the Rhine. These claims were filed by French and German fishermen, governments and business affected by the loss of tourism, and various businesses and industrial plants whose water supply had to be curtailed during the cleanup operations (Business Insurance, 1986).

In January 1987 Sandoz decided to substantially upgrade the water runoff containment capability at Schweizerhalle. They committed to building two catch basins, one with a water capacity of $15,000 \, m^3$ (4 million gal), and the other with a $2500 \, m^3$ (665 thousand gal) capacity. Both basins are connected to an on-site water treatment facility and to disposal tanks for temporary storage. Figure B.19 shows the locations of the catch basins and treatment facility between the Rhine and the Schweizerhalle plant sites of Sandoz, Ciba-Geigy, and Saurefabrik. The seven water-quality measuring stations shown in Figure B.18 are used to determine if and when the water should be discharged into the river.

Sandoz and Ciba-Geigy merged in 1996 to form Novartis. At the time, this was the largest corporate merger in history. Novartis is now at the forefront of European and international efforts to develop international guidelines for the safe warehousing of chemicals. These efforts mitigate but do not eliminate the long-lasting stigma of the Schweizerhalle fire and its environmental impact.

References

A Large Loss for K Mart: A Lesson for Others, *Record*, published by Factory Mutual Engineering Corporation, 1983.
Allan, R., The Great New York Telephone Fire, *IEEE Spectrum*, 34–40, June 1975.
Best, R., $100 Million Fire in K Mart Distribution Center, *Fire Journal*, 36–42, March 1983.
Best, R., Fire Walls that Failed: The K Mart Corporation Distribution Center Fire, *Fire Journal*, 74–86, May 1983.
Business Insurance, "Sandoz will be liable for Rhine pollution", *Business Insurance*, December 29 1986.
Chapman, E.F., 'Disaster at 'Ma Bell's'–Revisited, N.Y.F., published by the New York City Fire Department, 3rd Issue, 1978.
'Communication Disaster at Ma BELL's,' W.N.Y.F.: With New York Firefighters, published by the New York City Fire Department, 3rd Issue, 1975.
Factory Mutual Loss Report No. 77-F-125, 1978.
FTI Report, "Hinsdale Central office Fire," Joint Report of the State Five Marshall and Illinois Commerce Commission Staff, Forensics Technologies Int. Corp., 1989.
Harris, W.B., The Great Livonia Fire, *Fortune*, 132–135, 172–177, November 1953.
Hession, V.J., Fire Protection: A Burning Issue for Telcos, *Telephony*, March 14 1983.
How to Prevent Large-Loss Fires in Factories, *Engineering News Record*, 28–30, July 15 1954.
Isner, M. "Fire Investigation Report, Telephone Central office, Hinsdale, Illinois, May 8, 1988," NFPA, 1989.
Juillerat, E.E. and Gaudet, R.E., Chicago's McCormick Place Fire, *Fire Journal*, 14–22, May 1967.
Lathrop, J.K., World Trade Center Fire, *Fire Journal*, July 1975.
Lathrop, J.K., Telephone Exchange Fire, *Fire Journal*, July 1975.
Matter, B.E., 'The fire in Schweizerhalle: a review of current activities to evaluate potential health hazards and environmental damage, and some suggestions from a toxicological viewpoint, *Proceedings Conf. on Attitudes to Toxicology in the EEC* (P.L. Chombers, ed.), Wiley, 1987, 163–169.
Pryor, A.J., 'The Browns Ferry Nuclear Plant Fire,' SFPE Technical Paper, 1976.
Salzmann, J.-J., "Schweizerhalle and Its Consequences," Transcript of presentation to World Conference on Chemical Accidents, World Health Organization, July 1987.
"The Aftermath of the Sandoz Fire," Fire Prevention, pp. 13–20, No. 199, May 1987.
"Schweizerhalle, The Fire on 1 November 1986 and Its Aftermath," 1987 Sandoz Report received from Dr. George Suter, former Novartis Safety Director, December 2000.
Stein, L., *The Triangle Fire*, Lippincott Co., 1962.
'The General Motors Fire,' *Quarterly of the NFPA*, October 1953.

APPENDIX C: BLAST WAVES

The blast wave generated by an explosion consists of a shock front in which the pressure rises virtually instantaneously, followed by an expansion wave in which the pressure returns to its ambient value. At large distances from the explosion center, the pressure overexpands below ambient such that the blast wave has a negative phase as well as a positive phase. This is illustrated in Figure C.1 which shows the characteristic form of an ideal blast wave. The magnitudes of the pressure rise across the shock front, P_s, and the other blast wave parameters shown in Figure C.1 depend upon the explosion energy release characteristics and the distance from the effective center of the explosion. The shock pressure rise, P_s, generally decreases with distance from the explosion center. Eventually the blast wave decays to an acoustic wave traveling at the speed of sound in air (about 738 mph, 330 m/s).

The damage potential of a blast wave depends on both the shock front pressure rise, P_s, and the impulse, I_s, which is defined as the area under the positive phase of the pressure versus time curve. A particular object will suffer damage when the impulse and P_s both exceed damage threshold values. Damage threshold pressures and impulses corresponding to various categories of structural damage and injuries are listed in Table C.1 which is based primarily on results in Baker et al. (1983), Baker (1973), and Kinney and Graham (1985). They account for the static strength of the representative structures, their dynamic response times, ductility, and support boundary

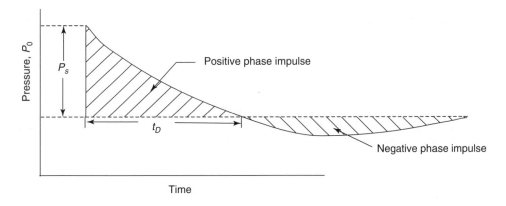

Figure C.1. Blast wave generated by an explosion

Table C.1. Blast damage and personal injury pressures and impulses

Structure/object	Pressure (psig)	Impulse (psi-msec)
Plate Glass Windows:		
20 ft² pane, 3/16" thick	0.3–0.6	–
10 ft² pane, 3/16" thick	0.6–1.0	–
10 ft² pane, 1/4" thick	1.1–1.6	–
Wood Roof Joist, 13 ft Span	0.5	–
Brick Wall – Minor Damage	0.7	16
Brick Wall – Major Damage	2.0	43
Wood Stud Wall, 7.5 ft high	1.0	1
Sheet Metal Panel Buckling	1.1–1.8	–
Wood Siding Failure	1.1–1.8	–
Cinder Block Wall Failure	1.8–2.9	–
Wood Frame Building Collapse	3.0–4.5	36
Oil Storage Tank Rupture	3.0–4.5	–
Structural Steel Building	4.5–7.3	–
Reinforced Concrete Wall	6.0–9.0	–
Total Destruction of Most Bldgs	10–12	–
Overturning of 10 ft high truck	0.3	110
Personal Injury	Pressure (psig)	Impulse (psi-msec)
Personnel Knock Down	0.5–1.5	–
Eardrum Rupture Threshold	5	7
50th Percentile Eardrum Rupture	15	22
Lung Damage Threshold	10	340
99% Lethal Lung Damage	50	1940

conditions. Structural response calculation methods that account for the dynamic character of the blast wave loads are also presented in a US Army Technical Manual (1987) for individual structural elements and in Whitney *et al.* (1988) for buildings of common construction.

C.1 Ideal blast waves

An ideal blast wave is one in which the energy is released rapidly compared to the time required for the blast wave to propagate to a particular target. This inevitably requires the explosive energy release to occur within a short distance compared to the distance to the target. Blast waves associated with condensed phase explosives such as TNT are effectively ideal blast waves. Many other types of explosions can be approximated as ideal blast waves in the far field away from the energy release site.

Mathematical models and data correlations for ideal blast waves (Baker, 1973) show that blast wave pressure and impulse both vary as distance divided by the one-third power of the blast wave energy. The nondimensionalized distance of a target from the energy release center is given by

$$\overline{R} = R(p_0/E)^{1/3} \qquad \text{[C.1]}$$

where R is the distance from energy source (m), $\overline{R}$ is the nondimensional distance, p_0 is the ambient pressure (kPa), and E is the blast wave energy (kJ).

APPENDIX C: BLAST WAVES

Often the blast wave energy is expressed in terms of the equivalent weight of TNT that would generate the same energy. The relationship is

$$W_{TNT} = E/4200 \, \text{kg} \quad \text{[C.2]}$$

where W_{TNT} is the equivalent weight of TNT (kg), E is the blast wave energy (kJ), and 4200 kJ/kg is the specific energy potential of TNT.

Blast wave calculations and correlations are geometry dependent. The simplest geometry is the spherically expanding ideal blast wave that occurs in the absence of confinement and/or structures between the target and the explosion center. Blast wave interactions with a structure often result in a complicated pattern of diffracted and reflected waves. If the target surface is situated parallel to the direction of blast wave propagation, the pressure exerted on the target is the incident pressure, also called the side-on pressure, P_s. If the target structure is situated perpendicular to the direction of blast wave propagation, the peak pressure exerted on the structure is the pressure just behind the reflected shock, P_r.

Ideal blast wave incident and reflected shock overpressures (pressure rise above atmospheric pressure) are shown as a function of nondimensional distance in Figure C.2. At values of $\overline{R}$ greater than about 2, the reflected shock overpressure is about twice the incident overpressure. Comparing the curves in Figure C.2, to the damage threshold given in Table C.1 for window breakage, windows would not expected to be damaged at values of $\overline{R}$ greater than about 10, unless they are very large and thin and situated perpendicular to the blast wave propagation direction.

Correlations for the ideal blast wave positive phase impulse, I_s, are given in Figure C.3. The impulses in Figure C.3 are normalized as:

$$\overline{I}_s = I_s a_0 / (p_0^{2/3} E^{1/3}) \quad \text{[C.3]}$$

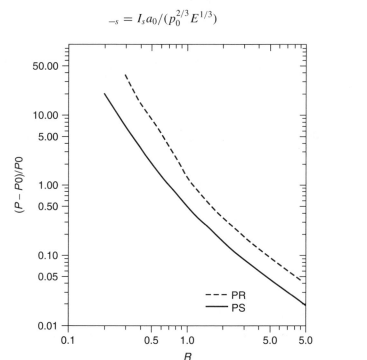

Figure C.2. Ideal blast wave incident and reflected shock overpressures as a function of nondimensional distance

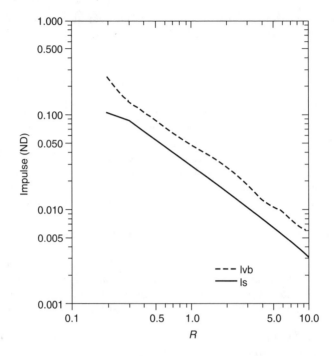

Figure C.3. Correlations for the ideal blast wave positive phase impulse I_s

where $\bar{I}_s$ is the nondimensional impulse, I_s is the impulse, a_0 is the speed of sound in ambient air (m/s), p_0 is the ambient pressure (kPa), and E is the blast wave energy (kJ).

The corresponding correlation for the positive phase impulse behind a normally reflected shock wave, I_r, can be calculated from

$$I_r = I_s(P_r/P_s) \quad [C.4]$$

where I_r is the reflected shock wave, I_s is the positive phase impulse, and P_r/P_s is the ratio of reflected shock pressure to incident shock pressure as determined from Figure C.2.

Application of the ideal blast wave correlation shown in Figures C.2 and C.3 requires knowledge of the blast wave energy. Energies generated in vessel rupture explosions are described in the next section.

C.2 Pressure vessel ruptures

The blast wave energy, E, generated during the rupture of a pressure vessel containing a compressed gas is given by (Baker et al., 1978)

$$E = V(P_1 - P_0)/(\gamma - 1) \quad [C.5]$$

where V is the vessel volume, P_1 is the gas pressure in the vessel, P_0 is ambient pressure, and γ is the gas ratio of specific heats.

Equation [C.5] is conservative in the sense that it gives energies slightly higher than energies calculated with more precise models of the gas expansion process. The accuracy of equation [C.5] should suffice for most engineering applications.

APPENDIX C: BLAST WAVES

If the vessel contains a liquefied gas that may partially condense upon sudden expansion, the energy released in expansion is equal to the change in internal energy, i.e.

$$E = m(u_1 - u_2) \qquad [C.6]$$

where u_1 and u_2 designate the fluid internal energy prior to and immediately after expansion to ambient pressure.

The expansion process is usually assumed to be isentropic since that usually provides a simple conservative estimate for E. Thermodynamic property data for liquefied gaseous fuels are available in Din (1967).

Blast wave pressures measured in vessel burst tests (Baker *et al.*, 1983, 1978) are usually lower than those given by TNT/ideal blast wave correlations for the same energy. The differences are significant near the ruptured vessel, but become almost negligible at nondimensional scaled distances greater than about 0.2.

Positive phase impulse correlations for vessel burst blast waves are given by the curve labeled I_{vb} in Figure C.3. For a given value of the energy normalized distance, $\overline{R}$, I_{vb} is significantly greater than I_s obtained from condensed phase blast wave correlations. This is due to the increased time scale for a vessel burst energy release compared to a condensed phase explosive blast wave.

C.3 Vapor cloud explosions

Within the vapor cloud, ideal blast wave correlations overestimate vapor cloud explosion peak pressures and underestimate impulses and positive phase durations. Non-ideal blast wave calculations have been reported (e.g. Strehlow, 1980), but they usually require empirical knowledge of flame speeds and/or accelerations. Far-field (outside the vapor cloud) blast waves can be approximated as ideal blast waves providing the effective energy, or yield, of the blast wave can be estimated. Most engineering estimates are based on effective yields in the range 1% to 5% depending on the reactivity of the flammable gas and the vapor release scenario. The size and shape of the dispersed vapor cloud, the vapor concentration distribution, and the presence of building or equipment obstructions (particularly repeated obstacles that induce rapid flame accelerations) are the primary considerations in the vapor release scenario.

C.4 Vented gas and dust explosions

The pressure rise in an enclosure during a deflagration type explosion is not a blast wave because the pressure is approximately uniform throughout the enclosure at any given time; i.e. there is a gradual pressure rise instead of a propagating shock wave. The magnitudes and rise times of enclosure pressures in gas and dust deflagrations are discussed in Zalosh, 1995.

If the enclosure is vented either deliberately or accidentally, blast waves form outside the enclosure and propagate into the surrounding atmosphere. Peak pressure data outside vented enclosures have been reported in Butlin and Tonkin (1974) and Hattwig (1978) for gas and dust explosions, respectively. The data obtained in a direction normal to the vented side of the enclosure can be represented by

$$\frac{P_d}{P_m} = \frac{k}{k + \frac{d}{\sqrt{A_v}}} \qquad [C.7]$$

where P_d is the peak pressure at a distance d from the vent, P_m is the maximum pressure developed within the enclosure, A_v is the explosion vent area in the enclosure, and k is an empirical constant.

The data in Butlin and Tonkin (1974) and Hattwig (1978) correspond to values of k in the range 0.8 to 1.7, with the value 1.7 providing a better fit to the data from the larger scale tests. Thus $k = 1.7$ would be the prudent value to use until a more comprehensive data base and correlation becomes available. The same value is also recommended for targets in other directions; the test data indicate that these targets can sometimes experience blast wave pressures as high as those in a direction normal to the vent.

References

Baker, W.E., *Explosions in Air*, University of Texas Press, 1973.

Baker, W.E., Kulesz, J.L., Richer, R.E., Bessey R.L., Westine, P.S., Parr, U.B. and Oldham, G.A., "Workbook for Estimating Effects of Accidental Explosions in Propellant Ground Handling and Transport Systems," NASA CR 3023, August 1978.

Baker, W.E., Cox, P.A., Westine, P.S., Kulesz, J.J. and Strehlow, R.A., *Explosion Hazards and Evaluation*, Elseveir, New York, 1983.

Butlin, R.N. and Tonkin, P.S., 'Pressures Produced by Gas Explosions in a Vented Compartment,' Fire Research Station Note No. 1019, September 1974.

Din, F., ed., *Thermodynamic Functions of Gases, Vol. 2, Air, Acetylene, Ethylene, Propane, and Arogon*, Butterworth, London, 1962.

Hattwig, M., Selected Aspects of Explosion Venting, *2nd Intl. Symposium on Loss Prevention and Safety Promotion in the Process Industries*, DECHEMA, 1978.

Kinney, G.F. and Graham, K.J., *Explosive Shocks in Air*, Springer-Verlag, 2nd Edition, 1985.

Strehlow, R.A., Accidental Explosions, *American Scientist*, **68**, 1980.

U.S. Army Technical Manual No. 5-1300, 'Structures to Resist the Effects of Accidental Explosions,' Department of Defense triservice document also available as NAVFAC P-397, and AFM 88-22, 1987.

Whitney, M.G., Ketchum, D.E. and Oswald, C.J., 'A Procedure to Assess Explosion Damage to Buildings of Common Construction,' Minutes of the 23rd Explosives Safety Seminar, Department of Defense Explosives Safety Board, 1988.

Zalosh, R.G., Explosion Protection, *SFPE Handbook of Fire Protection Engineering*, Chapter 3-16, National Fire Protection Association, 1995.

INDEX

Note: Page references followed by 'f' represents a figure and 't' represents a table.

actual delivered density (ADD)   158, 163, 164–165
aerosol cans   44, 46, 281, 353
aerosol products, fire protection   184–188
AFFF (aqueous film forming foam)   230, 236, 267, 276, 277
Ameritech Telephone Exchange Fire (1989)   20, 21
ammonium perchlorate   190, 191
arcing   302, 322
array tests, small palletized   140
arson   42
askarel transformers   329, 332
atmospheric absorption   342, 343
atmospheric pressure tanks   243
atmospheric transmissivity   342–343
automated storage and retrieval systems (ASRS)   125
automatic sprinklers   18, 270
automatic suppression systems   18–20, 43, 365

barriers
  fire   57
  passive   21, 43, 45, 51
  vapor   83, 84, 85
Basel 1986 fire   20, 21, 23, 271, 369–373
Bell Telephone Exchange Fire (1975)   20, 301, 354–358
blast waves   14, 33–36, 375–380
boiling liquid expanding vapor explosions (BLEVEs)   14, 247, 271, 276
Browns Ferry 1975 Nuclear Plant fire, Alabama   304, 354
bulk materials, storage of   171–199
buoyancy flux   31
buoyancy pressure differences   96–98
burner flames   202
bushings   328, 330
business interruption losses   8

cable fire incidents   300–304
cable fire suppression tests   314–317
cable fires, grouped   300
cable flammability, testing and classifications   304–309
cable tray fires
  coal-fired power plant Illinois (1998)   302
  heat release rates   311
cable trays   298, 300, 301, 302, 319, 321
  fire protection   320
  fire resistance   317
  Hinsdale fire simulation   313f
cables
  coatings   320
  coaxial   298, 300
  communication   297, 298, 299f, 311
  control tray   298
  grouped electrical   354
  heat release rates   307
  installation   300
  insulation   300, 309
  jacket material   309
  limited smoke (LS) rating   304
  limited use   304
  loadings   311
  optical fiber   298
  riser   304
cables and trays, protection guidelines and practices   319–322
calcium hypochlorite   188, 189, 190
carbon dioxide   95, 96, 198, 236–237, 316, 349
carbon monoxide   102, 196
ceiling jets   6
  warehouse fire plumes and   160–162
ceiling sprinklers   148–158, 174, 276, 281, 290, 352
  actuation time calculation   160
  discharge density   167
  location   161
Chernobyl nuclear reactor explosion   15t, 17
Chicago McCormack Place fire   20, 38, 350–351
chlorine vapor   30
chlorofluorocarbon propellants   184
cigarettes   17

circuit boards, flammability of electronic components 322
cold storage warehouse fire protection 167–168
combustion 3, 191
concrete 61, 70
cone calorimeters 134
cone roofs 250
configuration (view) factor 28, 29, 342
construction materials 57–61
containers
 failure times and modes 282–285
 small metal 285–289
 small plastic 289–293
 types 279–281
critical Froude numbers 104, 105, 106, 110
critical heat flux 29
critical temperature 58
cutting incidents 303

damage control measures 9
damage threshold pressure data 34
Dayton 1987 fire 78, 83, 113, 186, 281
differential scanning calorimetry (DSC) 192
differential thermal analysis (DTA) 192
diffusion flame emission/absorption coefficients 337
dilution ratio 226
discharge density 151
discharge rate 96
doors, smoke leakage 107
double carton tests 283, 284
downward flame spread rates 133
draft curtains 112, 114
drums
 designs and storage modes 268–270
 failure times and failure modes 271–276
 storage fire suppression systems 276–279
dry chemical suppression agents 234–236
dust, combustible 51
dust explosions 14, 196, 379–380

early warning fire detection systems 326
earthquakes, fire protection and 42
electrical cables
 and equipment 297–336
 fire resistant 23
 fire suppression testing 317
 flammability testing 304
 generic description 297–300
electrical current overloads, cable response 309
electrical faults 300, 350
electronic cabinets, flammability 323–324
electronic components, flammability 322
electronic equipment
 fire detection and suppression 326–327
 vulnerability 324–325
electrostatic discharge 51
electrostatic ignitions 209–215, 250
emergency evacuation requirements 30, 31
emergency exits 23, 24, 43, 46, 52–53, 107
environment, fire protection and 42–43
ethyl chloride, diked spill fire 32
evacuation zones 31
exhaust fans 100, 102
explosions 247, 375
 damage 8
 historical industrial 20–24
 multiple fatality 14–17
 plant layout protection 43–53
 prevention 14
 rapid phase transformation 247
 scenario identification 2–6
 vapor cloud 14, 379
exposure fires 29, 256, 303, 304

Factory Mutual Data Sheet 3–26, 36
Factory Mutual Data Sheet for Maximum Foreseeable Loss Fire Walls 73
Factory Mutual exposure fire categories, exposure fire separation distances 29
Factory Mutual Loss Prevention Sheet 1–28, 83
Factory Mutual standards 29
Factory Mutual water demand specifications 37
failure modes and effects analysis (FMEA) 2
FANRES computer model 103
fault tree analysis 9, 38
fire containment 43, 91
fire control criteria, ignition of wetted commodities 165–166
fire damage, environmental 8
fire doors 23, 78–83
fire emergency response procedures 351
fire exposure failure mode 270
fire exposure tests 282, 283
fire fighting, manual 43, 52, 107
fire fighting organizations 41–42
fire models, enclosure 103
fire plumes 6, 29–32, 107, 144
fire points
 flash points and 202–204
 time to reach 205–209
fire products collector data 141
fire products collector tests 146, 147, 164, 173–175, 179–180
fire propagation index 9
fire protection
 alternative protection evaluation 1, 2f, 8–9
 industrial 1–9

lessons learned 21–23
lessons not learned 23–24
plant layouts 43–53
siting considerations 27–43
fire resistance 42, 45, 57, 82
  calculations 61–67
  tests 67–72
fire resistant construction 57–89
fire scenarios
  identification 2–6
  loss experience 270–271
  storage tank loss history 247–251
fire spread 3, 247
fire suppression 42, 166–167, 321
fire tests 36, 135, 179–181
  automatic sprinkler 316
  large array sprinklered 147–149
  Plaza Hotel 102
  rack storage 149, 179, 181
  roll paper 173–176
  sprinklered 174, 180–181
fire walls 8, 21, 23, 73–78, 281, 349, 354, 361
fire-retarded acrylonitrile-butadiene-styrene
  (FRABS) 322
fires
  automatic detection 18–20, 24, 43
  electrical cables 14
  historical industrial 20–24, 347–373
  injuries and fatalities 7–8
  liquid burning rates of open roof 251
  multiple fatality 14–17
  nonthermal damage 7
  open tank 252, 254
  plastic storage 14
  smoldering 94
  yard storage 271
  see also warehouse fires
flame emissive powers 27, 29, 254, 337–341
flame height 27, 28, 29, 165, 341–342
flame propagation 308, 309
flame radiation 27–29, 87, 337–346
flame spread 6, 29, 85, 131, 134
flame spread rates
  free burn heat release rates and 160–161
  ventilation velocity 313
flame surface area 341
flame transmission 69
flame-target configuration factor 27, 28
flames 223, 253, 254
flammable liquids
  fires 72, 201
  ignitability and extinguishability 201–241
  incident data 201–202
  loss experience 281–282

problems caused by residue 24
in small containers 279–293
storage 24, 243–295
flash points 202, 204, 205, 224–228
flue space, effect of 124–129
fluoroprotein foam 267
FMRC clean room materials flammability test
  protocol 9
FMRC flammability tests 134
foam application systems 244
foam extinguishment 230–234
foam injection 267
foam systems 266, 270
foam-water systems 277, 287
foams
  high expansion 183, 230, 234, 357
  low expansion 198, 230–233
  medium expansion 230, 234
Ford Cologne, Germany warehouse fire 20, 21, 78, 118, 358–361
free burn calorimetry tests 284
free burn heat release rates 127, 160–161
fuel area limited fires 110
fuel fired equipment 17
furnace endurance times 70
furnace exposure tests 67–69
fusible links and detectors 81

gas leakage tests, high temperature 107
gaseous suppression agents 198
gaseous suppression system tests 316–317
gasoline fire data 227
General Motors Livonia fire 78, 83, 84, 347–350
general purpose cables 304
glass containers 281
GM Auto Transmission Plant Fire (1953) 20, 21, 24
Guidelines for Risk Management, EPA 49
Guidelines for Use of Vapor Cloud Dispersion Models 33
gypsum wallboard 72

Halon 1301 93, 94, 316
Halon replacement agents 95, 237–238
halon suppression, isolation and 91–96
hazard consequence models 53
hazardous materials, storage 23–24, 43–46
heat detectors 320
heat flux tests, flammable liquid fire exposure 256
heat release rate tests 9
heat release rates 27, 28, 29, 31, 102, 110, 126, 136, 307, 321
  cable tray fire 311
  double carton tests 284

heat release rates (*continued*)
  first ignited material 3
  laboratory measurements 134
  reduction 163–165
  roll paper 173
  spray fires 220
  storage tiers and 124, 127
heat transfer theory 86
high efficiency particle airfilter (HEPA) 91
Hinkley's model 113
Hinsdale Illinois Telephone Central Office fire 301–302, 363–369
horizontal cable trays 311–313, 314
horizontal flame spread rates 311
hot surfaces 201
HVAC (heating ventilating and air conditioning) system 96
hydrocarbon propellants 184
hydrocarbons, fluorinated 316
hydrochloric acid 368
hydrogen chloride 309, 326
hydrogen fluoride 238
hydrostatic pressure tests 281

ignition data 134
ignition sources 3, 4, 17–18, 43, 148, 202, 309, 311, 350
  electrical 17, 201
  isolation 46–51
ignition tests 131, 192–193, 205, 308
ignition time 208
ignition/damage threshold flux 28
incident heat flux 208
incident radiant flux 205
industrial explosions, historical 20–24
industrial fire protection, consequence analysis 1, 6–8
industrial fires, historical 20–24, 347–373
industrial fires and explosions, statistics 10–20
infrared systems 196, 198
ink and varnish manufacturing plant fire (1983) 211
insulation
  foamed 167
  halogenated 309
  steel 61
insulation materials 84
Insurance Services Office Fire Suppression Rating Schedule 41
intermediate bulk containers (IBC) 267
ionization smoke detectors 94, 357, 365
irritant substances, irritation thresholds 31

K Mart warehouse fire (1982) 20, 21, 78, 82, 113, 118, 185–186, 351–354
kerosene tank fires 254

laboratory flammability testing 131–135
lead-sheathed cable incidents 303
Life Safety Code 24, 53
lightning 247
liquids
  autoignition 205
  high flash point 224–226
  low flashpoint 227–228
loss expectancy 6

McCormick Place Convention Hall Fire (1967) 20, 38, 350–351
mass burning rates 124
mass flow rates 99, 110
maximum foreseeable losses 6, 21, 38
metal container tests 285, 288
metal containers 281
methane detectors 196
mineral-insulated metal-sheathed cables 298
monitors 196, 198
Moscow TV transmission tower fire (August 2000) 302, 303f

National Electrical Code 46–47
National Fire Protection Association (NFPA) 10, 29
National Institute of Standards and Technology (NIST) 114
New York City World Trade Center fire (1975) 304, 354
New York telephone exchange fire (1975) 20, 301, 354–358
nitrogen 198
Nonwoven Fabrics Handbook 178
nonwoven roll goods
  loss experience 179
  storage 178–182
nozzles
  agent delivery 235
  water spray tests 316
nuclear plant fires 15t, 17, 304, 354

open fired equipment 46
oxidizers, solid 188–191
oxygen concentration monitors 196

paint spray operations 219
pallet storage 120–122, 277, 293
panic loading tests 81
passive protection 317–319, 321

INDEX 385

PEPCON manufacturing facility fire Henderson Nevada (1987) 190–191
Piper Alpha oil production platform 10, 17
plant fire protection, major deficiencies 349
plants
  layout for fire/explosion protection 43–53
  site review 27–43
plastic tests 139
plastics, classification system 129
plenum cables 304
plenums 100, 101
plume theory 63
plumes
  dispersal 31, 32
  mass flow rates 110
point source approximation 343–346
polyethylene 135, 149, 272, 281
polyethylene terephthalate (PET) cups 141
polyisocyanurate insulated steel deck roofing 168
polypropylene 164, 180
polystyrene 139, 148, 168, 322
pool fires 222, 254
  heat release rates 215–219
  water miscible liquid 226–227
portable tanks, and intermediate bulk containers 267–268
power cables 297, 298, 311
pressure vessel ruptures 378–379
property damage 6–7

rack storage 122, 125, 141, 179, 182
rack storage fires, sprinkler water and 126
radiant heat fluxes 27, 28, 29, 68, 134, 253, 255
radiant view factor, calculation 254
radiation flame temperatures 338
radiation transmissivity 29
radiators 327–328
required delivered densities (RDD) 45, 126, 225
response time index (RTI) 144, 151
risk contour plots 53
roll loft factor (RLF) 179–180
roll paper
  loss experience 172–173
  storage 171–178
roof deck fire spread tests 21
roof loading 76
roof venting, heat and smoke 107–111
roof vents 349
roofing, insulated metal deck 83–86
room pressures 96
rubber tire storage 181–184

safe separation distances 27–36, 43
Sandoz Flammable Liquids Warehouse Fire (1986) 20, 21, 23, 271, 369–373
sanitizers 188
Setagaya telephone cable tunnel 1985 fire 304, 311
Sherwin-Williams warehouse fire (1987) 78, 83, 113, 186, 281
signal cables 297
small array tests 135–148
smoke, isolation and control 23
smoke control 91, 96–107
smoke damage 9, 114
smoke dampers 105, 107
smoke detection 322, 326
smoke detectors 81
  ionization 94, 357, 365
  location 106
smoke extractors 110
smoke generation rates 321
smoke isolation
  ventilation exhaust 100–103
  within rooms 96–107
smoke leakage, door and damper 107
smoke propagation, upstream 104–106
smoke release rate data 307, 310
smoldering fires 94, 196, 198
solid pile storage 118, 122, 293
soot release rate 310
soot yields 5, 9, 311, 313
spill fires 217–219, 287
spontaneous combustion 191
spontaneous ignition 17, 192, 193–196, 201, 250
spray fires 219–222, 228–230, 231, 233
spray ignitability tests 219
spray protection distances 333
spray-plume penetration model 163–164
Springfield MA (1988) fire 30, 188
sprinkler actuation model 162
sprinkler protection 117, 349, 361
  flammable liquids 24, 285–293
sprinkler systems 21, 42
  dry pipe 167, 168
  failure rates 41
  flow rates 36
  layout 51–52
sprinkler and water spray suppression tests 314–316
sprinklers 28, 29, 144, 154
  automatic 18, 45, 91
  early suppression fast response (ESFR) 23, 158, 184, 188
  flow rate requirements 148–159
  in-rack 122, 156–159, 181, 276, 281, 290

sprinklers (*continued*)
  large drop   150, 184
  protection guidelines   188
  protection scoping tests   267
  spray patterns   114
steel   57–61, 70
steel drums   268, 272
storage
  configurations   118–125
  on floor   179, 181
  on-side   172, 182, 183
  special commodities and bulk materials   171–199
storage heights   125–129, 156, 177, 182
storage racks   122
storage sprinkler protection   10
storage tanks   243–268
suppression systems   18–20, 24, 43, 237, 326–327, 365
swimming pool chemicals fires   30, 188

tank fires, suppression   266–267
tank truck fires   214
tanks
  burning rates and spacing criteria   251–255
  dome roof   250
  emergency venting   256–266
  fixed roof   243, 266
  floating roof   243, 250
  horizontal low pressure storage   245–246
  lifter-roof   244–245
  low pressure   245, 262
telephone exchange fires   20, 21, 301, 354–358
temperature monitors   196
temperatures
  ambient extreme   42–43
  ignitability   202–209
The Aerosol Handbook   185, 186
thermal blankets   317
thermal damage   325
thermal explosion theory   192
thermal insulation   65, 285, 317
thermal loads   6
thermal monitoring devices   196
thermal response parameters (TRP)   134, 308
thermal thrust force   70
thermisters   196
thermogravimetric analysis (TGA)   192
threshold limit value (TLV)   31
time adjusted quasi-steady (TASQS) formulation   161, 163
time-to-ignition test data   9, 134
transformers
  fire incidents   329–332

  fire protection   327–334
  fire scenarios   328–329
  generic description   327–328
  installation and fire protection guidelines   332
  water spray protection   332–334
Triangle Shirtwaist Company fire (1911)   20, 23, 52–53, 361–363
trichloroisocyanuric acid   30, 188, 189, 190
turbulence, effect on flame speeds   6
twin agent extinguishment, dry chemical   234–236

vapor clouds, flammable   33
vapor concentration   53
vapor explosions   14
vapor release rates   33
vapor venting   262, 264
ventilation   49, 50, 96, 106
venting, smoke isolation and   91–116
vents, location   112
vertical cable tray fire test data   309–311
vertical cable tray tests   307
vertical flame spread data   307
vertical tanks, vapor venting   262
very early warning fire detection (VEWFD) systems   326
volumetric expansion pressures   99–100

warehouse fires   72, 132, 186, 270–271, 282
  aerosol   185–186
  Cambridge MA (1965)   38, 39
  cold-storage   168
  losses   117–118
  MTM Partnership (1985)   185
  Worcester Massachusetts   168
warehouse storage   117–170
  commodity effects   128–149
  commodity generic classification   128–132
  effect of aisle widths   125–129
warehouses
  fire modeling   159–167
  sprinkler protection   10, 21–23
water application density   164
water application test results   181
water application times   226
water flow rates   36
water flux application rates   166
water runoff   23, 43
water spray
  credit factor   259
  densities   226
  discharge   198
  extinguishment   222–230
  nozzle tests   316
  protection   86, 87, 260

water supply   21, 36–41, 350
welding incidents   303
wet benches   92, 94, 96
white house tests   84–85
wind, flame diameter   216

wind tunnel tests   104
wood charring rate   72

yield strength   57, 58